Mapping and the Citizen Sensor

Edited by
Giles Foody, Linda See, Steffen Fritz, Peter Mooney, Ana-Maria Olteanu-Raimond, Cidália Costa Fonte and Vyron Antoniou

Published by
Ubiquity Press Ltd.
6 Windmill Street
London W1T 2JB

First published 2017

DOI: https://doi.org/10.5334/bbf

The full text of this book has been peer-reviewed to ensure high academic standards. For full review policies, see http://www.ubiquitypress.com/

Suggested citation:
Foody, G, See, L, Fritz, S, Mooney, P, Olteanu-Raimond, A-M, Fonte, C C and Antoniou, V (eds.) 2017 *Mapping and the Citizen Sensor.* London: Ubiquity Press. DOI: https://doi.org/10.5334/bbf. License: CC-BY 4.0

Contents

Supporting Institutions v

Chapter 1. Mapping and the Citizen Sensor 1

Giles Foody, Steffen Fritz, Cidália Costa Fonte, Lucy Bastin, Ana-Maria Olteanu-Raimond, Peter Mooney, Linda See, Vyron Antoniou, Hai-Ying Liu, Marco Minghini and Rumiana Vatseva

Chapter 2. Sources of VGI for Mapping 13

Linda See, Jacinto Estima, Andrea Pődör, Jamal Jokar Arsanjani, Juan-Carlos Laso Bayas and Rumiana Vatseva

Chapter 3. A Review of OpenStreetMap Data 37

Peter Mooney and Marco Minghini

Chapter 4. Production of Topographic Maps with VGI: Quality Management and Automation 61

Guillaume Touya, Vyron Antoniou, Sidonie Christophe, Andriani Skopeliti

Chapter 5. Motivating and Sustaining Participation in VGI 93

Steffen Fritz, Linda See and Maria Brovelli

Chapter 6. Considerations of Privacy, Ethics and Legal Issues in Volunteered Geographic Information 119

Peter Mooney, Ana-Maria Olteanu-Raimond, Guillaume Touya, Niels Juul, Seraphim Alvanides and Norman Kerle

Chapter 7. Assessing VGI Data Quality 137

Cidália Costa Fonte, Vyron Antoniou, Lucy Bastin, Jacinto Estima, Jamal Jokar Arsanjani, Juan-Carlos Laso Bayas, Linda See and Rumiana Vatseva

Chapter 8. The Impact of the Contribution Micro-environment on Data Quality: The Case of OSM 165

Vyron Antoniou and Andriani Skopeliti

Chapter 9. Visualisation and Communication of VGI Quality 197

Andriani Skopeliti, Vyron Antoniou and Temenoujka Bandrova

Chapter 10. The Relevance of Protocols for VGI Collection 223

Marco Minghini, Vyron Antoniou, Cidália Costa Fonte, Jacinto Estima, Ana-Maria Olteanu-Raimond, Linda See, Mari Laakso, Andriani Skopeliti, Peter Mooney, Jamal Jokar Arsanjani, Flavio Lupia

Chapter 11. Data and Metadata Management for Better VGI Reusability 249

Lucy Bastin, Sven Schade and Christian Schill

Chapter 12. Integrating Spatial Data Infrastructures (SDIs) with Volunteered Geographic Information (VGI) for creating a Global GIS platform 273

Demetris Demetriou, Michele Campagna, Ivana Racetin, Milan Konecny

Chapter 13. VGI in National Mapping Agencies: Experiences and Recommendations 299

Ana-Maria Olteanu-Raimond, Mari Laakso, Vyron Antoniou, Cidália Costa Fonte, Alexandra Fonseca, Magdalena Grus, Jenny Harding, Tobias Kellenberger, Marco Minghini, Andriani Skopeliti

Chapter 14. Opportunities for Volunteered Geographic Information Use in Spatial Planning 327

Matej Nikšič, Michele Campagna, Pierangelo Massa, Matteo Caglioni, Thomas Theis Nielsen

Chapter 15. Citizen Science and Citizens' Observatories: Trends, Roles, Challenges and Development Needs for Science and Environmental Governance 351

Hai-Ying Liu, Sonja Grossberndt and Mike Kobernus

Chapter 16. The Future of VGI 377

Vyron Antoniou, Linda See, Giles Foody, Cidália Costa Fonte, Peter Mooney, Lucy Bastin, Steffen Fritz, Hai-Ying Liu, Ana-Maria Olteanu-Raimond and Rumiana Vatseva

Supporting Institutions

This publication is based upon work from COST Action TD 1202 *Mapping and the Citizen Sensor*, supported by COST (European Cooperation in Science and Technology).

COST (European Cooperation in Science and Technology) is a funding agency for research and innovation networks. Our Actions help connect research initiatives across Europe and enable scientists to grow their ideas by sharing them with their peers. This boosts their research, career and innovation.

www.cost.eu

EUROPEAN COOPERATION IN SCIENCE & TECHNOLOGY

Funded by the Horizon 2020 Framework Programme of the European Union

CHAPTER 1

Mapping and the Citizen Sensor

Giles Foody*, Steffen Fritz†, Cidália Costa Fonte‡, Lucy Bastin§,¶, Ana-Maria Olteanu-Raimond‖, Peter Mooney**, Linda See†, Vyron Antoniou††, Hai-Ying Liu‡‡, Marco Minghini§§ and Rumiana Vatseva¶¶

*School of Geography, University of Nottingham, UK, giles.foody@nottingham.ac.uk

†International Institute for Applied Systems Analysis (IIASA), Schlossplatz 1, 2361 Laxenburg, Austria

‡Department of Mathematics, University of Coimbra, 3001-501 Coimbra, Portugal / INESC Coimbra, Rua Sílvio Lima, Pólo II, 3030-290 Coimbra, Portugal

§European Commission, Joint Research Centre, Ispra, Italy

¶Aston University, Birmingham UK

‖Paris-Est, LASTIG COGIT, IGN, ENSG, F-94160 Saint-Mande, France.

**Maynooth University, Maynooth, Ireland

††Hellenic Army General Staff, Geographic Directorate, PAPAGOU Camp, Mesogeion 227-231, Cholargos, 15561, Greece

‡‡Norwegian Institute for Air Research (NILU), Kjeller 2027, Norway

§§Department of Civil and Environmental Engineering, Politecnico di Milano, Piazza Leonardo da Vinci 32, 20133 Milano, Italy

¶¶National Institute of Geophysics, Geodesy and Geography, Bulgarian Academy of Sciences, Bulgaria

Abstract

The role of citizens in mapping has evolved considerably over the last decade. This chapter outlines the background to citizen sensing in mapping and sets the

How to cite this book chapter:

Foody, G, Fritz, S, Fonte, C C, Bastin, L, Olteanu-Raimond, A-M, Mooney, P, See, L, Antoniou, V, Liu, H-Y, Minghini, M and Vatseva, R. 2017. Mapping and the Citizen Sensor. In: Foody, G, See, L, Fritz, S, Mooney, P, Olteanu-Raimond, A-M, Fonte, C C and Antoniou, V. (eds.) *Mapping and the Citizen Sensor.* Pp. 1–12. London: Ubiquity Press. DOI: https://doi.org/10.5334/bbf.a.

scene for the chapters that follow, which highlight some of the main outcomes of a collaborative programme of work to enhance the role of citizens in mapping.

Keywords

Volunteered Geographic Information, mapping, citizens, sensors

1 Introduction

Accurate and timely maps are a fundamental resource for a vast array of applications. Maps are, for example, central to everyday activities ranging from route planning and the legal demarcation of space through to scientific undertakings such as the design of nature reserves for species conservation or the monitoring of terrestrial carbon pools in support of climate change policies. Maps, therefore, provide a range of services, including ones that support economic activity (e.g. location-based services) and enhance human health and well-being (e.g. damage maps for disaster relief and humanitarian aid programmes). Maps underpin popular location-based augmented reality mobile games such as Pokémon Go, and gaming activity can be used to help acquire geographic information for mapping (Antoniou and Schlieder, 2014). Map production and updating in a rapidly changing world is, however, a major scientific and practical challenge. The US National Academies, for example, highlight a key strategic question for the geographical sciences, which is: how can we better observe, analyse and visualise a changing world? (CSDGSND, 2010). This book is focused on the potential of citizen sensors, typically volunteers, to help in mapping activities. In the context of this book, we use the term mapping to refer to the process of creating maps. This term aims to be inclusive and thus covers any activity from the process of data gathering to the production of spatial and cartographic products.

Citizens have considerable potential as a source of geographic information and this activity is itself a further strategic priority identified by the US National Academies (CSDGSND, 2010). Citizens have been collecting georeferenced data of several types for some time (Boyd and Foody, 2014) but this activity, and its possible usefulness, is not well understood and therefore its potential remains unfulfilled. To help advance the role of citizens in mapping, a Cooperation in Science and Technology (COST) Action – where COST is a European framework to support research on topics of global relevance – called TD1202 Mapping and the Citizen Sensor[1] was launched. This book presents some of the work that has arisen from the Action's activities.

Mapping has a long history, and 'best practices' for authoritative mapping have been established and used for many years. For example, standards for topographic mapping have been defined and used by major government agencies (Olteanu-Raimond et al., 2017). Similarly, in relation to thematic mapping from remote sensing, best practices for map validation have been defined (Strahler et al., 2006; Olofsson et al., 2014). The various bodies engaged in authoritative mapping, however, often cannot meet mapping requirements or 'best practices', which can be impractical to implement (Rahmatizadeh et al., 2016) – for example, data collection that follows a strict probabilistic sample design or the need for large sample sizes for thematic map validation. In this situation there are a variety of ways in which mapping activity could progress. The problems of authoritative mapping could simply be recognised and standards lowered. This rather negative approach would appear to be a retrograde step. It would, for example, leave thematic maps unvalidated, representing no more than one possible representation, one untested hypothesis, of contestable value (Strahler et al., 2006; McRoberts, 2011). Alternatively, and more constructively, techniques that require only relatively limited amounts of reference data could be used. For example, semi-supervised techniques that can make use of unlabelled information could be used in the production of thematic maps from remote sensing (Bruzzone et al., 2006) and model-based rather than standard design-based inference could be adopted in map evaluation (McRoberts, 2010; Foody, 2012). A further alternative is to utilise the enormous potential of citizen sensors. For example, data from citizen observations have already been used as a cost effective alternative to collect reference data for hybrid map generation (Schepaschenko et al., 2015; See et al., 2015).

The role of citizens has been noted in a variety of subjects, from astronomy to zoology (Raddick and Szalay, 2010; Dickinson et al., 2010; Wiersma, 2010; Muller et al., 2015; Rossiter et al., 2015). Citizens have also already contributed greatly to mapping activities, including, for example, to major programmes such as bird species distribution mapping (Dickinson et al., 2010; Wiersma, 2010) and to the pioneering production of national land cover datasets such as the first land utilisation survey of the UK in the 1930s (Parece and Campbell, 2015). The role of citizens in mapping has, however, benefited greatly from recent advances in geoinformation technologies. Technological advancement has fostered the emerging role of the citizen as a source of data. Due to the proliferation of location aware devices and the opportunities of Web 2.0, it is now possible for citizens to easily acquire, share and use geographical information. This activity has been named or described in a variety of ways, notably as crowdsourcing, volunteered geographic information (VGI), user generated spatial content, neogeographies and the pervasive media (See et al., 2016). These various terms are often used to help differentiate between activity that is passive or active, and between information that is truly volunteered or that is being provided for a modest, and possibly non-financial, reward. In this book, there is no particular desire to distinguish between the different approaches, although the detail can sometimes be important, and the focus is simply on citizen-derived geographical data. The citizens contributing data may be anyone: they could be children or adults, they may be amateurs or experts, they may have differing motivations and may even be contributing without knowing so.

Citizen sensing has dramatically affected mapping and map use, impacting on routine daily life activities such as gaming and tourism as well as on science and technology more generally. Resources such as Google Earth, Bing Maps and even maps that are citizen-generated through projects such as OpenStreetMap (OSM) are now widely and routinely used by diverse amateur and professional communities. Furthermore, possibly radical impacts on mapping activity are likely to occur (Olteanu-Raimond et al., 2017) and some argue that a new data-rich paradigm is emerging with VGI (Jiang and Thill, 2015; Li et al., 2016). These future developments should arise from the trend for continued technological advances but also from an increased provision of free, or at least inexpensive, remote sensing data and increasing access to official government data resources. These tremendous opportunities do, of course, come with challenges. In the big data era, there is now, paradoxically, so much data that problems in mapping may arise. The curse of data volume can be likened to the widely encountered Hughes phenomenon, in which map accuracy declines as data dimensionality increases for a fixed ground dataset (Richards, 2013). Immense volumes of data from future remote sensing will amount to a deluge; for example, Sentinel 2 satellites alone will produce 1.6 TB of data per day, and yet they are just one pair of the over 350 Earth observing satellites that are to be launched by 40 different countries by 2023 (Foody et al., 2015). There are also clear challenges with citizen-derived data. These datasets can be voluminous, as with other components of the developing field of big geospatial data, and their size and dynamic nature may need to be recognised explicitly if they are to be used efficiently and effectively (Herrera et al., 2015; Li et al., 2016). Citizen-derived data are also often of varied (and typically unknown) quality and trust levels (Goodchild and Glennon, 2010). Moreover, the data generated may be poorly described and associated with little if any metadata. To realise the full potential of citizen sensing, there is a need to establish good practices and perhaps even protocols for some activities (Schade and Tsinaraki, 2016). This will be a challenging task, not least due to issues such as the diversity of datasets generated, the range of devices used and sensitivities to error and uncertainty, which are often application-specific. Additionally, there are a suite of other major considerations in the use of VGI, including ownership rights, as well as privacy, legal and ethical issues (Granell and Ostermann, 2016). As a further complication, there may be tensions between different parts of the community, with, for example, some calling for anonymity and privacy as an essential feature (Mozas-Calvache, 2016) while others want information on volunteers to be available to aid assessments of trust (Zhao et al., 2016). There is also clearly a strong desire to not 'kill off the golden goose' by laying down strict rules and procedures that end up making volunteering an onerous task and ultimately deter the provision of citizen-derived data. A variety of priorities have been identified that must be addressed in order to facilitate citizen sensing, including issues such as standardisation and interoperability (Brown et al., 2013), and groups are working on defining good practices to encourage mapping-related applications (Pocock et al., 2014a; 2014b). This book reports on some of the activities of one group, the participants of COST Action TD1202. This Action has addressed a wide range of issues connected with citizen sensing in mapping, from advice on photography that might be uploaded to social media sites (Antoniou et al., 2016) to informing the activities of European national mapping agencies (NMAs) (Olteanu-Raimond et al., 2017). The production of the book involved considerable input from the Action and beyond. We are grateful to all who helped bring this book to fruition from authors to publishers but we wish to also highlight here the significant inputs from Bénédicte Bucher who reviewed the manuscript for publication and Nourane Clostre who copyedited it.

2 Outline of the Book

This book is intended to closely reflect the main research themes of COST Action TD1202. One of the first themes addressed was how VGI is acquired, managed, stored and disseminated. Building upon a review that systematically evaluated VGI websites and mobile applications to characterise VGI (See et al., 2016), Chapter 2 provides an overview of different sources of VGI for mapping. The sources are first distinguished by (i) whether the VGI can be considered as framework data (i.e. of the type generally collected by NMAs) or whether they fall into 'other' types of data (e.g. weather and traffic data) and (ii) whether the VGI is actively or passively collected. The chapter then provides a range of examples that illustrate these four types of citizen-contributed data, as well as a brief discussion on 3D VGI. Chapter 3 then discusses one of the most successful VGI projects, which is OSM, and provides a comprehensive introduction to this data source, including how it is being used in a range of services and applications in education, mapping, visualisation and research. The current status and positioning of OSM as a VGI project is also evaluated. The chapter then closes with discussions on future issues that need to be considered by contributors to and users of OSM in order for it to continue its success and growth. In Chapter 4, the emphasis shifts to exploring automated mapmaking with the use of OSM data. The chapter starts by examining why traditional automated mapping processes are not adapted to VGI and describes attempts to solve this problem. The focus then turns towards the level of detail of OSM features and how it can be inferred and harmonised for different features, which aims to aid map generalisation. How other VGI sources, such as geotagged photographs, can help to evaluate the quality of OSM prior to the application of any automatic mapmaking processes is also presented. Finally, issues related to advanced map stylisation with VGI are discussed.

Another prominent theme of the Action has been to gain a better understanding of the motivations of contributors to VGI, and this theme is outlined in Chapter 5. This chapter reviews the literature on motivation and incentives for participation in VGI projects and then presents case studies to reflect on what motivations and incentives have worked well, including how to sustain participation in VGI activities in the longer term. When considering citizens as part of the VGI equation, legal issues and issues such as data privacy and the ethics of data use and reuse immediately come to the forefront. These are discussed in detail in Chapter 6 with specific reference to VGI as a unique source of information.

The quality of citizen-sensor-derived VGI is often a problem, as sources range from naïve, poorly trained citizens to authoritative experts and may even include people contributing erroneous data maliciously. Hence another major theme of the Action has been data quality. It is important to note that VGI can be as good as, if not better than, authoritative datasets in terms of quality (Antoniou and Skopeliti, 2015; See et al., 2013; Dorn et al., 2015). However, even if the data collected could be trusted in terms of features such as their accuracy, there are a variety of other concerns, relating to issues such as the spatial sampling and bias of data collection (Brown, 2017) and the ability to repeat and replicate studies, that may limit the scientific value of the data (Ostermann and Granell, 2017). Much VGI is collected opportunistically and is spatially biased, for instance by digital divides between urban and rural regions or between developed and developing countries (Estima et al., 2014; Neis and Zielstra, 2014). There are also social divides, with most contributions made by young citizens who are technologically savvy (Haworth et al., 2015). Some of the Action's work has focused on how VGI could be usefully used in map validation (Fonte et al., 2015), taking quality considerations into account. In this book, Chapters 7 to 9 all deal with quality-related issues of VGI. Chapter 7 is dedicated to the assessment of VGI quality, and presents the challenges that are raised by this type of data for quality assessment. It provides an overview of how the data quality elements included in the ISO 19157 standard can be applied to VGI as well as of the limitations of these elements. A description of additional indicators that can be used to assess VGI quality is then made. Efforts developed to establish workflows to assess VGI data quality are then presented and discussed, as well as efforts to combine data quality indicators to assess VGI fitness-for-use.

Returning back to OSM, Chapter 8 discusses the evolution of OSM quality from a novel point of view; the chapter deviates from the more traditional quality measurements or quality statistics used in most OSM quality studies and examines the evolution of OSM data quality as a function of the OSM micro-environment, such as OSM specifications and OSM editors. The evolution of OSM specifications, taking into account a number of different factors that directly affect the quality of contributions, is examined. The evolution of OSM editors is also presented, as they are literally the entry point for all OSM contributions. Finally, the combined impact of these two factors on the overall OSM quality is discussed. In Chapter 9, a framework for VGI quality visualisation is presented that supports both the communication and the exploration of VGI quality. This framework is based on four factors: the available methods for quality visualisation of spatial data; the nature of VGI data quality; user profiles; and the visualisation environment. The chapter then discusses how the framework can be implemented with VGI data.

One critical issue related to the diversity and quality of spatial data is the need to develop good practices. Here, there is a tension between the desire to encourage volunteers without constraining their involvement and the desire to acquire useful data. The latter could be aided by the specification of best practices or even protocols, but if these become too onerous they may actually act to deter volunteers. Since, for example, much current VGI is derived from geotagged photographs and from vector data, such as in the OSM project, the proposal of good practices for key mapping-related activities is one major way in which the Action has helped contribute to the development of the subject. Thus, Chapter 10 explores the role of protocols as tools to guide data collection in VGI projects with the purpose of increasing the quality of user contributions. With the help of technology, protocols should balance the opposing needs of providing VGI contributors with detailed instructions and keeping intact their enthusiasm and motivation. With this in mind, a general protocol is formalised, and specific, real-world applications of the protocol are presented. In Chapter 11, the means by which citizen-generated data may be published and documented to make these datasets discoverable and reusable for robust and reproducible science is investigated. The current state of the art is assessed, with particular attention to the role and adoption of Data Management Plans for citizen science initiatives and observatories. The relevance and availability of existing data and metadata standards, vocabularies and tools which can be employed to support interoperable storage and dissemination of VGI are evaluated, and reference is made to examples of good practice from existing infrastructures. Finally, in Chapter 12, the challenges of integrating VGI with the Infrastructure for Spatial Information in the European Community (INSPIRE) directive are discussed, contrasting Spatial Data Infrastructures (SDIs) with VGI. This is followed by a discussion of the set of critical issues that arise when integrating INSPIRE and VGI and of what the prospects for integration are, providing illustrative examples. Finally, a conceptual framework is presented for what an SDI-VGI integrated GIS platform could look like.

A final theme in the Action has been the role of citizen sensing in map production. The research undertaken was aimed at defining the needs of the map producing community, identifying the sensitivity and tolerance of mapping methods to different types of error and uncertainty in VGI, and assessing the potential role of current VGI efforts as well as of active citizen sensing in the activities of NMAs. A survey of key map producers, notably European NMAs, was undertaken to establish their current and potential future use of VGI to inform their work (Olteanu-Raimond et al., 2017). Chapter 13 builds upon this work and provides an overview of the experiences of some European NMAs in engaging with VGI. It also provides recommendations to support wider engagement with the VGI community and to help ensure that the potential of VGI in mapping is fully exploited and used in the workflows of NMAs in the future. Switching to another public stakeholder, i.e. urban planners, Chapter 14 discusses the value and opportunities of VGI, and of its more passive equivalent, social media geographic information (SMGI), for urban planning. A number of examples are provided to illustrate how this new source of information can be used to improve visualisation, planning processes, evaluation of plans and decision-making. The use of VGI and SMGI in smart cities initiatives is also examined. One recent trend has been towards the development of citizen observatories and hence Chapter 15 discusses their increasing role in engaging citizens in science, environmental monitoring and policy-making. The chapter provides an overview of existing and planned citizen observatories and of where further developments are happening at the European front. The chapter closes with a discussion of the key challenges and development needs for policy- and decision-makers in the future.

The term VGI has been in existence for only a decade, yet the number of new applications and the increased involvement of citizens in mapping and environmental monitoring has literally exploded. The final chapter of the book examines what the future trends in VGI might be and the increasing role that smart cities and society will play in this innovative area. It is clear that the future for VGI is very bright; the key is to not waste these valuable citizen-based resources but to find ways to maximise the synergies between stakeholders across multiple levels of society.

Notes

[1] http://www.citizensensor-cost.eu/

Reference list

Antoniou, V., Fonte, C., See, L., Estima, J., Arsanjani, J., Lupia, F., Minghini, M., Foody, G., Fritz, S., 2016. Investigating the feasibility of geo-tagged photographs as sources of land cover input data. *ISPRS International Journal of Geo-Information* 5, 64. DOI: https://doi.org/10.3390/ijgi5050064

Antoniou, V., Schlieder, C., 2014. Participation patterns, VGI and gamification, in: Proceedings of AGILE 2014. Presented at the AGILE 2014, Castellón, Spain, pp. 3–6.

Antoniou, V., Skopeliti, A., 2015. Measures and indicators of VGI quality: An overview, in: ISPRS Annals of the Photogrammetry, Remote Sensing and Spatial Information Sciences. Presented at the ISPRS Geospatial Week 2015, ISPRS Annals, La Grande Motte, France, pp. 345–351.

Boyd, D.S., Foody, G., 2014. Volunteered geographic information. *Geography* 99, 157–160.

Brown, G., 2017. A review of sampling effects and response bias in Internet participatory mapping (PPGIS/PGIS/VGI). *Transactions in GIS* 21, 39–56. DOI: https://doi.org/10.1111/tgis.12207

Brown, M., Sharples, S., Harding, J., Parker, C.J., Bearman, N., Maguire, M., Forrest, D., Haklay, M., Jackson, M., 2013. Usability of geographic information: Current challenges and future directions. *Applied Ergonomics* 44, 855–865. DOI: https://doi.org/10.1016/j.apergo.2012.10.013

Bruzzone, L., Chi, M., Marconcini, M., 2006. A novel transductive SVM for semisupervised classification of remote-sensing images. *IEEE Transactions on Geoscience and Remote Sensing* 44, 3363–3373. DOI: https://doi.org/10.1109/TGRS.2006.877950

Committee on Strategic Directions for the Geographical Sciences in the Next Decade (CSDGSND), 2010. *Understanding the Changing Planet: Strategic Directions for the Geographical Sciences*. National Academies Press, Washington, DC, USA.

Dickinson, J.L., Zuckerberg, B., Bonter, D.N., 2010. Citizen science as an ecological research tool: Challenges and benefits. *Annual Review of Ecology, Evolution, and Systematics* 41, 149–172. DOI: https://doi.org/10.1146/annurev-ecolsys-102209-144636

Dorn, H., Törnros, T., Zipf, A., 2015. Quality evaluation of VGI using authoritative data—A comparison with land use data in Southern Germany. *ISPRS International Journal of Geo-Information* 4, 1657–1671. DOI: https://doi.org/10.3390/ijgi4031657

Estima, J., Fonte, C.C., Painho, M., 2014. Comparative study of Land Use/Cover classification using Flickr photos, satellite imagery and Corine Land Cover database, in: Proceedings of the 17th AGILE International Conference on Geographic Information Science. Presented at the 17th AGILE International Conference on Geographic Information Science, Castellon, Spain.

Fonte, C.C., Bastin, L., See, L., Foody, G., Lupia, F., 2015. Usability of VGI for validation of land cover maps. *International Journal of Geographical Information Science* 1–23. DOI: https://doi.org/10.1080/13658816.2015.1018266

Foody, G., 2012. Latent class modeling for site- and non-site-specific classification accuracy assessment without ground data. *IEEE Transactions on Geoscience and Remote Sensing* 50, 2827–2838. DOI: https://doi.org/10.1109/TGRS.2011.2174156

Foody, G., Mooney, P., See, L., Kerle, N., Olteanu-Raimond, A.-M., Fonte, C.C., 2015. *Current Status and Future Trends in Crowd-Sourcing Geographic Information*. Working Paper of COST Action TD1202. University of Nottingham, Nottingham, UK.

Goodchild, M.F., Glennon, J.A., 2010. Crowdsourcing geographic information for disaster response: a research frontier. *International Journal of Digital Earth* 3, 231–241. DOI: https://doi.org/10.1080/17538941003759255

Granell, C., Ostermann, F.O., 2016. Beyond data collection: Objectives and methods of research using VGI and geo-social media for disaster management. *Computers, Environment and Urban Systems* 59, 231–243. DOI: https://doi.org/10.1016/j.compenvurbsys.2016.01.006

Haworth, B., Bruce, E., Middleton, P., 2015. Emerging technologies for risk reduction: assessing the potential use of social media and VGI for increasing community engagement. *Australian Journal of Emergency Management* 30, 36–41.

Herrera, F., Sosa, R., Delgado, T., 2015. *GeoBI and big VGI for crime analysis and report*. IEEE, pp. 481–488. DOI: https://doi.org/10.1109/FiCloud.2015.112

Jiang, B., Thill, J.-C., 2015. Volunteered Geographic Information: Towards the establishment of a new paradigm. *Computers, Environment and Urban Systems*, Special Issue on Volunteered Geographic Information 53, 1–3. DOI: https://doi.org/10.1016/j.compenvurbsys.2015.09.011

Li, S., Dragicevic, S., Castro, F.A., Sester, M., Winter, S., Coltekin, A., Pettit, C., Jiang, B., Haworth, J., Stein, A., Cheng, T., 2016. Geospatial big data handling theory and methods: A review and research challenges. *ISPRS Journal of Photogrammetry and Remote Sensing* 115, 119–133. DOI: https://doi.org/10.1016/j.isprsjprs.2015.10.012

McRoberts, R.E., 2011. Satellite image-based maps: Scientific inference or pretty pictures? *Remote Sensing of Environment* 115, 715–724. DOI: https://doi.org/10.1016/j.rse.2010.10.013

McRoberts, R.E., 2010. Probability- and model-based approaches to inference for proportion forest using satellite imagery as ancillary data. *Remote Sensing of Environment* 114, 1017–1025. DOI: https://doi.org/10.1016/j.rse.2009.12.013

Mozas-Calvache, A.T., 2016. Analysis of behaviour of vehicles using VGI data. *International Journal of Geographical Information Science* 0, 1–20. DOI: https://doi.org/10.1080/13658816.2016.1181265

Muller, C. l., Chapman, L., Johnston, S., Kidd, C., Illingworth, S., Foody, G., Overeem, A., Leigh, R. R., 2015. Crowdsourcing for climate and atmospheric sciences: current status and future potential. *International Journal of Climatology* 35, 3185–3203. DOI: https://doi.org/10.1002/joc.4210

Neis, P., Zielstra, D., 2014. Recent developments and future trends in volunteered geographic information research: The case of OpenStreetMap. *Future Internet* 6, 76–106. DOI: https://doi.org/10.3390/fi6010076

Olofsson, P., Foody, G., Herold, M., Stehman, S.V., Woodcock, C.E., Wulder, M.A., 2014. Good practices for estimating area and assessing accuracy of land change. *Remote Sensing of Environment* 148, 42–57. DOI: https://doi.org/10.1016/j.rse.2014.02.015

Olteanu-Raimond, A.-M., Hart, G., Foody, G., Touya, G., Kellenberger, T., Demetriou, D., 2017. The scale of VGI in map production: A perspective of European National Mapping Agencies. *Transactions in GIS* 21, 74–90. DOI: https://doi.org/10.1111/tgis.12189

Ostermann, F.O., Granell, C., 2017. Advancing science with VGI: Reproducibility and replicability of recent studies using VGI: Reproducibility and replicability of VGI studies. *Transactions in GIS* 21, 224–237. DOI: https://doi.org/10.1111/tgis.12195

Parece, T.E., Campbell, J.B., 2015. Land use/land cover monitoring and geospatial technologies: An overview, in: Younos, T., Parece, T.E. (Eds.), *Advances in Watershed Science and Assessment*. Springer International Publishing, Cham, pp. 1–32.

Pocock, M.J.O., Chapman, D.S., Sheppard, L.J., Roy, H.E., 2014a. *A Strategic Framework to Support the Implementation of Citizen Science for Environmental Monitoring. Final Report to SEPA*. Centre for Ecology & Hydrology, UK.

Pocock, M.J.O., Chapman, D.S., Sheppard, L.J., Roy, H.E., 2014b. *Choosing and Using Citizen Science: a guide to when and how to use citizen science to monitor biodiversity and the environment*. Centre for Ecology & Hydrology, UK.

Raddick, M.J., Szalay, A.S., 2010. The universe online. *Science* 329, 1028–1029. DOI: https://doi.org/10.1126/science.1186936

Rahmatizadeh, S., Rajabifard, A., Kalantari, M., 2016. A conceptual framework for utilising VGI in land administration. *Land Use Policy* 56, 81–89. DOI: https://doi.org/10.1016/j.landusepol.2016.04.027

Richards, J.A., 2013. *Remote Sensing Digital Image Analysis*. Springer-Verlag, Berlin, Heidelberg.

Rossiter, D.G., Liu, J., Carlisle, S., Zhu, A.-X., 2015. Can citizen science assist digital soil mapping? *Geoderma* 259–260, 71–80. DOI: https://doi.org/10.1016/j.geoderma.2015.05.006

Schade, S., Tsinaraki, C., 2016. *Survey Report: Data Management in Citizen Science Projects, EUR 27920 EN*. European Commission, Luxembourg. DOI: https://doi.org/10.2788/539115.

Schepaschenko, D., See, L., Lesiv, M., McCallum, I., Fritz, S., Salk, C., Moltchanova, E., Perger, C., Shchepashchenko, M., Shvidenko, A., Kovalevskyi, S., Gilitukha, D., Albrecht, F., Kraxner, F., Bun, A., Maksyutov, S., Sokolov, A., Dürauer, M., Obersteiner, M., Karminov, V., Ontikov, P., 2015. Development of a global hybrid forest mask through the synergy of remote sensing, crowdsourcing and FAO statistics. *Remote Sensing of Environment* 162, 208–220. DOI: https://doi.org/10.1016/j.rse.2015.02.011

See, L., Comber, A., Salk, C., Fritz, S., van der Velde, M., Perger, C., Schill, C., McCallum, I., Kraxner, F., Obersteiner, M., 2013. Comparing the quality of crowdsourced data contributed by expert and non-experts. *PLoS ONE* 8, e69958. DOI: https://doi.org/10.1371/journal.pone.0069958

See, L., Fritz, S., Perger, C., Schill, C., McCallum, I., Schepaschenko, D., Duerauer, M., Sturn, T., Karner, M., Kraxner, F., Obersteiner, M., 2015. Harnessing the power of volunteers, the internet and Google Earth to collect and validate global spatial information using Geo-Wiki. *Technological Forecasting and Social Change* 98, 324–335. DOI: https://doi.org/10.1016/j.techfore.2015.03.002

See, L., Mooney, P., Foody, G., Bastin, L., Comber, A., Estima, J., Fritz, S., Kerle, N., Jiang, B., Laakso, M., Liu, H.-Y., Milčinski, G., Nikšič, M., Painho, M., Pődör, A., Olteanu-Raimond, A.-M., Rutzinger, M., 2016. Crowdsourcing, citizen science or Volunteered Geographic Information? The current state of crowdsourced geographic information. *ISPRS International Journal of Geo-Information* 5, 55. DOI: https://doi.org/10.3390/ijgi5050055

Strahler, A.H., Boschetti, L., Foody, G., Friedl, M.A., Hansen, M.C., Herold, M., Mayaux, P., Morisette, J., Stehman, S.V., Woodcock, C.E., 2006. *Global Land Cover Validation: Recommendations for Evaluation and Accuracy Assessment of Global Land Cover Maps (No. EUR 22156 EN).* European Commission.

Wiersma, Y.F., 2010. Birding 2.0: Citizen science and effective monitoring in the Web 2.0 world. *Avian Conservation and Ecology* 5. DOI: https://doi.org/10.5751/ACE-00427-050213

Zhao, Y., Zhou, X., Li, G., Xing, H., 2016. A spatio-temporal VGI model considering trust-related information. *ISPRS International Journal of Geo-Information* 5, 10. DOI: https://doi.org/10.3390/ijgi5020010

CHAPTER 2

Sources of VGI for Mapping

Linda See*, Jacinto Estima†, Andrea Pődör‡, Jamal Jokar Arsanjani§, Juan-Carlos Laso Bayas* and Rumiana Vatseva¶

*International Institute for Applied Systems Analysis (IIASA), Schlossplatz 1, 2361 Laxenburg, Austria, see@iiasa.ac.at

†NOVA IMS, Universidade Nova de Lisboa, 1070-312, Lisbon, Portugal

‡Institute of Geoinformatics, Óbuda University Alba Regia Technical Faculty, Székesfehérvár 8000, Hungary

§Department of Planning and Development, Aalborg University Copenhagen, A.C. Meyers Vænge 15, DK-2450 Copenhagen, Denmark

¶National Institute of Geophysics, Geodesy and Geography, Bulgarian Academy of Sciences, Bulgaria

Abstract

The concept of Volunteered Geographic Information (VGI) is often exemplified by the mapping of features in OpenStreetMap (OSM), yet there are many other sources of VGI available. Some VGI is very focused on the creation of map-based products, while in other applications location is simply one attribute that is routinely collected, due to the proliferation of Global Positioning System (GPS) enabled devices, e.g. mobile phones and tablets. This chapter aims to provide an overview of the variety of sources of VGI currently available, categorised according to whether they can contribute to framework data (i.e. the type of data that are commonly part of the spatial data infrastructure of national mapping agencies and governments) or not and whether the data have been actively or passively collected. A range of examples are presented to illustrate the different types of VGI in each of

How to cite this book chapter:

See, L, Estima, J, Pődör, A, Arsanjani, J J, Bayas, J-C L and Vatseva, R. 2017. Sources of VGI for Mapping. In: Foody, G, See, L, Fritz, S, Mooney, P, Olteanu-Raimond, A-M, Fonte, C C and Antoniou, V. (eds.) *Mapping and the Citizen Sensor*. Pp. 13–35. London: Ubiquity Press. DOI: https://doi.org/10.5334/bbf.b.

these main categories. Finally, the chapter discusses some of the main issues surrounding the use of VGI and points to chapters in the book where these issues are described in more detail.

Keywords

Volunteered Geographic Information, framework data, active data collection, passive data collection, crowdsourcing

1 Introduction

Crowdsourced mapping and citizen-driven spatial data collection are radically changing the relationship between traditional map production and those individuals and organisations that consume the data. In the past, authoritative maps such as road networks and building footprints were firmly in the domain of national mapping agencies (NMAs), where the maps were created by professionals. Today NMAs still fulfil this role but they face a relatively new, citizen mapping community, armed with online mapping tools, open access to very-high-resolution satellite imagery/aerial photography and mobile devices with GPS (Global Positioning System) for geotagging features. The result has been an abundance of maps that are created by citizens and a blurring of the traditional boundaries between map producers and consumers, as citizens take on the dual role of production and consumption (Coleman et al., 2009; See et al., 2016b).

At the same time, citizens have become empowered to collect and map features and objects that are not traditionally mapped by NMAs, such as sentiments and hiking/biking routes, among many others. OpenStreetMap (OSM) is one of the most successful and most commonly cited examples (e.g. Fan et al., 2016; Hagenauer and Helbich, 2012; Haklay, 2010; Jokar Arsanjani et al., 2015b; Mooney and Corcoran, 2013) of this new phenomenon, referred to in the geographical literature as Volunteered Geographic Information (VGI), a term originally coined by Goodchild (2007). Numerous other terms have been proposed that refer to similar phenomena, all of which have citizens and citizen participation at their core. In the field of geography and urban planning, public participation in Geographic Information Systems (PPGIS) appeared in the late 1990s, as a way of improving the public consultation experience and fostering public engagement (Kingston et al., 2000; Sieber, 2006) and can be thought of as a precursor to VGI, when Web 2.0 technologies and online mapping were still in their infancy. In other fields, for example in ecology, conservation and biodiversity monitoring, there has been a long tradition of citizen involvement in science, such as the Audubon Society's Christmas Bird Count, which started in the 1900s (LeBaron, 2007). In these domains, citizen involvement has commonly been referred to as public participation in scientific research

(PPSR) (Bonney et al., 2009a) and more recently as citizen science (Bonney et al., 2009b), where data collection, often geotagged, is only one component of citizen participation. In yet another domain, i.e. that of the business world, the term crowdsourcing has emerged to refer to the outsourcing of tasks to the crowd (Howe, 2006). Crowdsourcing can be used for financial remuneration (Buhrmester et al., 2011) or for other, more altruistic reasons, e.g. searching for the remains of the Malaysian Airways plane that went missing in 2014 (Whittaker et al., 2015) or providing hotel and restaurant reviews on sites like TripAdvisor; other initiatives can be found in Sester et al. (2014).

Many other terms exist and the reader is referred to a recent review by See et al. (2016b) for a broader overview. For the purpose of this book, we use the term VGI to mean geotagged data contributed by citizens, whether map-based or where location is simply an attribute in a much larger dataset. The term covers many different domains of activities, from monitoring the weather to species identification and georeferencing old historical maps contained in digital libraries. This chapter aims to provide an overview of the variety of sources of VGI currently available, categorised according to whether they are framework data (i.e. the type of data that are commonly part of the spatial data infrastructure of national mapping agencies and governments) or not and whether the data have been actively or passively collected, as outlined in Section 2 below. A range of examples is then presented in Section 3 to illustrate the different types of VGI in each of these main categories. Finally, the main issues that currently surround VGI are highlighted, providing a link to different chapters in the book that describe these issues in more detail.

2 Categorisation of VGI Sources for Mapping

To help organise the diverse range of VGI sources available for mapping, we have categorised them based on two main criteria. The first one is whether the data fall into the territory of NMAs; we refer here to such data as 'framework data'. Framework data are typically data that are collected by government agencies, and which can be organised into the following themes: geodetic control, orthoimagery, elevation, transportation, hydrography, governmental units and cadastre, and comprise the basic components of a government's spatial data infrastructure (SDI; Elwood et al., 2012). These data will be collected by professionals and have minimum levels of error specified in their production, with update cycles that depend on national budgets but will generally range from one to five years. Depending on the country, the content of these datasets may also vary; for example, some countries do not have cadastres, while others may include a gazetteer as part of their SDI. In the European Union, the INSPIRE (Infrastructure for Spatial Information in Europe) Directive specifies the types of framework data that all EU member states should collect (EC, 2007); the type of data specified in the Directive's Annexes I and II corresponds to the types of data outlined in Elwood et al. (2012), but Annex II additionally includes land cover and geology, and Annex III contains much more detail in terms of land use and socio-economic data. For the purpose of this chapter, however, we take framework data to mean the most basic components of an SDI as outlined by Elwood et al. (2012).

The second criterion is whether the data have been contributed actively or passively (Harvey, 2013). Active data collection includes campaigns that call for participation or where people sign up to complete micro-tasks with the full knowledge that they are contributing the data for a specific purpose, e.g. the active mapping of features in OSM. In passive mode, participants may be providing geotagged information willingly, e.g. through social media, but the data may then be used for purposes, such as for behavioural studies or marketing purposes, that contributors are unaware of since they did not read the terms of participation in detail or modify their privacy settings (if available). Examples of this are geotagged tweets from Twitter, geotagged photographs from Flickr and Instagram, etc. There is a tradeoff between the two data sources; active data are often easier to process since they were collected with a specific purpose in mind and often with some type of protocol or minimum data requirements, while passive data may not meet the minimum requirements of an application. In addition, passive data can be 'big data' in terms of volume and complexity, but may thus also require considerable post-processing before use. Regardless of how the data are collected, the importance of this new wave of data collection, i.e. VGI, for the public and private sectors and for scientific research is yet to be truly exploited.

Using these two criteria to categorise VGI, i.e. framework vs. non-framework data and active vs. passive data collection, there are four categories in which VGI can fall. The first category is VGI that can contribute to framework data and that is actively contributed by volunteers. In this category fall projects that can be used to update or correct the types of data routinely collected by NMAs; the category is represented by the upper right quadrant of Figure 1. The second category is non-framework data (or data that are not routinely collected by NMAs but are useful for other agencies and scientific research) where active participation by volunteers is evident; it is located in the bottom right quadrant of Figure 1. The left half of Figure 1 contains the other two categories, i.e. framework and non-framework data that are passively collected, e.g. through social media or sensors such as the GPS of a mobile phone. The four quadrants in Figure 1 are then populated with different sources of VGI; examples of these sources are provided in Section 3. Note that the exact location of the VGI examples within each quadrant has no significance – they are simply arranged for optimal readability. A fifth category has been added to consider three-dimensional VGI; although this type of VGI could also be characterised by the two criteria introduced in this section, we provide a separate discussion of it, focused on height data, OSM and publicly available sources of elevation, in Section 3.5, since this is a new area of VGI.

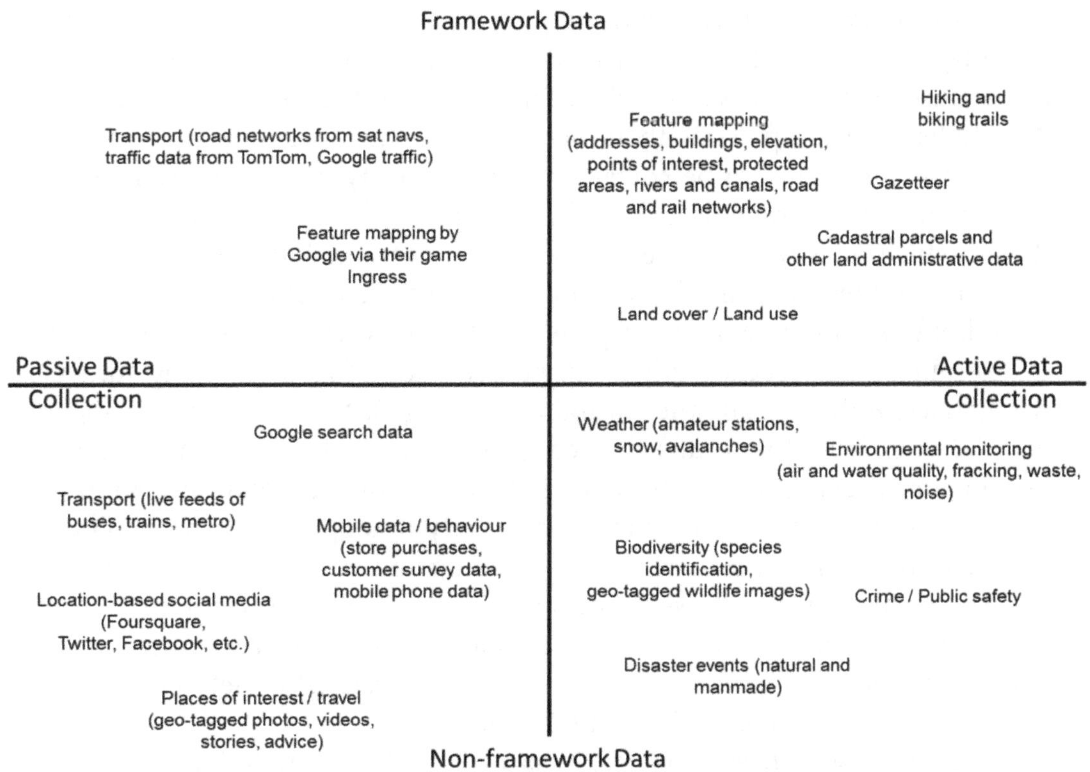

Fig. 1: Categorisation of VGI based on whether it consists of framework or non-framework data and whether the data have been actively or passively collected. This figure is modified from See et al. (2016b).

3 Examples of VGI Sources for Mapping

3.1 Active Framework Data

OSM, as already mentioned, is one of the most successful and commonly cited examples of VGI sources, and aims at creating a world map freely available to anyone (Jokar Arsanjani et al., 2015b). OSM is a prime example of feature mapping and covers data types often found in topographic databases and transportation networks; an extensive overview of this initiative is provided in Chapter 3 of this book (Mooney and Minghini, 2017). Google Map Maker[1] is another example of an application that allows volunteers to map features such as roads and points of interest (POI). These are then displayed on Google Maps in certain countries where the review process is well developed enough to ensure a minimum level of quality.

A second example of active framework data contributed by citizens is the mapping of cadastral boundaries and properties (Kalantari and La, 2015). This is particularly relevant for developing countries where land rights are not well documented. This is also relevant in places where surveying is very expensive and time-consuming and so has not been carried out in all areas, which leads to a stagnation in the property market. An example from Greece is outlined by

Basiouka and Potsiou (2012), who conducted an experiment in the rural part of the village of Tsoukalades, on the island of Lefkada, where fifteen volunteer land owners used a handheld GPS to delineate their land parcel boundaries. When the results were compared with an official survey, the locations and shapes of all parcels were found to be correct and the majority of the parcels had area calculations that were within the tolerance limits of the specifications set by the Hellenic Cadastre. Moreover, the land owners wanted to be involved in the collection of these data and hence motivation was high. Thus, citizen involvement holds great potential for helping to gather this type of framework information. In a more recent study by Basiouka et al. (2015), surveying students were tasked with assessing the feasibility of using OSM for cadastral mapping in Athens, Greece. The results showed good accuracy, low costs, and ease-of-use for non-experts, indicating that OSM is one possible solution for crowdsourcing land parcels and features, particularly if adopting a hybrid solution in which surveying experts are used in training and quality assurance. Mobile phones can also be used for securing land rights; GeoODK (Geographic Open Data Kit) is an Android-based mobile phone app for spatial and attribute data collection that is being used by the Cadasta Foundation[2] to help people map their lands and resources and assert their rights.

In the area of gazetteers, Wikimapia[3] is a very well known initiative that aims to describe places in the world (Goodchild, 2007). It is freely available and all the content is provided by volunteers. Users can mark places, add descriptions with links and upload and categorise photos. Entries are then voted on by a group of peers. To access the raw data, the Wikimapia API and Motomapia[4] are available. GeoNames[5] is another gazetteer, containing over 10 million geographical names and available to download free of charge: volunteers can contribute by editing existing names or adding new names through the GeoNames website.

Mapping of land cover and land use is another area of framework data. Some of the current authoritative products have been created globally, e.g. GlobeLand30 (Chen et al., 2015); regionally, such as CORINE land cover[6] for EU countries or AFRICOVER for some African countries (FAO, 1998); and nationally by NMAs, e.g. the land cover map of Great Britain produced by the Centre for Ecology and Hydrology (Fuller et al., 2002). These authoritative products use satellite and aerial imagery in combination with different types of classification algorithms, and there is often a long period of time between updates due to the difficulty of the task. One problem that has been highlighted by researchers is that when these maps are compared spatially, there are often areas where they disagree (Fritz et al., 2011). Several efforts have been undertaken to tackle this problem, with a promising contribution from VGI. For example, the Geo-Wiki tool[7] for crowdsourcing land cover data asks volunteers to interpret very-high-resolution satellite imagery from Google Earth and Bing to increase the amount of in-situ data for producing and validating land cover products (Fritz et al.,

2012; See et al., 2015). One of the latest Geo-Wiki applications is called Foto-Quest Austria[8], and, in contrast to the online Geo-Wiki applications, encourages volunteers to go out into the field and collect land cover and land use information using a mobile app. The idea behind the project is to see whether volunteers can collect in-situ data based on the Land Use and Coverage Area frame Survey (LUCAS) protocol (Eurostat, 2015) and complement this authoritative data source. LUCAS is currently the only official validation dataset for products such as CORINE land cover and the very-high-resolution (VHR) layers produced as part of the Copernicus land monitoring service (Büttner and Eiselt, 2013; Gallego, 2011). Thus, any additional in-situ data have great value for calibration and validation of products from Earth Observation, especially in terms of density and frequency of updating (See et al., 2016a). Initial results from a comparison of land cover and land use data collected from the app with the authoritative LUCAS data indicate that volunteers are able to identify basic land cover and land use types on the ground but that more detailed land cover types will require some training (Laso Bayas et al., 2016). The app is currently being rolled out to other EU countries. Similar tools to Geo-Wiki have been developed by other research teams. For example, the VIEW-IT application (Clark and Aide, 2011) is a collaborative effort to record reference information on land use and land cover, while Google Earth Grids (Jacobson et al., 2015) allows users to create an interactive and user-specified grid over Google Earth imagery and identify the land cover in each square of the grid.

As shown in Figure 1, a final area where VGI has been used to actively map framework data is that of biking and hiking trails (which may or may not appear in the topographic databases of NMAs; thus this category could also be included in active non-framework data). An example of such an initiative is MapMyFitness[9], which is a suite of mobile apps and websites that provide interactive tools to map and share fitness activities including running, walking, cycling and hiking[10]. Each of these provide paths and trails that could be incorporated into the topographic database of an NMA. Bikemap[11] and Bikely[12] are other examples of initiatives to map bike routes, with many more examples to be found online. Bikemap has more than 2.8 million cycling routes available, where the routes are accessible via the web interface and also through the API, while routes in Bikely can be accessed via the web interface or downloaded in GPX and KML formats. Finally, there are many hiking sites available. An example is AllTrails[13], which is a platform for sharing geotagged user-generated travel content. Travel experiences are shared through an interactive map and can include photographs plotted along the trip route; mobile apps and a developer API are available to access the platform and manage the data. Wikiloc[14], with more than 2 million users, around 5 million outdoor trails and 8 million photographs, is very popular for discovering and sharing the best trails for outdoor activities, and offers routes and waypoints (POIs) along with elevation profiles, distances and images taken.

3.2 Active Non-framework Data

In contrast to active framework data, there are many diverse examples of initiatives for active non-framework data. It is not possible to comprehensively list all of them or even touch upon every domain in which these initiatives are emerging, as this is a very dynamic area: the reader is referred to sites such as those of SciStarter[15] and the Citizen Science Alliance[16], which are portals to many other citizen science projects. Not all are spatially-oriented but location is usually a key attribute collected by citizens. Here we have chosen to focus on five main areas shown in Figure 1: weather, biodiversity, environment, disasters and crime.

Amateur weather stations are a prime example of active data contributions and have become important sources of information for applications in hydrology, drought, agriculture, engineering and architecture, among others (Doesken and Reges, 2010). The US National Weather Service Cooperative Observer Program is a weather and observing network of more than 8,700 volunteers who provide observations from farms, urban areas, national parks, coastlines and mountaintops within the US (Leeper et al., 2015). There are other similar initiatives, such as the Citizen Weather Observer Program[17], which collects data from more than 7,000 stations in North America and sends around 50,000 to 75,000 observations every hour, and Weather Underground[18], which is a weather service that provides real-time weather information for free over the Internet and incorporates data from more than 200,000 personal weather stations around the world. Other notable initiatives include CoCoRaHS, which is a community-based network of volunteers who measure and map precipitation in the form of rain, hail and snow, and a mobile app called mPING[19], which allows users to contribute weather reports. As of mid-2015, CoCoRaHS volunteers have submitted over 31 million daily precipitation reports and tens of thousands of reports of hail, heavy rain and snow (Reges et al., 2016), while the data collected through mPING are used to fine-tune weather forecasts.

Biodiversity monitoring is the second area where volunteers have been actively contributing non-framework data. There are hundreds of different citizen science projects in this area, mainly because there is a long history of citizen involvement in conservation, as mentioned previously. Some of these are local projects, collecting data on a small scale, while others have more global reach. An example of a more local project is the Invaders of Texas Program, where citizen scientists are trained to detect the arrival and dispersal of invasive species and report them using the online mapping database (Gallo and Waitt, 2011). iSpot[20] and iNaturalist[21] are initiatives with global reach and both have mobile apps for data collection, where the data collected by citizens have been used in scientific research (e.g. Silvertown et al., 2015).

Citizens are also active in monitoring the environment. Global Water Watch[22], which is a voluntary network that monitors surface waters for the improvement of both water quality and public health, is a prime example of such monitoring. Another example is the Global Learning and Observations to Benefit the Environment (GLOBE) Program, which aims to increase environmental awareness and to actively involve schools in science; there, students perform measurements that are of research quality and report their observations to archives designed for the study of the Earth. Since 1995, the GLOBE network has grown to include representatives from 112 countries. One of the environmental parameters measured in the framework of the GLOBE Program is air pollution in terms of aerosols. In addition to creating awareness about aerosols and their role in climate and air quality, the measurements can be of significant value for validation of satellite products (Brooks and Mims, 2001; Boersma and de Vroom, 2006). More recently, the EU has funded four citizen observatories[23] covering different aspects of citizen-based environmental monitoring: Citi-Sense (air pollution); Omniscentis (odours); CobWeb (land cover and land use); and WeSenseIt (flooding).

Another environmental issue in cities, especially in dense urban areas, is noise, which can become a public health issue in extreme cases. NoiseWatch[24] is a citizen science project supported by the European Environment Agency that integrates noise data from official scientific sources with noise data collected from crowdsourced observations. A mobile application can be used by citizens to measure the level of noise in their location, which is automatically uploaded to a central database. These data can then be used to develop noise maps for decision-making. Finally, in the area of light pollution, the Cities at Night[25] initiative is a citizen science project to help georeference photographs of cities taken by astronauts on the International Space Station at night. Using these images, it is possible to compare the efficiency of lighting across different cities on the planet as well as study their light pollution, which can have a negative effect on ecosystems and health (Falchi et al., 2011).

The fourth area of active non-framework data collection is in disaster mapping. The Humanitarian OpenStreetMap Team (HOT)[26] is an initiative that rallies a huge network of volunteers when disaster strikes to create maps that enable responders to reach those in need. HOT was launched after the January 12, 2010 Haiti earthquake, when 600 remotely located volunteer mappers built a base layer map to support the aid effort (Soden and Palen, 2014). HOT volunteers were also effectively mobilised during the November 8, 2013 Typhoon Yolanda in the Philippines (Palen et al., 2015). Going back to earthquakes, Did You Feel It?[27] is an initiative from the United States Geological Survey (USGS) that maps where earthquakes were experienced by individuals and the severity of the damage. Any citizen who feels an earthquake can report it online by selecting the earthquake from a real-time map of earthquakes and filling in a survey with detailed questions on their experiences as well as their location.

The final area being considered here is crime and public safety. Citizens are willing to contribute especially when they feel threatened. Alertos[28] is a citizen observation platform to report crime and similar events to the legal authorities in Guatemala, Latin America. An interactive map showing reported events by category and time is also available on the website. WikiCrimes[29] is a collaborative wiki-type initiative to report crime events of different categories through the website. Such events can then be visualised and filtered using an interactive map. Mobile apps are also available to provide users with information on the safety of a place based on the analysis of the reported events. CrimeReports[30] and SpotCrime[31] are examples of similar initiatives for reporting data on different types of crimes in the US, Canada and the UK. Emotional and perception mapping is another area where initiatives have emerged to understand the level of security perceived by citizens and their spatial distribution. Measuring the fear of crime has been undertaken as part of a research project developed at Óbudai University Alba Regia Technical Faculty Institute of Geoinformatics: contributors are asked to fill an online survey[32] and draw a red or grey polygon to report that they are feeling respectively unsafe or safe. Finally, the Ushahidi platform[33] has been used to map reports of violence in Kenya after the post-election violence in 2008. Since then several initiatives have used this platform to empower citizens to report different events, e.g. the Map it. End it[34] initiative to map technology-related violence against women and the Egyptian Zabatak[35] initiative.

3.3 Passive Framework Data

There are not many examples of passive framework data collection but such collection does exist, e.g. through the Google Traffic application: through a smartphone with the Google Maps app installed and the location functionality activated, users continuously send Google anonymous data on how fast they are moving. Google then analyses the data coming in from the same location and sends back accurate information on traffic conditions. Such information on traffic volumes and hotspots can be used to improve road planning (see e.g. Barth, 2009) as well as road mapping (Ekpenyong et al., 2009). Satellite navigation companies also gather traffic and travel data from their customers' devices in a passive mode. In addition, the TomTom satellite navigation company has developed the Map Share Reporter[36] as a way of allowing customers to make active changes to the map and share these with other TomTom users. Thus, they are crowdsourcing improvements to their product.

Another example is the crowdsourcing of features using gamification via the Google Ingress game[37] to improve Google Maps. The idea behind the game is to find a portal and capture it. In the process of doing this, players are asked to travel on specific routes and photograph locations or features along their way to the portal. In this way Google gathers information from the players. The main goal of the players is to gain control over the portals and have fun, so the data collection has been seamlessly integrated into the game.

This is an example of a very cleverly disguised way of updating map features through crowdsourcing.

3.4 Passive Non-framework Data

Several examples can be found in the category of non-framework data contributed passively by citizens, and can be mapped and analysed for different applications. The Google search engine is used approximately 3.5 billion times per day[38], where Google collects the search terms along with other data such as the location where the search has been made. This allows Google to analyse a vast amount of data, e.g. trends in influenza based on frequency of searching (Ginsberg et al., 2009). To allow researchers to analyse the data using their own queries, Google has developed some online tools. For example, Google Trends[39] is a tool that shows the frequency of a particular search term relative to the total search volume across various regions of the world, and in various languages. Choi and Varian (2012) demonstrated how Google trends can help to predict current phenomena much quicker than the usual reporting process in diverse areas such as motor vehicles and parts, initial claims for unemployment benefits or travel planning. Another tool called Google Correlate[40] works in the reverse way. Users upload a time series or spatial pattern of interest and the software returns the queries that best mimic the data (Mohebbi et al., 2011): Google calculates a correlation coefficient between the uploaded time series and the time series of every query in their database, and the results displayed are those queries that generate the highest correlation with the uploaded data.

Another big-data source of passively collected non-framework data is real-time transport information such as live feeds from buses, metro stations, bike scheme data, trains, etc. APIs are available to retrieve the data and can be brought together in dashboard type applications that provide information on the status of different transportation systems in real-time, the weather, air pollution, electricity demand, etc. For example, the CityDashboard project[41] was developed by the Centre for Advanced Spatial Analysis at UCL, London, and is available for a number of UK cities. The CityDashboard data have also been used to extract useful information for other purposes such as generating insights into sustainable transport systems (O'Brien et al., 2014) or the health impact of bicycle sharing systems (Woodcock et al., 2014); for example, the Bike Share Map[42] shows the status of biking system docks in real-time for several cities around the world. Uniman et al. (2010) used data from the Oyster Smart Card (public transport card for the London Underground) to determine the reliability of the Underground system. Using data on the entries and exits to/from London Underground stations, they developed metrics based on the travel time of passengers. This type of big data (where there are more than 1.3 billion metro and 2.4 billion bus journeys annually in London; Transport for

London, 2015), has great potential for improving passenger experiences and for planning future transport projects.

Mobile phone data from communication network operators represent another big-data source of passively collected non-framework data. These data have been analysed to investigate applications in areas such as transportation planning (Di Lorenzo et al., 2016), user behaviour (Bianchi et al., 2016), public health (Oliver et al., 2015), the spatial spread of diseases such as cholera (Bengtsson et al., 2015) or population displacement after a major disaster (Wilson et al., 2016).

A fourth area of passively collected non-framework data is travel websites and travel blogs, where all of the information provided is attached to a location and can therefore be mapped. TripAdvisor is the world's largest travel site, where users rate their accommodation, restaurants and attractions, providing their collective intelligence to the system. Any users can then access this information for free to make informed decisions. There are many examples of booking sites that draw upon TripAdvisor or have their own rating system based upon user feedback, e.g. Booking.com and Trivago, among many others.

Social media websites such as Facebook and Twitter are also prime examples that fall within this category of passive non-framework data collection; information can be shared with location data, depending on whether users enable this option in the application. Geotagged tweets are now being used in a number of applications, mostly related to crisis events and disaster management. For example, Twitter was used during the 2010 Pakistan floods (Murthy and Longwell, 2013) and tweets were an active source of information during flooding in Jakarta, allowing for the creation of open source flood maps through the Peta Jakarta initiative[43].

Finally, websites that allow users to share geotagged photographs are included in this category. Panoramio, Flickr and Instagram are a few examples of such initiatives. Users upload their photographs along with additional information such as date and time, textual tags and geotags, among others, making it possible to map the photographs. Research has been conducted to explore ways to use such data for different applications including land cover and land use mapping (Estima and Painho, 2014; Antoniou et al., 2016).

3.5 3D VGI

The third dimension in geospatial data is height or elevation. Height is now being added by volunteers to mapping initiatives such as OSM, e.g. the heights of buildings and roof geometry, which means that 3D models of cities can be created from VGI (Goetz and Zipf, 2013). Height values of GPS traces in OSM also show a promising way of retrieving 3D information for elaborating height information from SRTM and ASTER DEM models (John et al., 2016). A 3D model of a city can be generated using a GIS package or via OSM-3D, which allows OSM to be visualised as a 3D model on a virtual globe (Over et al., 2010). However, height information is still not commonly added to buildings on OSM, with less than 1.5% of buildings having height information available in November 2011 (Goetz and Zipf, 2013). If more height data were added to OSM, it would open up many possibilities for urban planning, transportation planning, navigation and disaster management, among others, particularly in locations where an SDI is currently lacking.

Elevation data are publicly available through the NASA's Shuttle Radar Topography Mission (SRTM) at a resolution of 30m. A new source of higher resolution elevation data, which are being collected by volunteers, is Unmanned Aerial Vehicles (UAVs). When DEMs generated using UAVs were compared with DEMs from LIDAR in the context of hydrological modelling (Leitão et al., 2016), the results were promising and UAVs represented an affordable option for 3D mapping. UAVs are also used in mapping damages after a disaster event (Adams and Friedland, 2011). To accommodate the growing source of aerial imagery from UAVs and other freely available satellite imagery, Development Seed and HOT have developed OpenAerialMap[44], which is a new service for contributing to and accessing this new source of data from volunteers.

4 Issues Related to VGI for Mapping

One of the main issues that is always raised with VGI, and is often perceived as a barrier to its further use, is the quality of the data. For this reason a considerable quantity of literature has appeared on this topic (see e.g. Antoniou and Skopeliti, 2015; Bordogna et al., 2015; Flanagin and Metzger, 2008; Jokar Arsanjani et al., 2015a). There is an ISO standard for spatial quality that can be applied to VGI, but additional quality indicators are required due to the characteristics that are specific to VGI. This ISO framework, along with additional quality indicators, is discussed in more detail in Chapter 7 by Fonte et al. (2017). Quality is of particular interest to NMAs, some of which see the possibility of using VGI as a way to potentially update maps that would otherwise only be re-surveyed professionally every few years, or view VGI as a complementary source of information of a richer nature, e.g. footpaths and cycle paths that may not be mapped. NMA experiences of VGI for these purposes is documented in Chapter 13 by Olteanu-Raimond et al. (2017), including the barriers to the adoption of this source of information. Demetriou et al. (2017) in Chapter 12 consider the broader question of integrating VGI with SDIs and how this might be achieved in the future.

Another key issue that is commonly discussed in relation to VGI, in particular active VGI projects, is how to recruit participants, keep them motivated and sustain the project in the future (see e.g. Coleman et al., 2009; Nov et al., 2010; Reed et al., 2013). However, more research is still needed that looks into what constitutes effective incentives for participation and how citizens can be mobilised to participate in ways that are mutually beneficial to them while contributing VGI. These aspects of recruitment, motivation and sustainability are covered in detail in Chapter 5 by Fritz et al. (2017), where the authors review a series of crowdsourcing initiatives in a comparative analysis on recruitment strategies, techniques for motivation and, more generally, issues of sustainability.

The involvement of citizens in VGI immediately raises critical questions regarding copyright, ownership, data privacy and licensing of the data, particularly when the data contributed by citizens are then integrated with third party base layers (see e.g. the work by Saunders et al. (2012) within a Canadian context). There are also ethical issues with VGI data use with respect to health and disease surveillance (Blatt, 2015). The chapter by Mooney et al. (2017) on privacy, ethics and legal issues tackles these concerns in more detail.

Finally there is a new trend in the development of citizen observatories, which are defined as a framework that combines participatory community monitoring (including policy-makers, scientists and other stakeholders) with technology such as web portals, mobile devices and low-cost sensors (Liu et al., 2014). This new trend is the subject of Chapter 15 by Liu et al. (2017).

5 Conclusions

This chapter provided an overview of sources of VGI for mapping, categorised according to whether the data are collected by government agencies as part of an SDI (i.e. framework data) or in other domains (e.g. weather or ecology, among others), as well as according to the mode of data collection, i.e. active or passive. A range of examples were then provided to illustrate the different types of VGI that fall into these categories. 3D VGI was discussed as a special case. With advances in technology, e.g. 3D mobile phones, and the increasing interest in UAVs, many new, low-cost solutions will emerge, from biomass mapping to hydrological modelling to smart cities applications. Finally, the chapter introduced some of the main issues surrounding the use of VGI, including, among others, quality, participant recruitment and motivation and the trend toward citizen observatories, which are the subjects of different chapters throughout the book. New advances in data mining and knowledge discovery techniques may also help to improve the quality of VGI in the future.

The wide range of VGI as a data source for mapping illustrates the growing interest in collecting and using these data for many different purposes. VGI has the potential to complement but also rival more traditional mapping sources in both quality and richness. What has been presented here is only the start of a growing citizen-based contribution to many different domains. Many of the sources listed in this chapter will disappear, only to be replaced by many other projects and initiatives in the future. For NMAs, the key will be the successful engagement of citizens in helping to update and correct the more authoritative sources in such a way that both entities benefit in the long run.

Acknowledgements

This work was supported by the EU FP7-funded ERC grant Crowdland (No. 617754).

Notes

1 https://www.google.com/mapmaker
2 http://cadasta.org/
3 http://wikimapia.org/
4 http://www.motomapia.com/
5 http://www.geonames.org/
6 http://land.copernicus.eu/pan-european/corine-land-cover
7 http://www.geo-wiki.org
8 http://fotoquest.at
9 http://www.mapmyfitness.com/
10 Respectively through MapMyRun (http://www.mapmyrun.com/), MapMyWalk (http://www.mapmywalk.com/), MapMyRide (http://www.mapmyride.com/) and MapMyHike (http://www.mapmyhike.com/).
11 https://www.bikemap.net/
12 http://www.bikely.com/
13 https://www.alltrails.com/
14 http://www.wikiloc.com/
15 https://scistarter.com/
16 http://www.citizensciencealliance.org/
17 http://wxqa.com/
18 https://www.wunderground.com
19 http://www.nssl.noaa.gov/projects/ping/
20 http://www.ispotnature.org/communities/global
21 http://www.inaturalist.org/
22 http://www.globalwaterwatch.org/
23 http://www.citizen-obs.eu/
24 http://discomap.eea.europa.eu/map/NoiseWatch/
25 http://www.citiesatnight.org/
26 https://hotosm.org/
27 http://earthquake.usgs.gov/data/dyfi/
28 http://alertos.org/
29 http://wikicrimes.org
30 https://www.crimereports.com
31 http://spotcrime.com
32 http://bunmegelozes.amk.uni-obuda.hu/MainPageEng.php?ln=1
33 https://www.ushahidi.com/
34 https://www.takebackthetech.net/mapit/

[35] http://zabatak.com/
[36] http://www.tomtom.com/mapshare/tools
[37] https://www.ingress.com/
[38] http://www.internetlivestats.com/google-search-statistics/
[39] https://www.google.com/trends/
[40] https://www.google.com/trends/correlate/
[41] http://citydashboard.org/
[42] http://bikes.oobrien.com/
[43] https://petajakarta.org/banjir/en/
[44] https://openaerialmap.org/

Reference list

Adams, S.M., Friedland, C.J., 2011. A survey of Unmanned Aerial Vehicle (UAV) usage for imagery collection in disaster research and management. Presented at the Ninth International Workshop on Remote Sensing for Disaster Response, Stanford, CA.

Antoniou, V., Fonte, C., See, L., Estima, J., Arsanjani, J., Lupia, F., Minghini, M., Foody, G., Fritz, S., 2016. Investigating the feasibility of geo-tagged photographs as sources of land cover input data. *ISPRS International Journal of Geo-Information* 5, 64. DOI: https://doi.org/10.3390/ijgi5050064

Antoniou, V., Skopeliti, A., 2015. Measures and indicators of VGI quality: An overview, in: ISPRS Annals of the Photogrammetry, Remote Sensing and Spatial Information Sciences. Presented at the ISPRS Geospatial Week 2015, ISPRS Annals, La Grande Motte, France, pp. 345–351.

Barth, D., 2009. The bright side of sitting in traffic: Crowdsourcing road congestion data. Google Official Blog.

Basiouka, S., Potsiou, C., 2012. VGI in Cadastre: a Greek experiment to investigate the potential of crowd sourcing techniques in Cadastral Mapping. *Survey Review* 44, 153–161. DOI: https://doi.org/10.1179/1752270611Y.0000000037

Basiouka, S., Potsiou, C., Bakogiannis, E., 2015. OpenStreetMap for cadastral purposes: an application using VGI for official processes in urban areas. *Survey Review* 47, 333–341. DOI: https://doi.org/10.1179/1752270615Y.0000000011

Bengtsson, L., Gaudart, J., Lu, X., Moore, S., Wetter, E., Sallah, K., Rebaudet, S., Piarroux, R., 2015. Using mobile phone data to predict the spatial spread of Cholera. *Scientific Reports* 5, 8923. DOI: https://doi.org/10.1038/srep08923

Bianchi, F.M., Rizzi, A., Sadeghian, A., Moiso, C., 2016. Identifying user habits through data mining on call data records. *Engineering Applications of Artificial Intelligence* 54, 49–61. DOI: https://doi.org/10.1016/j.engappai.2016.05.007

Blatt, A.J., 2015. Data privacy and ethical uses of Volunteered Geographic Information, in: *Health, Science, and Place*. Springer International Publishing, Cham, pp. 49–59.

Boersma, K.F., de Vroom, J.P., 2006. Validation of MODIS aerosol observations over the Netherlands with GLOBE student measurements. *Journal of Geophysical Research* 111. DOI: https://doi.org/10.1029/2006JD007172

Bonney, R., Ballard, H., Jordan, R., McCallie, E., Phillips, T., Shirk, J., Wilderman, C.C., 2009a. *Public Participation in Scientific Research: Defining the Field and Assessing its Potential for Informal Science Education (A CAISE Inquiry Group Report)*. Center for Advancement of Informal Science Education (CAISE), Washington DC.

Bonney, R., Cooper, C.B., Dickinson, J., Kelling, S., Phillips, T., Rosenberg, K.V., Shirk, J., 2009b. Citizen science: A developing tool for expanding science knowledge and scientific literacy. *BioScience* 59, 977–984. DOI: https://doi.org/10.1525/bio.2009.59.11.9

Bordogna, G., Carrara, P., Criscuolo, L., Pepe, M., Rampini, A., 2015. A user-driven selection of VGI based on minimum acceptable quality levels. *ISPRS Annals of Photogrammetry, Remote Sensing and Spatial Information Sciences II-3/W5*, 277–284. DOI: https://doi.org/10.5194/isprsannals-II-3-W5-277-2015

Brooks, D.R., Mims, F.M., 2001. Development of an inexpensive handheld LED-based Sun photometer for the GLOBE program. *Journal of Geophysical Research* 106, 4733–4740. DOI: https://doi.org/10.1029/2000JD900545

Buhrmester, M., Kwang, T., Gosling, S.D., 2011. Amazon's Mechanical Turk: A new source of inexpensive, yet high-quality, data? *Perspectives on Psychological Science* 6, 3–5. DOI: https://doi.org/10.1177/1745691610393980

Büttner, G., Eiselt, B., 2013. *LUCAS and CORINE Land Cover*. [pdf] Luxembourg: EEA and Eurostat. Available at: http://ec.europa.eu/eurostat/documents/205002/274769/LUCAS_UseCase-CLC.pdf [Last accessed 16 May 2017]

Chen, J., Chen, J., Liao, A., Cao, X., Chen, L., Chen, X., He, C., Han, G., Peng, S., Lu, M., Zhang, W., Tong, X., Mills, J., 2015. Global land cover mapping at 30 m resolution: A POK-based operational approach. *ISPRS Journal of Photogrammetry and Remote Sensing, Global Land Cover Mapping and Monitoring* 103, 7–27. DOI: https://doi.org/10.1016/j.isprsjprs.2014.09.002

Choi, H., Varian, H., 2012. Predicting the present with Google Trends. *Economic Record* 88, 2–9. DOI: https://doi.org/10.1111/j.1475-4932.2012.00809.x

Clark, M.L., Aide, T.M., 2011. Virtual Interpretation of Earth Web-Interface Tool (VIEW-IT) for collecting land-use/land-cover reference data. *Remote Sensing* 3, 601–620. DOI: https://doi.org/10.3390/rs3030601

Coleman, D.J., Georgiadou, Y., Labonte, J., 2009. Volunteered geographic information: The nature and motivation of produsers. *International Journal of Spatial Data Infrastructures Research* 4, 332–358.

Demetriou, D, Campagna, M, Racetin, I, Konecny, M. 2017. Integrating Spatial Data Infrastructures (SDIs) with Volunteered Geographic Information (VGI) for creating a Global GIS platform. In: Foody, G, See, L, Fritz, S, Mooney, P, Olteanu-Raimond, A-M, Fonte, C C and Antoniou, V. (eds.) *Mapping and the Citizen Sensor*. Pp. 273–297. London: Ubiquity Press. DOI: https://doi.org/10.5334/bbf.l.

Di Lorenzo, G., Sbodio, M., Calabrese, F., Berlingerio, M., Pinelli, F., Nair, R., 2016. AllAboard: Visual exploration of cellphone mobility data to optimise public transport. *IEEE Transactions on Visualization and Computer Graphics* 22, 1036–1050. DOI: https://doi.org/10.1109/TVCG.2015.2440259

Doesken, N., Reges, H., 2010. The value of the citizen weather observer. *Weatherwise* 63, 30–37.

Ekpenyong, F., Palmer-Brown, D., Brimicombe, A., 2009. Extracting road information from recorded GPS data using snap-drift neural network. *Neurocomputing*, 73, 24–36. DOI: https://doi.org/10.1016/j.neucom.2008.11.032

Elwood, S., Goodchild, M.F., Sui, D.Z., 2012. Researching volunteered geographic information: Spatial data, geographic research, and new social practice. *Annals of the Association of American Geographers* 102, 571–590. DOI: https://doi.org/10.1080/00045608.2011.595657

Estima, J., Painho, M., 2014. Photo based Volunteered Geographic Information initiatives: A comparative study of their suitability for helping quality control of Corine Land Cover. *International Journal of Agricultural and Environmental Information Systems* 5, 73–89. DOI: https://doi.org/10.4018/ijaeis.2014070105

European Commission (EC), 2007. *The INSPIRE Directive*. Available from: http://inspire.ec.europa.eu/

Eurostat, 2015. *LUCAS 2015 (Land Use / Cover Area Frame Survey). Technical reference document C1 Instructions for Surveyors*. Eurostat.

Falchi, F., Cinzano, P., Elvidge, C.D., Keith, D.M., Haim, A., 2011. Limiting the impact of light pollution on human health, environment and stellar visibility. *Journal of Environmental Management* 92, 2714–2722. DOI: https://doi.org/10.1016/j.jenvman.2011.06.029

Fan, H., Yang, B., Zipf, A., Rousell, A., 2016. A polygon-based approach for matching OpenStreetMap road networks with regional transit authority data. *International Journal of Geographical Information Science* 30, 748–764. DOI: https://doi.org/10.1080/13658816.2015.1100732

Flanagin, A., Metzger, M., 2008. The credibility of volunteered geographic information. *GeoJournal* 72, 137–148.

Fonte, C C, Antoniou, V, Bastin, L, Estima, J, Arsanjani, J J, Bayas, J-C L, See, L and Vatseva, R. 2017. Assessing VGI Data Quality. In: Foody, G, See, L, Fritz, S, Mooney, P, Olteanu-Raimond, A-M, Fonte, C C and Antoniou, V. (eds.) *Mapping and the Citizen Sensor*. Pp. 137–163. London: Ubiquity Press. DOI: https://doi.org/10.5334/bbf.g.

Food and Agriculture Organization of the United Nations (FAO), 1998. *Land Cover and Land Use: The FAO AFRICOVER Programme*.

Fritz, S., McCallum, I., Schill, C., Perger, C., See, L., Schepaschenko, D., van der Velde, M., Kraxner, F., Obersteiner, M., 2012. Geo-Wiki: An online platform for improving global land cover. *Environmental Modelling & Software* 31, 110–123. DOI: https://doi.org/10.1016/j.envsoft.2011.11.015

Fritz, S, See, L and Brovelli, M. 2017. Motivating and Sustaining Participation in VGI. In: Foody, G, See, L, Fritz, S, Mooney, P, Olteanu-Raimond, A-M, Fonte, C C and Antoniou, V. (eds.) *Mapping and the Citizen Sensor*. Pp. 93–117. London: Ubiquity Press. DOI: https://doi.org/10.5334/bbf.e.

Fritz, S., See, L., McCallum, I., Schill, C., Obersteiner, M., van der Velde, M., Boettcher, H., Havlík, P., Achard, F., 2011. Highlighting continued uncertainty in global land cover maps for the user community. *Environmental Research Letters* 6, 44005. DOI: https://doi.org/10.1088/1748-9326/6/4/044005

Fuller, R.M., Smith, G.M., Sanderson, J.M., Hill, R.A., Thomson, A.G., 2002. The UK Land Cover Map 2000: Construction of a parcel-based vector map from satellite images. *The Cartographic Journal* 39, 15–25. DOI: https://doi.org/10.1179/caj.2002.39.1.15

Gallego, F.J., 2011. Validation of GIS layers in the EU: getting adapted to available reference data. *International Journal of Digital Earth* 4, 42–57. DOI: https://doi.org/10.1080/17538947.2010.512746

Gallo, T., Waitt, D., 2011. Creating a successful citizen science model to detect and report invasive Species. *BioScience* 61, 459–465. DOI: https://doi.org/10.1525/bio.2011.61.6.8

Ginsberg, J., Mohebbi, M.H., Patel, R.S., Brammer, L., Smolinski, M.S., Brilliant, L., 2009. Detecting influenza epidemics using search engine query data. *Nature* 457, 1012–1014. DOI: https://doi.org/10.1038/nature07634

Goetz, M., Zipf, A., 2013. The evolution of geo-crowdsourcing: Bringing Volunteered Geographic Information to the third dimension, in: Sui, D., Elwood, S., Goodchild, M. (Eds.), *Crowdsourcing Geographic Knowledge*. Springer Netherlands, Dordrecht, pp. 139–159.

Goodchild, M.F., 2007. Citizens as sensors: the world of volunteered geography. *GeoJournal* 69, 211–221. DOI: https://doi.org/10.1007/s10708-007-9111-y

Hagenauer, J., Helbich, M., 2012. Mining urban land-use patterns from volunteered geographic information by means of genetic algorithms and artificial neural networks. *International Journal of Geographical Information Science* 26, 963–982. DOI: https://doi.org/10.1080/13658816.2011.619501

Haklay, M., 2010. How good is volunteered geographical information? A comparative study of OpenStreetMap and Ordnance Survey datasets. *Environment and Planning B: Planning and Design* 37, 682–703. DOI: https://doi.org/10.1068/b35097

Harvey, F., 2013. To volunteer or to contribute locational information? Towards truth in labeling for crowdsourced geographic information, in: Sui, D., Elwood, S., Goodchild, M. (Eds.), *Crowdsourcing Geographic Knowledge*. Springer Netherlands, Dordrecht, Netherlands, pp. 31–42.

Howe, J., 2006. The rise of crowdsourcing. *Wired Magazine* 14, 1–4.

Jacobson, A., Dhanota, J., Godfrey, J., Jacobson, H., Rossman, Z., Stanish, A., Walker, H., Riggio, J., 2015. A novel approach to mapping land conversion using Google Earth with an application to East Africa. *Environmental Modelling & Software* 72, 1–9. DOI: https://doi.org/10.1016/j.envsoft.2015.06.011

John, S., Hahmann, S., Rousell, A., Löwner, M.-O., Zipf, A., 2016. Deriving incline values for street networks from voluntarily collected GPS traces. *Cartography and Geographic Information Science* 1–18. DOI: https://doi.org/10.1080/15230406.2016.1190300

Jokar Arsanjani, J., Mooney, P., Zipf, A., Schauss, A., 2015a. Quality assessment of the contributed land use information from OpenStreetMap versus authoritative datasets, in: Jokar Arsanjani, J., Zipf, A., Mooney, P., Helbich, M. (Eds.), *OpenStreetMap in GIScience, Lecture Notes in Geoinformation and Cartography*. Springer International Publishing, Cham, pp. 37–58.

Jokar Arsanjani, J., Zipf, A., Mooney, P., Helbich, M. (Eds.), 2015b. *OpenStreetMap in GIScience, Lecture Notes in Geoinformation and Cartography*. Springer International Publishing, Cham.

Kalantari, M., La, V., 2015. Assessing OpenStreetMap as an open property map, in: Jokar Arsanjani, J., Zipf, A., Mooney, P., Helbich, M. (Eds.), *OpenStreetMap in GIScience*. Springer International Publishing, Cham, pp. 255–272.

Kingston, R., Carver, S., Evans, A., Turton, I., 2000. Web-based public participation geographical information systems: an aid to local environmental decision-making. *Computers, Environment and Urban Systems* 24, 109–125. DOI: https://doi.org/10.1016/S0198-9715(99)00049-6

Laso Bayas, J.-C., See, L., Fritz, S., Sturn, T., Perger, C., Duerauer, M., Karner, M., Moorthy, I., Schepaschenko, D., Domian, D., McCallum, I., 2016. Crowdsourcing in-situ data on land cover and land use using gamification and mobile technology. *Remote Sensing* 8(11), 905. DOI: https://doi.org/10.3390/rs8110905.

LeBaron, G., 2007. Audubon's Christmas Bird Count from 19th Century Conservation Action to 21st Century Citizen Science. Presented at the Citizen Science Toolkit Conference, Cornell Lab of Ornithology, Ithaca, New York, USA, p. 10

Leeper, R.D., Rennie, J., Palecki, M.A., 2015. Observational perspectives from U.S. Climate Reference Network (USCRN) and Cooperative Observer Program (COOP) Network: Temperature and Precipitation comparison. *Journal of Atmospheric and Oceanic Technology* 32, 703–721. DOI: https://doi.org/10.1175/JTECH-D-14-00172.1

Leitão, J.P., Moy de Vitry, M., Scheidegger, A., Rieckermann, J., 2016. Assessing the quality of digital elevation models obtained from mini unmanned aerial vehicles for overland flow modelling in urban areas. *Hydrology and Earth System Sciences* 20, 1637–1653. DOI: https://doi.org/10.5194/hess-20-1637-2016

Liu, H-Y, Grossberndt, S and Kobernus, M. 2017. Citizen Science and Citizens' Observatories: Trends, Roles, Challenges and Development Needs for

Science and Environmental Governance. In: Foody, G, See, L, Fritz, S, Mooney, P, Olteanu-Raimond, A-M, Fonte, C C and Antoniou, V. (eds.) *Mapping and the Citizen Sensor*. Pp. 351–376. London: Ubiquity Press. DOI: https://doi.org/10.5334/bbf.o.

Liu, H.-Y., Kobernus, M., Broday, D., Bartonova, A., 2014. A conceptual approach to a citizens' observatory – supporting community-based environmental governance. *Environmental Health* 13. DOI: https://doi.org/10.1186/1476-069X-13-107

Mohebbi, M., Vanderkam, D., Kodysh, J., Schonberger, R., Choi, H., Kumar, S., 2011. *Google Correlate Whitepaper*. Google, San Francisco, CA.

Mooney, P., Corcoran, P., 2013. Analysis of interaction and co-editing patterns amongst OpenStreetMap contributors. *Transactions in GIS* 18, 633–659. DOI: https://doi.org/10.1111/tgis.12051

Mooney, P and Minghini, M. 2017. A Review of OpenStreetMap Data. In: Foody, G, See, L, Fritz, S, Mooney, P, Olteanu-Raimond, A-M, Fonte, C C and Antoniou, V. (eds.) *Mapping and the Citizen Sensor*. Pp. 37–59. London: Ubiquity Press. DOI: https://doi.org/10.5334/bbf.c.

Mooney, P, Olteanu-Raimond, A-M, Touya, G, Juul, N, Alvanides, S and Kerle, N. 2017. Considerations of Privacy, Ethics and Legal Issues in Volunteered Geographic Information. In: Foody, G, See, L, Fritz, S, Mooney, P, Olteanu-Raimond, A-M, Fonte, C C and Antoniou, V. (eds.) *Mapping and the Citizen Sensor*. Pp. 119–135. London: Ubiquity Press. DOI: https://doi.org/10.5334/bbf.f.

Murthy, D., Longwell, S.A., 2013. Twitter and Disasters. *Information, Communication & Society* 16, 837–855. DOI: https://doi.org/10.1080/1369118X.2012.696123

Nov, O., Arazy, O., Anderson, D., 2010. Technology-mediated citizen science participation: A motivational model, in: *Proceedings of the Fifth International AAAI Conference on Weblogs and Social Media*. ACM Press, pp. 249–255. DOI: https://doi.org/10.1145/1772690.1772766

O'Brien, O., Cheshire, J., Batty, M., 2014. Mining bicycle sharing data for generating insights into sustainable transport systems. *Journal of Transport Geography* 34, 262–273. DOI: https://doi.org/10.1016/j.jtrangeo.2013.06.007

Oliver, N., Matic, A., Frias-Martinez, E., 2015. Mobile network data for public health: Opportunities and challenges. *Frontiers in Public Health* 3, 1–12. DOI: https://doi.org/10.3389/fpubh.2015.00189

Olteanu-Raimond, A-M, Laakso, M, Antoniou, V, Fonte, C C, Fonseca, A, Grus, M, Harding, J, Kellenberger, T, Minghini, M, Skopeliti, A. 2017. VGI in National Mapping Agencies: Experiences and Recommendations. In: Foody, G, See, L, Fritz, S, Mooney, P, Olteanu-Raimond, A-M, Fonte, C C and Antoniou, V. (eds.) *Mapping and the Citizen Sensor*. Pp. 299–326. London: Ubiquity Press. DOI: https://doi.org/10.5334/bbf.m.

Over, M., Schilling, A., Neubauer, S., Zipf, A., 2010. Generating web-based 3D City Models from OpenStreetMap: The current situation in Germany.

Computers, Environment and Urban Systems 34, 496–507. DOI: https://doi.org/10.1016/j.compenvurbsys.2010.05.001

Palen, L., Soden, R., Anderson, T.J., Barrenechea, M., 2015. Success & scale in a data-producing organization: The socio-technical evolution of OpenStreetMap in response to humanitarian events, in: *Proceedings of the 33rd Annual ACM Conference on Human Factors in Computing Systems*. ACM Press, New York, NY, USA, pp. 4113–4122. DOI: https://doi.org/10.1145/2702123.2702294

Reed, J., Raddick, M.J., Lardner, A., Carney, K., 2013. *An exploratory factor analysis of motivations for participating in Zooniverse, a collection of virtual citizen science projects*. IEEE, pp. 610–619. DOI: https://doi.org/10.1109/HICSS.2013.85

Reges, H.W., Doesken, N., Turner, J., Newman, N., Bergantino, A., Schwalbe, Z., 2016. COCORAHS: The evolution and accomplishments of a volunteer rain gauge network. *Bulletin of the American Meteorological Society*. DOI: https://doi.org/10.1175/BAMS-D-14-00213.1

Saunders, A., Scassa, T., Lauriault, T.P., 2012. Legal issues in maps built on third party base layers. *Geomatica* 66, 279–290. DOI: https://doi.org/10.5623/cig2012-054

See, L., Fritz, S., Perger, C., Schill, C., McCallum, I., Schepaschenko, D., Duerauer, M., Sturn, T., Karner, M., Kraxner, F., Obersteiner, M., 2015. Harnessing the power of volunteers, the internet and Google Earth to collect and validate global spatial information using Geo-Wiki. *Technological Forecasting and Social Change* 98, 324–335. DOI: https://doi.org/10.1016/j.techfore.2015.03.002

See, L., Fritz, S., Dias, E., Hendriks, E., Mijling, B., Snik, F., Stammes, P., Vescovi, F., Zeug, G., Mathieu, P.-P., Desnos, Y.-L., Rast, M., 2016a. A new generation of tools for crowdsourcing and citizen science to support Earth Observation calibration and validation. *IEEE Geosciences and Remote Sensing Magazine* 4, 38–50. DOI: https://doi.org/10.1109/MGRS.2015.2498840.

See, L., Mooney, P., Foody, G., Bastin, L., Comber, A., Estima, J., Fritz, S., Kerle, N., Jiang, B., Laakso, M., Liu, H.-Y., Milčinski, G., Nikšič, M., Painho, M., Pődör, A., Olteanu-Raimond, A.-M., Rutzinger, M., 2016b. Crowdsourcing, citizen science or Volunteered Geographic Information? The current state of crowdsourced geographic information. *ISPRS International Journal of Geo-Information* 5, 55. DOI: https://doi.org/10.3390/ijgi5050055

Sester, M., Jokar Arsanjani, J., Klammer, R., Burghardt, D., Haunert, J.-H., 2014. Integrating and generalising Volunteered Geographic Information, in: Burghardt, D., Duchêne, C., Mackaness, W. (Eds.), *Abstracting Geographic Information in a Data Rich World*. Springer International Publishing, Cham, pp. 119–155.

Sieber, R., 2006. Public participation geographic information systems: A literature review and framework. *Annals of the Association of American Geographers* 96, 491–507. DOI: https://doi.org/10.1111/j.1467-8306.2006.00702.x

Silvertown, J., Harvey, M., Greenwood, R., Dodd, M., Rosewell, J., Rebelo, T., Ansine, J., McConway, K., 2015. Crowdsourcing the identification of organisms: A case-study of iSpot. *ZooKeys* 480, 125–146. DOI: https://doi.org/10.3897/zookeys.480.8803

Soden, R., Palen, L., 2014. From crowdsourced mapping to community mapping: The post-earthquake work of OpenStreetMap Haiti, in: Rossitto, C., Ciolfi, L., Martin, D., Conein, B. (Eds.), *COOP 2014 – Proceedings of the 11th International Conference on the Design of Cooperative Systems, 27–30 May 2014, Nice (France)*. Springer International Publishing, pp. 311–326.

Transport for London, 2015. Record passenger numbers on London's transport network. Available at: http://www.tfl.gov.uk/ [Last accessed 13 April 2017]

Uniman, D., Attanucci, J., Mishalani, R., Wilson, N., 2010. Service reliability measurement using automated fare card data: Application to the London Underground. *Transportation Research Record: Journal of the Transportation Research Board* 2143, 92–99. DOI: https://doi.org/10.3141/2143-12

Whittaker, J., McLennan, B., Handmer, J., 2015. A review of informal volunteerism in emergencies and disasters: Definition, opportunities and challenges. *International Journal of Disaster Risk Reduction* 13, 358–368. DOI: https://doi.org/10.1016/j.ijdrr.2015.07.010

Wilson, R., zu Erbach-Schoenberg, E., Albert, M., Power, D., Tudge, S., Gonzalez, M., Guthrie, S., Chamberlain, H., Brooks, C., Hughes, C., Pitonakova, L., Buckee, C., Lu, X., Wetter, E., Tatem, A., Bengtsson, L., 2016. Rapid and near real-time assessments of population displacement using mobile phone data following disasters: The 2015 Nepal earthquake. *PLoS Currents*. DOI: https://doi.org/10.1371/currents.dis.d073fbece328e4c39087bc086d694b5c

Woodcock, J., Tainio, M., Cheshire, J., O'Brien, O., Goodman, A., 2014. Health effects of the London bicycle sharing system: health impact modelling study. *British Medical Journal* 348, g425–g425. DOI: https://doi.org/10.1136/bmj.g425

CHAPTER 3

A Review of OpenStreetMap Data

Peter Mooney* and Marco Minghini†

*Department of Computer Science, Maynooth University, Maynooth, Co. Kildare, Ireland, Peter.mooney@nuim.ie

†Department of Civil and Environmental Engineering, Politecnico di Milano, Piazza Leonardo da Vinci 32, 20133 Milano, Italy

Abstract

While there is now a considerable variety of sources of Volunteered Geographic Information (VGI) available, discussion of this domain is often exemplified by and focused around OpenStreetMap (OSM). In a little over a decade OSM has become the leading example of VGI on the Internet. OSM is not just a crowdsourced spatial database of VGI; rather, it has grown to become a vast ecosystem of data, software systems and applications, tools, and Web-based information stores such as wikis. An increasing number of developers, industry actors, researchers and other end users are making use of OSM in their applications. OSM has been shown to compare favourably with other sources of spatial data in terms of data quality. In addition to this, a very large OSM community updates data within OSM on a regular basis. This chapter provides an introduction to and review of OSM and the ecosystem which has grown to support the mission of creating a free, editable map of the whole world. The chapter is especially meant for readers who have no or little knowledge about the range, maturity and complexity of the tools, services, applications and organisations working with OSM data. We provide examples of tools and services to access, edit, visualise and make quality assessments of OSM data. We also provide a number of examples of applications, such as some of those

How to cite this book chapter:

Mooney, P and Minghini, M. 2017. A Review of OpenStreetMap Data. In: Foody, G, See, L, Fritz, S, Mooney, P, Olteanu-Raimond, A-M, Fonte, C C and Antoniou, V. (eds.) *Mapping and the Citizen Sensor.* Pp. 37–59. London: Ubiquity Press. DOI: https://doi.org/10.5334/bbf.c.

used in navigation and routing, that use OSM data directly. The chapter finishes with an indication of where OSM will be discussed in the other chapters in this book, and we provide a brief speculative outlook on what the future holds for the OSM project.

Keywords

OpenStreetMap, geodata, open data, Volunteered Geographic Information (VGI)

1 Introduction

The OpenStreetMap (OSM) project was founded in 2004 and has now positioned itself as the most famous example of Volunteered Geographic Information (VGI) on the Internet (Jokar Arsanjani et al., 2015). While OSM is only one of many well established and well known VGI projects (See et al., 2016), it holds a dominant position in the VGI landscape. Chapter 2 of this book, by See et al. (2017), gives an overview of different sources of VGI in the context of its usage and characteristics. In recent years OSM has attracted very significant research attention (Mooney, 2015) and could almost be considered a field of research in its own right (Jokar Arsanjani et al., 2015); given the influence of OSM on the VGI and citizen sensor research landscape, this chapter will provide an introduction to and overview of the OSM project.

OSM was founded in 2004 by then MSc student Steve Coast, who created the idea as part of a thesis dissertation. Around that time the concept of crowdsourcing, collaboration and Web-based co-production or creation of knowledge was beginning to gain momentum. Coast's idea was simple: if I collect geographic data about my area – where I have local knowledge – and you collect geographic data about your area – where you have local knowledge – then these can be combined, and we can begin to build a spatial database of a region. If this scales up to a larger *crowd* of people, then it is very possible to crowdsource the mapping of the entire world. The OSM mission statement grew out of this simple idea, which was to be a collaborative project that created a free editable map of the world. Rather than the focus being on outputs in the form of cartographic products and maps, the core of OSM is a spatial database, which contains geographic data and information from all over the world. Many authors and commentators have speculated on the ingredients for the rapid and sustained success of OSM since 2004. A number of factors are seen as having been influential in OSM's development. In the first instance one of these factors is Web 2.0, or the interactive web (O'Reilly, 2007), which facilitates the development of large scale collaborative projects that can see hundreds or thousands of people contributing simultaneously – the most famous example of this is Wikipedia. Secondly the availability of low-cost, high-quality and high-accuracy Global Positioning System (GPS) means that consumers or citizens can now collect geographic information using smart devices such as their smartphones or dedicated GPS units; these geographic data can then be uploaded and contributed to OSM. The third factor is related to the citizen contributors: the OSM project welcomes anyone to register and take part as a contributor. Contributors can span the entire spectrum of geographic and Information Technology expertise: from beginner or newcomer to expert level geographer or software developer.

1.1 How Does One Contribute to OSM?

The OSM data model is very straightforward to understand. There are three primitive data types or objects: nodes, ways (polygons and polylines) and relations (logical collections of ways and nodes). A way is made up of at least two nodes (for polylines) or three nodes (for closed polygons). A node represents a geographic point feature and its coordinate is usually expressed as latitude and longitude. Within OSM, every object must have at least one attribute or tag (a key/value pair) assigned to it to describe its characteristics. There are many guides and tutorial documents on how one begins to map with OSM; recently the company Mapbox provided an updated set of documentation for this[1]. The OSM Map Features pages on the OSM wiki (OpenStreetMap, 2016) represent the reference document describing the officially adopted OSM tags. These tags have been agreed upon over the years and there are wiki pages written to describe the likely usage and use case scenarios of each tag. OSM follows a folksonomy approach to tagging, and, in theory, any tag can be associated with any object (Ballatore and Mooney, 2015). Contributors are free to create their own tags. As several authors have shown (Ballatore and Mooney, 2015; Ballatore and Zipf, 2015), this can lead to disagreements amongst contributors or confusion on how to use specific tags in certain geographic scenarios (for example tagging an object representing an unpaved pedestrian footpath). Services such as taginfo[2] allow exploration and visualisation of the most frequently used tags and their keys for the entire OSM database. The taginfo service is particularly useful for understanding the style or structure of tags used on specific object types, conceptualising the very wide range of values some keys are assigned in tags and the spatial distribution of tags. Taginfo is constantly updated in near real-time and stores the tags from every object in the global OSM database. There is no theoretical limit on the number of tags that can be assigned to any object. Nodes that have a tag with a key name are usually called Points of Interest (POI) and usually represent the position of some object or structure of general interest. Keys in OSM can be internationalised to accommodate languages other than English, which, due to OSM's origins, has established itself as the *lingua franca* of the project (Ballatore and Mooney, 2015).

There are many software tools available to automate the process of contributing data or editing existing data. The most widely used and popular is the JOSM (Java for OSM) tool[3], followed by the Web-based iD editor[4]; JOSM is acknowledged as being a software tool more suited to more experienced OSM contributors while the iD editor is very straightforward to use and is integrated into the OSM map homepage. New data submitted to OSM or existing data edited within the OSM database are available for access almost immediately, and the OSM map on the OSM homepage will render changes quickly (within 30 minutes). As we shall discuss in Section 2, there are many ways in which one can access and download OSM data for other uses. On a more technical level, every object within the OSM database (nodes, ways or relations) has several data attributes including: a globally unique ID; a version number, which indicates how many times the object has been edited; a timestamp of the most recent edit; and the user ID and the username of the contributor who created (or last edited) the object.

Anyone can sign up and register for free as a contributor to OSM. In July 2016, there were over 2.7M registered contributors, as outlined on the OSM wiki[5]; upon sign-up, a contributor can begin contributing or mapping new data in OSM or editing existing data stored in the OSM spatial database. However, it is not easy to automatically access attribute or demographic information about these user contributors from the OSM database or associated services. Several researchers (Neis et al., 2013 and references therein) have attempted to classify and understand *who* the contributors are to OSM through analysis of their editing and contribution patterns over a long period of time.

There are multiple ways users can contribute data to OSM. The simplest one is through the digitisation of objects (such as buildings, roads and rivers) that are visible on openly licensed satellite imagery. The most used imagery, available by default in the OSM iD editor, is the one provided under a compatible licence by Microsoft (Coast, 2010). While this way of contributing data allows volunteers to map places even when remote from the mapped place, other instruments, such as GPS receivers and paper-based tools like Field Papers[6], allow users to physically survey an area and then upload or insert the information into the OSM database. One of the more controversial methods of contributing data to the OSM database is through the bulk import of suitably licensed geographic data. The pros and cons of taking a geographic dataset produced outside of OSM and importing it into the OSM database have been discussed by many authors (Zielstra et al., 2013), and the issue remains a contentious one amongst the OSM community. One of the most powerful arguments against this bulk import is that it goes against the very ethos of OSM that data be collected or mapped by OSM contributors based on an ability to verify the quality of the data, ability itself founded on local knowledge, physical collection of the data or geographic expertise. Many examples of bulk import are available on the OSM wiki website[7], with the TIGER data import of roads and highways into OSM United States and the CORINE LandCover map import into OSM France amongst the most well known and controversial.

The remainder of this chapter is organised as follows: in the next section, we provide an overview of how OSM is accessed, visualised and used in research, software development and other applications. In the final section of the chapter, we provide some concluding remarks and points for discussion on OSM; we also outline where the reader will find more discussion of and information on OSM in the proceeding chapters of this volume. The overall purpose of this chapter is to introduce readers unfamiliar with OSM to the project and the types of applications it is currently used for. We let other chapters in this volume to describe specific aspects of OSM (data quality, visualisation of OSM, motivations of contributors, etc.) in more technical detail.

2 Applications Using OSM Data

In the introductory section of this chapter, we mentioned that, while much of the focus of OSM is on the maps and cartographic products derived from the OSM data, the core product of OSM is the spatial database. This second section will provide a comprehensive list of a number of projects, organisations, services, software and applications that make direct use of OSM data, with references and links provided at the end of the chapter. A number of such lists and descriptions are available on the Internet (e.g. on the OSM wiki[8]), but, to the authors' knowledge, this is the first list provided in an academic paper. Due to the free and open availability of OSM data and the increasing popularity of OSM worldwide, it would be impossible to list all of the existing projects and applications. Making use of OSM data has become so easy and immediate that new tools are created almost every day. Some of these applications become very popular and well known while other applications are limited to single languages or user groups. Therefore we limit the items on this list to what we consider from our knowledge of OSM to be the most popular, up-to-date and successful applications based on OSM data. The description of each item on the list serves as a reference and starting point for readers having no or limited experience in OSM.

We understand that links to online services and websites change over time and can become obsolete or broken. However, with this in mind, the list itself serves as a commentary on the diversity of application areas where OSM is used. We organise the list under the following headings: Data Download Applications and Services, Education and Research Use of OSM, Disaster and Humanitarian OSM, Government and Industry Usage, Visualisation of OSM Data, Software (OSM Editors, Routing Services, Vector Rendering, other services), Quality Assurance for OSM, and Games and Leisure. For more applications and services, a very extensive list is maintained on the OSM wiki[9].

2.1 Data Download Applications and Services

Regardless of the types of applications and visualisations that can be produced with OSM, the applications and services that provide access to the data within the OSM database are arguably the most important part of the OSM's data architecture. Geofabrik is one of the best known providers of access to OSM data and provides access to continental-, national- and regional-sized data extracts[10]; the data are uploaded very frequently (at least hourly) and are provided in a number of different formats. The OSM wiki provides access to the so-called Planet.osm file[11], which is the entire OSM database contained in one very large XML or compressed format file. This file is updated every few days. The wiki page lists many mirror servers providing access to the Planet.osm file, with many of these servers providing the file updated on an hourly basis. OSM also provides an API[12] that allows extracting and saving raw data from/to the OSM database. There are API calls to create, read, update and delete map data for OSM, and this provides software developers and applications with the most up-to-date data available. However, queries for very large amounts of data (such as city- or country-sized) are discouraged and disallowed. The Overpass API service[13], with its popular frontend Overpass Turbo[14], is a read-only API that allows access to selected parts of the OSM map database; clients send queries using a special API query language or using the graphical interface provided by Overpass Turbo. The Overpass API also allows programmatic calls for data extracts of arbitrary geographic size. The commercial company Mapzen provides OSM data for download in city- or region-based extract sizes from their Metro Extracts[15] service: a number of data formats are provided and their data extracts are updated on a weekly basis. A simple and popular way to download small amounts of OSM data is provided on the OSM homepage and consists in using its 'export' feature[16]. This allows users to browse the OSM map and select small regions using a bounding rectangle, which can then download OSM data to the calling device. All of the services mentioned so far provide, as standard, OSM data in the default OSM XML data format[17]. As most types of XML, OSM XML requires special software tools in order to be processed, and there are many options available for this task[18]. Data providers such as Geofabrik[19] and Mapzen[20] also provide OSM data in common formats, such as SHP files: this allows users to process and visualise the data using desktop GIS tools.

2.2 Education and Research Use of OSM

The ability to access the entire OSM spatial database on an hourly basis or even more frequently has proved a great attraction for the research community over the past number of years (Jokar Arsanjani et al., 2015). There has been a steady increase year-on-year of the number of papers being produced by the academic community in the domain of VGI, and OSM forms a major component of this work. In 2015, one of the first edited volumes on OSM as a research topic was published (Jokar Arsanjani et al., 2015); the volume considered OSM's role in GIScience and contained a very wide range of research topics, from navigation and routing to data quality and visualisation. Similarly, two EU COST Actions focused on VGI that ran from 2012 to 2016, TD1202 'Mapping and the Citizen Sensor' (from where this volume comes)[21] and IC1203 'ENERGIC'[22], have produced some excellent research around OSM. In other educational settings, a repository such as TeachOSM[23] provides a set of community- contributed resources for teachers, trainers, educators and instructors who want to bring OSM into their classrooms. The classroom can be a very important setting for educating the next generation of OSM mappers or contributors. There are many examples, including 'a world-record humanitarian mapathon that took place at the Politecnico di Milano in northern Italy in March 2016'[24]: This mapathon event involved over two hundred children from six elementary schools in the Milan province. This mapathon resulted in the mapping of over 5000 buildings in Swaziland (Ebrahim et al., 2016). More information can also be found in Chapter 5 of this book, by Fritz et al. (2017).

2.3 Disaster and Humanitarian OSM

OSM data and mapping has been used extensively in recent disaster and humanitarian emergencies and operations all over the world. The Humanitarian OpenStreetMap Team (HOT)[25] is a nonprofit organisation leading the international efforts in community mapping projects. Through its open source Tasking Manager[26], HOT coordinates online collaborative mapping based on OSM when major disaster strikes anywhere in the world, such as during the Nepal earthquake in 2015 and the Japan and Ecuador earthquakes in 2016; in regions such as Nepal, OSM very often is the only available source of mapping data and cartography that rescuers and aid agencies can use. The Missing Maps project[27] is an open, collaborative humanitarian project aiming to map the most vulnerable places in the developing world. Missing Maps founders and members are mainly humanitarian organisations (e.g. the American Red Cross and Doctors Without Borders) and NGOs; the project's volunteered mapping is again based on OSM data and the HOT Tasking Manager. The University of Heidelberg hosts the disastermappers project[28], which aims to educate and train university students about mapping in OSM for humanitarian purposes. Reaction time is often very quick and successful with OSM. Examples include a 5-day period of mapping where the Humanitarian OSM Team and volunteers mapped over 100,000 buildings and hundreds of miles of roads in Guinea when Ebola broke out in 2014[29]. The efforts of the OSM community in times of humanitarian crisis are easy to visualise, as snapshots of OSM data can be extracted to show the effects of mapping before and after a particular event. HOT shows the changes[30] in the OSM map that occurred after the city of Tacloban in the Philippines was devastated by the super typhoon Haiyan in 2013.

2.4 Government and Industry Usage

OSM is being used in industry and by government agencies around the world. Indeed there is a large number of companies listed on the OSM wiki[31] who provide consultancy based on OSM data. This consultancy has a wide range of applications, including Web-based mapping, Web GIS, data analysis, routing and navigation, and data extraction. There are several leading companies in this domain including: Mapbox[32], MapQuest[33], Stamen[34], Mapzen[35], CampTo-Camp[36] and Geofabrik[18]. Most of these companies also provide OSM services back to the OSM user community, including OSM data extracts, web-map layers for online mapping and specialist visualisation.

Government usage of OSM is more difficult to track unless it is advertised and highlighted by the government agencies involved. From the opposite direction, there has been significant use of government data in OSM, with several high-profile data imports having been performed over the years. These imports are based on the imported data having an acceptable open data licence allowing the corresponding geodata to be inserted into the OSM database. The imports include: the TIGER (the Topologically Integrated Geographic Encoding and Referencing system) data, produced by the US Census Bureau, in the USA; plan.at in Austria; GeoBase as a complete map of Canada; and the CORINE Land Cover map in France.

In 2013, New York City opened up many 'high-value datasets to the public, making it possible to use these data to improve OSM'[37], facilitated and assisted by Mapbox[30]. 'In return, New York City's GIS team is informed of changes made in OSM related to their datasets, which helps keep their map data current.' This effectively made the New York City municipality a participant and contributor to OSM in the United States. MapGive[38] is an initiative of the US Department of State's Humanitarian Information Unit, 'mak[ing] it easy for new volunteers to learn to map and get involved in online tasks'. Portland's TriMet traffic authority uses OSM to power their multi-modal traffic planner[39]. The Gendarmerie Nationale (one of the national police forces in France) uses OSM maps inside their police cars[40]. The CROWDGOV report by Haklay et al. (2014) has a number of examples of governmental use of OSM around the world. There is still some reluctance by government agencies to use VGI and OSM as a complement to their own sources of spatial data (Olteanu-Raimond et al., 2017b); however, examples do exist, such as the French National Address Database (BAN), which 'associates each address listed on the French territory (25 million addresses) with its geographic coordinates' (the database 'does not contain any nominative data'). BAN is the result of 'an innovative collaboration model between public authorities' in France and OSM France 'to build an essential reference for the economy, society and public services'[41].

2.5 Visualisation of OSM Data

From anecdotal evidence, visualisation of OSM data is certainly one of the most popular applications of OSM data. Visualisation of OSM data is facilitated by the flexible availability of the OSM data (see Section 2.1) and the very wide range of visualisation tools available, which can natively process OSM data directly or from a spatial database. There is a vast number of examples, and we provide a small selection here for the purposes of illustrating the breadth of applications.

OpenTopoMap[42] provides a topographic visualisation of OSM data combined with SRTM elevation data. The map tiles in OpenTopoMap are available for use as a web-map layer in other applications. OpenCycleMap[43] is an OSM rendering 'primarily aimed at showing information useful to cyclists'. The OpenCycleMap global cycling map is based on data from OSM and is updated frequently. The OpenCycleMap website indicates that 'at low zoom levels, it is intended for overviews of national cycling networks; at higher zoom levels, it should help with planning which streets to cycle on, where cyclists can park their bikes, etc.' It is also available for use as a web-map layer in other applications. In a similar fashion, the Hike & Bike Map[44] visualisation of OSM data highlights hiking and biking routes by using a specific cartographic style to highlight these routes. The OpenSnowMap[45] is an OSM-based map rendering of ski slopes and lifts. It integrates OSM data, MODIS/Terra Snow Cover 8-Day Global data[46] and SRTM 90m Digital Elevation data. As of December 2016, over 100,000 km of skiing trails have already been mapped. OsmHydrant[47] is a special map showing the position of hydrants, water tanks and suction points, with the purpose of assisting local authorities and fire departments. While there is an emphasis on visualisation, it allows OSM contributors to map new hydrants and edit the existing ones. As of July 2016, almost 45000 hydrants had been added. OpenFireMap[48] is an OSM rendering, highlighting 'fire stations, hydrants, water tanks, and ponds used for firefighting (suction points)'. It does not provide editing facilities directly. The Stamen company in the United States provides several cartographic variations on the standard OSM map representations. These are available for use as web-map layers in other applications. Three of the most popular web-maps provided by Stamen are the terrain representation[49], the black and white representation[50] and the very artistic watercolor representation[51]. There is also a good deal of visualisation of OSM in 3D: one of the best examples is the OSM Buildings[52] JavaScript library for visualising OpenStreetMap building geometry on 2D and 3D maps. F4map[53] is a French company providing cartography and visualisation services: one of its products is a 3D visualisation of the world using OSM data. In other types of visualisation, Kothic JS[54] is an in-development new technology that renders OSM data 'on the fly' using HTML5 without the need for raster tile images. Mapbox Studio[55] is a suite of free and paid-for tools to produce 'vector tiles', which can be rendered either server-side or client-side, with many different customisations available according to the OSM data being used.

2.6 *OSM-based Software*

As mentioned above, the OSM community has created a vast ecosystem of software tools and services. As is the case with the visualisation of OSM data, it is not possible to give an in-depth list of software. We have organised this section into three subsections: OSM data editors, OSM-based routing services and other services.

2.6.1 OSM Data Editors

OSM is an openly accessible spatial database which any contributor can supply geodata to and whose existing data any contributor can also edit. It is therefore very important that software tools be available to support this editing work for contributors. The OSM wiki contains an extensive list of OSM data editing tools[56] and a comparison of their characteristics. In this section we outline five of the most famous and well known OSM editors. The iD editor[57] is a Web-based editor for OSM and is the editor that is integrated into the OSM homepage. The JOSM editor[3] is a Java editor for OSM and is considered an editor for skilled OSM contributors. It 'supports loading GPX tracks, background imagery and OSM data from local sources as well as from online sources and allows' direct editing of the OSM data; a number of plugins provide other advanced functions. Potlatch[58] is a flash-based web editor for OSM. Vespucci[59] is the first OSM editor specifically developed for small and large Android-based devices; it provides a reasonably extensive set of editing functionalities, which makes it usable on the field by novice and experienced OSM contributors. Merkaartor[60] is a desktop-based software editor for OSM that is available for installation and use on most operating systems; similarly to JOSM and Vespucci, Merkaartor provides a wide range of functionalities.

2.6.2 OSM-based Routing Services

OSM-based routing services are software-based solutions that use the data in the OSM database for the purposes of generating routing and navigation solutions. Routing and navigation is possible when objects in OSM have attributes (tags) that are helpful in solving these problems. The ability to apply attributes from different thematic areas on the same object (such as a road or a street) means that different routing applications can be easily developed.

The Open Source Routing Machine (OSRM)[61] is a C++ routing engine for finding 'shortest paths in road networks'. It supports car, bicycle and walk modes and is 'easily customized through profiles'. GraphHopper[62] is a company based in Germany focused on delivering the 'fastest possible routing algorithms' and 'privacy protection' using open source software for their customers. Their open source routing library and server includes elevation data and allows routing for several difficult vehicle types. The MapQuest Directions API[63] is offered by the US company MapQuest and calculates 'point-to-point, multipoint, and optimized routes'. The API can be used by any application, and the directions are based on OSM data. OpenRouteService[64] is a routing service developed by the GIScience Research Group at Heidelberg University (Germany); it provides routing capabilities for different categories (including wheelchairs users), features an advanced graphic interface and is also available in a mobile version. Kurviger[65] is a specialised routing service for motorcyclists, which computes optimal paths considering the topography of the terrain. It is only available in German. Cruiser for Android[66] is an Android-based mapping and navigation application. Wheelmap.org[67] is an open and free online map of wheelchair-accessible places. While it is not actually a routing application per se, it provides information on the wheelchair-accessibility of public places, which is very useful for wheelchair users, by allowing contributors to directly edit OSM to provide accessibility information. ViaMichelin[68] is a 'wholly owned subsidiary of the Michelin Group'[69]; it 'designs, develops and markets digital travel assistance products and services for road users in Europe', and the German version of their route planner uses an OSM Outdoor Layer visualisation[70]. INRIX Traffic[71] is a commercial product for navigation and traffic information that uses OSM data; the application learns the preferences and daily routines of the user, and, based on the learned activities, makes a daily personalised itinerary with the anticipated tours and frequently used routes.

2.6.3 Other Services

In this section, we provide some links to other services that use OSM but do not necessarily fit neatly inside our classifications. In OSM, nodes that have specific tags are often called POI amongst contributors and users of OSM. There is no absolute set of tags that qualify as indicating a POI, but usually a POI will have tags related to amenities, such as buildings, shopping, education or buildings with cultural and historical significance. The OpenPoiMap[72] provides a map-based visualisation of all POI in OSM for any part of the world: POI are presented as individual layers, which can be turned on or off, and, based on what visualisation information the map provides, contributors can then edit the POI data directly in OSM using the links provided on the interface. The

Places! service[73] attempts to present a visualisation of the analysis of patterns in place names within given countries based on the OSM database for those countries. For example, Places! tries to find patterns in the spatial distribution of places in Switzerland containing the term 'berg' or places in the United Kingdom containing the term 'hill' in their name. The analysis is performed offline and updated regularly.

The OSM Analytics[74] application recently launched by HOT provides interactive functionality to analyse how specific OSM features are mapped in a specific region. This tool allows the user to select the geographic region of interest and shows a graph of the mapping activity in that region. It is possible to select a specific time interval to view the number of newly mapped or edited features in that period; the map will highlight the matching buildings, as related to this time interval. This tool is a very useful way to obtain a high-level view of how OSM developed in a particular region. Finally, the Show-Me-The-Way application[75] is an interactive web application that displays near real-time edits performed by contributors to OSM. The application loads recent edits and displays them by jumping to the particular region where the edit was made. This type of visualisation is possible owing to the fact that very recent edits submitted to OSM by contributors are immediately available for access by anyone who connects to the OSM API or other services listed in Section 2.6.

2.7 Quality Assurance for OSM

The quality of OSM data is under constant scrutiny by the scientific community. The quality of data in OSM is one of the major concerns that industry and authoritative agencies such as National Mapping Agencies (NMAs), Land and Cadastral Agencies and other types of government agencies have about OSM (Olteanu-Raimond et al., 2017b). In practice, there is no single set of metrics or criteria against which OSM can be measured that will satisfy all users for the myriad of possible end applications. The quality of the OSM data and suitability for a particular application, purpose or use case is very much dependent on the characteristics of the problem being tackled. The OSM community recognises the importance of data quality, and a very wide range of tools and applications have been developed to tackle this issue. In this section, we provide some introduction to a small number of these. A comprehensive list is maintained on the OSM wiki[76].

BBBike and Geofabrik deliver the OSM Map Compare tool[77], which allows visual comparison of OSM map layers with other popular mapping systems such as Google, Bing, HERE, ESRI, etc. The web map interface allows users to visually compare any region in OSM with the corresponding mapping in the other popular systems. IGN France (French National Institute of Geographic and Forest Information) provides a very similar system to Map Compare with their Ma Visionneuse[78] application, which allows OSM to be compared with IGN layers, amongst others; this is particularly useful for comparison between French web map layers. The OSM Inspector[79], also by Geofabrik, provides an overlay of potential errors or data quality problems onto an OSM map. These problems include: very long ways (polylines); self-intersecting ways, polygons or polylines, which are represented by only one node; and polygons or polylines that have duplicate nodes contained within them.

Taginfo[2] is a very popular Web-based application that displays up-to-date statistics about the tags used in the OSM database, e.g. which tags are used, how many times they are used, where a certain tag occurs, etc. Taginfo is particularly useful for finding problems with the keys or values in tags, the popularity of tags, where specific tags are used and which other tags are used in combination with them. The use of taginfo to find problems with tagging relates to its very comprehensive listing of the ranking of popularity/application of values to specific keys in tags. This can quickly allow an OSM expert to identify instances of an incorrect assignment of values in tags that has an overall effect on tag data quality. Taginfo does not provide any information on errors relating to geometry or topology. Osmose[80], an acronym for OpenStreetMap Oversight Search Engine, is a quality assurance tool available to detect issues in OSM data; it is also useful for integrating third-party datasets. It tries to detect anomalies in the data and then display them on an OSM map, from which contributors can fix or update them. Keep Right[81] is one of the oldest quality assurance tools in OSM. It displays automatically detected errors on the OSM map or in a list format, and it detects a very wide set of error types, including geometry errors, topological errors, attribution errors and other general OSM errors.

MapRoulette[82] is a Web-based application that proposes challenges to fix errors in OSM. Each challenge represents a set of tasks, and OSM contributors can fix the errors by performing edits in OSM in the usual way. The challenges vary in difficulty, allowing contributors to choose the types of errors that they feel confident about fixing. The fixing is very heavily focused on the contributors' interpretation of information from aerial imagery. DeepOSM[83] attempts to detect problems in OSM road networks using neural networks. The system downloads satellite imagery and the corresponding OSM data that show roads/features for that area. This allows DeepOSM to generate training and evaluation data for the neural networks, which then calculate predictions of misregistered roads in OSM.

The Grass&Green project (Ali et al., 2016) asks OSM contributors to correct tagging or classification of land use features involving grass or green areas. This application provides a two-screen interface, where an OSM feature is highlighted on the standard OSM web-map layer and in aerial imagery. The user (who needs to have an OSM account) must then provide an appropriate classification for this entity by choosing what he/she believes is correct from the list of classifications: grass, park, garden, forest and meadow. The JOSM

Validator[84] 'is a core feature of JOSM which checks and fixes invalid data' that have been contributed to OSM or are being contributed for the first time. The validator checks and fixes a wide variety of problems, including topological errors, unclosed polygons and overlapping areas.

Academic research has produced a wide range of quality assessment and comparison tools for OSM (Ostermann and Granell, 2017). One of the most recently published is that of Brovelli et al. (2017): this open source software tool provides an automated comparison of street network data in OSM with that in an authoritative dataset. Users of the tool must provide the authoritative dataset for comparison.

2.8 Games, Leisure and General Public Information

In this final section of applications for OSM, we describe a mixture of applications that use OSM for the purposes of games, leisure or general public information.

'Collapse – The Division Game'[85] is a simulation game based on open datasets (including OSM data), created by Ubisoft to introduce the environment upon which the new online action game 'TomClancy's The Division' (for Windows, Playstation and Xbox)[86] is based. The user is the first person in the world infected with a virus, and the game realistically simulates the diffusion of the virus until the collapse of society; OSM data relating to health facilities, societal infrastructure and transportation are used in the simulation. The OSM game Kort[87] is very similar to MapRoulette[79], with the exception that Kort drives a gamification approach to OSM error fixing. Kort was developed for usage mainly on mobile devices but also works well on most browsers. For both solving tasks and checking existing solutions, points (so-called Koins) can be earned. The goal is to continually rise through the ranks of the high-score list. Additionally, players are also awarded medals for their efforts. At the time of writing, there are over 2,000 active players having solved almost 50,000 tasks. The solutions to tasks must be evaluated and accepted by other users before they are submitted to the OSM database.

In a YouTube video[88], an OSM contributor provides a video-based visualisation of the contribution of nodes to OSM over the period 2004–2016. Nodes in OSM that have had more editing activity on them are coloured using a heatmap approach. This timelapse video and many others listed on the OSM wiki[89] provide a very good high-level overview of how OSM has developed since its inception. The node density map by tyrasd[90] provides a static visual overview of how many nodes are mapped within any OSM region. Lukas Martinelli[91] produced a Global Noise Pollution map based on the urban infrastructure data in OSM for cities and urban areas. GoodCityLife is a group of freelance researchers in urban dynamics who use OSM to produce visualisations. One such visualisation is their Smelly Maps[92], which uses the underlying OSM data for a city or region to calculate if there is likely to be nasty odours or smells in a locality. Bahnhof.de[93] is the website providing information about railway stations in Germany; OSM is used as the base layer for the mapping on this information website. The flight simulation software World2XPlane by X-Plane[94,95] is also worth mentioning; this software takes OSM data and converts the data into scenery for X-Plane. It uses as much information as possible to generate highly realistic scenery.

3 Conclusions and Discussion

In this chapter, we have provided an overview of the OSM project. As mentioned in the introduction, OSM is probably the most famous example of VGI on the Internet today. Even at the time of writing (during the summer of 2016), the project continued to grow and expand, with over 2.7M registered contributors/users and almost 3.4B nodes of data, which made up almost 350M polygons and polylines. Around 37,000 contributors are active in OSM during a typical month. OSM can certainly claim to be the largest freely and openly accessible database of geographic data in the world. Indeed its rate of growth in terms of geographic data and frequency of contributions and editing brings OSM into the realm of geographic big data (Leonelli, 2014). When one considers the extended OSM ecosystem of open source software, data download services, data visualisation services, wiki help systems, mailing lists and forums, OSM serves as a very suitable starting point for any discussion on VGI. Indeed one could speculate on how VGI would have developed if OSM had been absent from this space. This chapter has attempted to give the reader who is new to OSM an introduction to the OSM ecosystem while providing the reader familiar with OSM an overview of where OSM currently stands in the world of VGI.

In the remaining chapters of this book, OSM will be mentioned and discussed in many different ways. In Chapter 4, Touya et al. (2017) address the challenges of automated mapmaking using VGI as the input data, and the authors consider OSM as a key source, but not the only source, of this VGI data. Chapter 2, See et al. (2017) has already indicated that there are many sources of VGI available today. While OSM is open data and is licensed under the Open Data Commons Open Database License (ODbL), there are privacy and ethical issues around the reuse of OSM data. In OSM, one is free to copy, distribute, transmit and adapt OSM data, as long as credit is provided to OSM and its contributors. If one alters or builds upon the data, then the resultant data must also be distributed under the same licence. Chapter 6 tackles some of these issues for OSM and VGI in general (Mooney et al., 2017). In Chapter 8, Antoniou and Skopeliti (2017) consider how the concept of quality has evolved in OSM over time through the analysis of the evolution of OSM data specifications and of OSM editors. The very evolution and changes over time to the OSM ecosystem can influence the quality of OSM data. Related to this theme, Chapter 9, by Skopeliti et al. (2017), considers how quality in VGI can be visualised and communicated effectively, with significant research work having already been carried out on this topic using OSM as the case-study. As discussed earlier in this chapter, OSM has a very flexible and easy-to-understand approach to the contribution of new geographic data or editing of existing data in the OSM database. Chapter 10 considers best practices for VGI data collection, and Minghini et al. (2017) propose in that chapter that the lack of protocols and the flexibility of contribution is not necessarily a good thing in terms of producing consistently high-quality VGI data. Chapter 11 (Bastin et al., 2017) considers VGI data management and suggests ways in which OSM can be integrated into the so-called Semantic Web, where all OSM's data would be converted to Linked Data. Finally, Chapter 13 (Olteanu-Raimond et al., 2017a) discusses VGI and the role of NMAs, with OSM often seen as a rival or competitor to the geographic data services provided by these agencies. As is obvious from this overview of the remaining chapters of the book, a deep scientific discussion of VGI is impossible without reflecting on and considering the impact and influence of OSM. This is certainly very likely to continue for many years to come.

3.1 The Future of OSM

OSM's greatest strength will always be its huge pool of contributors. Thousands of these contributors have collected and generated some of the world's best street and topographic data without expensive teams of professional surveyors or world-class equipment. As the world and the urban and natural environment change every day, OSM contributors have the ability to depict this changing world in a map and a database that belong to them. OSM may not yet have the advanced types of features that Google Maps has – street-view images, multi-modal navigation, social recommendations, etc. – but it may soon have. Mapillary[96,97], which is a service for crowdsourcing street-level photographs using smartphones and computer vision, has almost 70 million geotagged street-level photographs at the time of writing. Mapillary shares the open data ethos of OSM and they can work well together (Juhász and Hochmair, 2016). Very similarly, efforts are in place to link OSM elements with their corresponding Wikipedia pages and Wikidata items. As an example, the WTOSM[98] (Wikipedia To OSM) service developed by the Italian OSM community automatically identifies Wikipedia pages that can be linked (by means of tags) to OSM elements. Mature services such as OpenRouteService provide navigation services based wholly on OSM's database. One of the factors in the evolution of OSM over the past decade or so has been the ability of the project to adapt and expand in the face of technological advancements in other areas of ICT and Open Source Software. Web service access to the OSM database or its mirrors has improved and is very stable, allowing developers to build an array of applications using the data directly from the database.

There are some challenges for OSM going forward. These challenges are a mixture of factors based on the social and technological aspects of VGI (Mooney, 2015). Contributors can make edits to the OSM global database without any real controls or moderation at the point of contribution. Despite the fact that there are many applications available for an a posteriori quality check (see Section 2.7), as long as edits can be made without initial controls the issue of OSM data quality will remain a contentious one. Relatively *unknown* contributors from an *unknown* crowd supplying geospatial data is a concern to end users and stakeholders such as NMAs, government agencies and commercial companies. There have been many instances in the past where large amounts of OSM data have been deleted by new or inexperienced contributors. Some authors have considered the problem of automated detection of instances of vandalism and of the purposeful deletion of data in OSM (Neis et al., 2012). Many local OSM communities have long debated the wish and need to implement tools for checking and approving contributions (e.g. by more experienced contributors or by the community itself). However, such an implementation would be clearly against the very same nature of the OSM project, and no formal actions are yet in place in this regard.

Several academic studies have shown that for specific regions of the world, OSM has reached a very high and mature level of completeness and spatial accuracy compared to data from sources such as NMAs (Dorn et al., 2015). One of the major challenges will be to sustain the contributor motivation for editing and maintaining the OSM database into the future (Budhathoki and Haythornthwaite, 2012). Every day sees less *white space* or empty places on the OSM map. Similar scenarios are being observed in Wikipedia (Jankowski-Lorek et al., 2016). The task of being an OSM contributor is changing from that of being the contributor of brand new geodata to OSM to that of *map gardening* (McConchie, 2016; Sinton, 2016); in this latter case, contributors are not necessarily involved in contributing new material to OSM but are attending to the upkeep and update of the existing geometry and attribute data (tags) in the database.

As geolocation is further embedded into social media, user-generated content on the Internet, etc., issues of privacy and ethics can be raised (Blatt, 2015), and the work outlined in Chapter 6 of this book (Mooney et al., 2017), highlighting these problems in relation to VGI, will become critical; currently, very little work has been undertaken by the research community into privacy and ethics in VGI. In the final chapter of one of the first edited volumes dedicated to OSM, Mooney (2015) advises that the academic community has a significant role to play in the future of OSM; through scientific research and investigation, the academic community is encouraged to feed its results and experiences back directly into the OSM community and become more closely involved in the day-to-day workings of the OSM ecosystem. This model has been very successful in the open source software community, and this can extend to the OSM world.

Notes

[1] https://www.mapbox.com/blog/redesigned-osm-mapping-guides/
[2] https://taginfo.openstreetmap.org
[3] https://josm.openstreetmap.de
[4] http://wiki.openstreetmap.org/wiki/ID
[5] http://wiki.openstreetmap.org/wiki/Stats
[6] http://fieldpapers.org
[7] http://wiki.openstreetmap.org/wiki/Import/Catalogue
[8] https://wiki.openstreetmap.org/wiki/Main_Page
[9] http://wiki.openstreetmap.org/wiki/List_of_OSM-based_services
[10] http://download.geofabrik.de/
[11] http://wiki.openstreetmap.org/wiki/Planet.osm
[12] http://wiki.openstreetmap.org/wiki/API
[13] http://wiki.openstreetmap.org/wiki/Overpass_API
[14] https://overpass-turbo.eu/
[15] https://mapzen.com/data/metro-extracts/
[16] http://wiki.openstreetmap.org/wiki/Export
[17] http://wiki.openstreetmap.org/wiki/OSM_XML
[18] http://wiki.openstreetmap.org/wiki/Software/Desktop
[19] https://www.geofabrik.de/
[20] https://mapzen.com/products/#data
[21] http://www.citizensensor-cost.eu
[22] http://vgibox.eu
[23] http://teachosm.org/en/
[24] https://hotosm.org/updates/2016-03-09_200_kids_map_swaziland_for_malaria_elimination
[25] https://hotosm.org
[26] http://tasks.hotosm.org
[27] http://www.missingmaps.org
[28] https://disastermappers.wordpress.com/
[29] https://www.mapbox.com/blog/ebola-mapping-progress/
[30] http://pierzen.dev.openstreetmap.org/hot/leaflet/OSM-Compare-before-after-philippines.html#12/11.2197/124.9925
[31] http://wiki.openstreetmap.org/wiki/Commercial_OSM_Software_and_Services
[32] https://www.mapbox.com
[33] http://www.mapquest.com
[34] http://maps.stamen.com
[35] https://mapzen.com
[36] http://www.camptocamp.com/en/
[37] https://www.mapbox.com/blog/nyc-and-openstreetmap-cooperating-through-open-data/
[38] http://mapgive.state.gov/

39 http://trimet.org/#/planner
40 https://twitter.com/Gendarmerie/status/691947889103392768
41 http://www.modernisation.gouv.fr/sites/default/files/fichiers-attaches/ban_cp_150415_en.pdf
42 http://opentopomap.org
43 http://opencyclemap.org/
44 http://hikebikemap.org
45 http://www.opensnowmap.org
46 https://nsidc.org/data/MOD10A2
47 https://www.osmhydrant.org
48 http://openfiremap.org
49 http://maps.stamen.com/terrain
50 http://maps.stamen.com/toner
51 http://maps.stamen.com/watercolor
52 http://osmbuildings.org/
53 http://www.f4map.com/
54 https://github.com/kothic/kothic-js
55 https://www.mapbox.com/mapbox-studio/
56 http://wiki.openstreetmap.org/wiki/Editors
57 http://wiki.openstreetmap.org/wiki/ID
58 http://wiki.openstreetmap.org/wiki/Potlatch_2
59 http://wiki.openstreetmap.org/wiki/Vespucci
60 http://wiki.openstreetmap.org/wiki/Merkaartor
61 http://project-osrm.org
62 https://graphhopper.com
63 https://developer.mapquest.com/products/directions
64 http://www.openrouteservice.org
65 https://kurviger.de
66 https://wiki.openstreetmap.org/wiki/Cruiser
67 http://wheelmap.org/en/map#/?zoom=14
68 http://www.viamichelin.de/
69 https://en.wikipedia.org/wiki/Michelin
70 http://www.thunderforest.com/maps/outdoors/
71 https://www.engadget.com/2016/03/30/inrix-traffic-app-uses-ai-to-learn-your-driving-habits
72 http://openpoimap.org
73 http://bgrsquared.com/places
74 http://osm-analytics.org
75 https://osmlab.github.io/show-me-the-way/
76 http://wiki.openstreetmap.org/wiki/Quality_assurance
77 http://mc.bbbike.org/mc/
78 http://mavisionneuse.ign.fr/visio.html?lon=3.46539&lat=46.044673&zoom=15&num=4&mt0=ign-cartes&mt1=ign-scexstandard&mt2=google-map&mt3=osmfr

[79] http://tools.geofabrik.de/osmi
[80] http://wiki.openstreetmap.org/wiki/Osmose
[81] http://wiki.openstreetmap.org/wiki/Keep_Right
[82] http://maproulette.org
[83] https://libraries.io/github/trailbehind/DeepOSM
[84] http://wiki.openstreetmap.org/wiki/JOSM/Validator
[85] http://collapse-thedivisiongame.ubi.com
[86] http://tomclancy-thedivision.ubi.com/game/en-us/home/
[87] http://play.kort.ch
[88] https://www.youtube.com/watch?v=FdRO-QZaWX8
[89] http://wiki.openstreetmap.org/wiki/Timelapse_videos
[90] http://tyrasd.github.io/osm-node-density/
[91] http://lukasmartinelli.ch/gis/2016/04/03/openstreetmap-noise-pollution-map.html
[92] http://goodcitylife.org/smellymaps/
[93] http://www.bahnhof.de/bahnhof-de/Karlsruhe_Hbf.html?hl=karlsruhe
[94] http://www.flightsim.com/vbfs/content.php?16301-OpenStreetMap-Tutorial
[95] http://www.x-plane.com/desktop/home/
[96] http://blog.mapillary.com/update/2016/05/24/use-mapillary-editing-OSM.html
[97] http://www.openstreetmap.org/user/Blackbird27/diary/38711
[98] http://geodati.fmach.it/gfoss_geodata/osm/wtosm/en_US/index_2.html

Reference list

Ali, A.L.; Sirilertworakul, N.; Zipf, A.; Mobasheri, A. 2016. Guided Classification System for Conceptual Overlapping Classes in OpenStreetMap. *ISPRS International Journal of Geo-Information* 5, no. 6: 87.

Antoniou, V and Skopeliti, A. 2017. The Impact of the Contribution Microenvironment on Data Quality: The Case of OSM. In: Foody, G, See, L, Fritz, S, Mooney, P, Olteanu-Raimond, A-M, Fonte, C C and Antoniou, V. (eds.) *Mapping and the Citizen Sensor*. Pp. 165–196. London: Ubiquity Press. DOI: https://doi.org/10.5334/bbf.h.

Ballatore, A., Mooney, P., 2015. Conceptualising the geographic world: the dimensions of negotiation in crowdsourced cartography. *International Journal of Geographical Information Science* 29, 2310–2327. DOI: https://doi.org/10.1080/13658816.2015.1076825

Ballatore, A., Zipf, A., 2015. A conceptual quality framework for Volunteered Geographic Information, in: Fabrikant, S.I., Raubal, M., Bertolotto, M., Davies, C., Freundschuh, S., Bell, S. (Eds.), *Spatial Information Theory*. Springer International Publishing, Cham, pp. 89–107.

Bastin, L, Schade, S and Schill, C. 2017. Data and Metadata Management for Better VGI Reusability. In: Foody, G, See, L, Fritz, S, Mooney, P, Olteanu-Raimond, A-M, Fonte, C C and Antoniou, V. (eds.) *Mapping and the Citizen Sensor*. Pp. 249–272. London: Ubiquity Press. DOI: https://doi.org/10.5334/bbf.k.

Blatt, A.J., 2015. Data privacy and ethical uses of Volunteered Geographic Information, in: *Health, Science, and Place*. Springer International Publishing, Cham, pp. 49–59.

Brovelli, M.A., Minghini, M., Molinari, M., Mooney, P., 2017. Towards an automated comparison of OpenStreetMap with authoritative road datasets. *Transactions in GIS* 21, 191–206. DOI: https://doi.org/10.1111/tgis.12182

Budhathoki, N.R., Haythornthwaite, C., 2012. Motivation for open collaboration: Crowd and community models and the case of OpenStreetMap. *American Behavioral Scientist* 57, 548–575. DOI: https://doi.org/10.1177/0002764212469364

Coast, S., 2010. Microsoft Imagery details. OpenStreetMap blog. Available at https://blog.openstreetmap.org/2010/11/30/microsoft-imagery-details. [Last accessed 7 July 2016].

Dorn, H., Törnros, T., Zipf, A., 2015. Quality evaluation of VGI using authoritative data—A comparison with land use data in Southern Germany. *ISPRS International Journal of Geo-Information* 4, 1657–1671. DOI: https://doi.org/10.3390/ijgi4031657

Ebrahim, M., Minghini, M., Molinari, M., Torrebruno, A., 2016. MiniMapathon: Mapping the world at 10 years old. In: Proceedings of the 8th Annual International Conference on Education and New Learning Technologies (EDULEARN 2016), Barcelona, Spain, 4–6 July 2016, pp. 4200–4208. DOI: https://doi.org/10.21125/edulearn.2016.2018

Fritz, S, See, L and Brovelli, M. 2017. Motivating and Sustaining Participation in VGI. In: Foody, G, See, L, Fritz, S, Mooney, P, Olteanu-Raimond, A-M, Fonte, C C and Antoniou, V. (eds.) *Mapping and the Citizen Sensor*. Pp. 93–117. London: Ubiquity Press. DOI: https://doi.org/10.5334/bbf.e.

Haklay, M., Antoniou, V., Basiouka, S., Soden, R., Mooney, P., 2014. *Crowdsourced Geographic Information Use in Government*. Report to GFDRR (World Bank). UCL-Geomatics, London.

Jankowski-Lorek, M., Jaroszewicz, S., Ostrowski, Ł., Wierzbicki, A., 2016. Verifying social network models of Wikipedia knowledge community. *Information Sciences* 339, 158–174. DOI: https://doi.org/10.1016/j.ins.2015.12.015

Jokar Arsanjani, J., Zipf, A., Mooney, P., Helbich, M., 2015. An introduction to OpenStreetMap in Geographic Information Science: Experiences, research, and applications, in: Jokar Arsanjani, J., Zipf, A., Mooney, P., Helbich, M. (Eds.), *OpenStreetMap in GIScience, Lecture Notes in Geoinformation and Cartography*. Springer International Publishing, Cham, pp. 1–18.

Juhász, L., Hochmair, H.H., 2016. User contribution patterns and completeness evaluation of Mapillary, a crowdsourced street level photo service:

Contribution patterns of Mapillary. *Transactions in GIS* 20, 925–947. DOI: https://doi.org/10.1111/tgis.12190

Leonelli, S., 2014. What difference does quantity make? On the epistemology of Big Data in biology. *Big Data & Society* 1, 2053951714534395. DOI: https://doi.org/10.1177/2053951714534395

McConchie, A., 2016. OpenStreetMap past(s), OpenStreetMap future(s). Stamen blog. Available at https://hi.stamen.com/openstreetmap-past-s-openstreetmap-future-s-cafddc2a4736#.hklbicd24 [Last accessed 22 March 2017]

Minghini, M., Antoniou, V., Fonte, C.C., Estima, J., Olteanu-Raimond, A.-M., Minghini, M, Antoniou, V, Fonte, C C, Estima, J, Olteanu-Raimond, A-M, See, L, Laakso, M, Skopeliti, A, Mooney, P, Arsanjani, J J, Lupia, F. 2017. The Relevance of Protocols for VGI Collection. In: Foody, G, See, L, Fritz, S, Mooney, P, Olteanu-Raimond, A-M, Fonte, C C and Antoniou, V. (eds.) *Mapping and the Citizen Sensor*. Pp. 223–247. London: Ubiquity Press. DOI: https://doi.org/10.5334/bbf.j.

Mooney, P., 2015. Quality assessment of the contributed land use information from OpenStreetMap versus authoritative datasets, in: Jokar Arsanjani, J., Zipf, A., Mooney, P., Helbich, M. (Eds.), *OpenStreetMap in GIScience, Lecture Notes in Geoinformation and Cartography*. Springer International Publishing, Cham, pp. 319–324.

Mooney, P, Olteanu-Raimond, A-M, Touya, G, Juul, N, Alvanides, S and Kerle, N. 2017. Considerations of Privacy, Ethics and Legal Issues in Volunteered Geographic Information. In: Foody, G, See, L, Fritz, S, Mooney, P, Olteanu-Raimond, A-M, Fonte, C C and Antoniou, V. (eds.) *Mapping and the Citizen Sensor*. Pp. 119–135. London: Ubiquity Press. DOI: https://doi.org/10.5334/bbf.f.

Neis, P., Goetz, M., Zipf, A., 2012. Towards automatic vandalism detection in OpenStreetMap. *ISPRS International Journal of Geo-Information* 1, 315–332. DOI: https://doi.org/10.3390/ijgi1030315

Neis, P., Zielstra, D., Zipf, A., 2013. Comparison of volunteered geographic information data contributions and community development for selected world regions. *Future Internet* 5, 282–300. DOI: https://doi.org/10.3390/fi5020282

Olteanu-Raimond, A-M, Laakso, M, Antoniou, V, Fonte, C C, Fonseca, A, Grus, M, Harding, J, Kellenberger, T, Minghini, M, Skopeliti, A. 2017a. VGI in National Mapping Agencies: Experiences and Recommendations. In: Foody, G, See, L, Fritz, S, Mooney, P, Olteanu-Raimond, A-M, Fonte, C C and Antoniou, V. (eds.) *Mapping and the Citizen Sensor*. Pp. 299–326. London: Ubiquity Press. DOI: https://doi.org/10.5334/bbf.m.

Olteanu-Raimond, A.-M., Hart, G., Foody, G., Touya, G., Kellenberger, T., Demetriou, D., 2017b. The scale of VGI in map production: A perspective of European National Mapping Agencies. *Transactions in GIS* 21, 74–90. DOI: https://doi.org/10.1111/tgis.12189

OpenStreetMap, 2016. The OpenStreetMap Wiki Homepage. Available at https://wiki.openstreetmap.org/wiki/Main_Page [Last accessed 01 July 2016].

O'Reilly, T., 2007. What is Web 2.0: Design patterns and business models for the next generation of software. *Communications & Strategies* 65, 17–37.

Ostermann, F.O., Granell, C., 2017. Advancing science with VGI: Reproducibility and replicability of recent studies using VGI. *Transactions in GIS* 21, 224–237. DOI: https://doi.org/10.1111/tgis.12195

See, L, Estima, J, Pőd.r, A, Arsanjani, J J, Bayas, J-C L and Vatseva, R. 2017. Sources of VGI for Mapping. In: Foody, G, See, L, Fritz, S, Mooney, P, Olteanu-Raimond, A-M, Fonte, C C and Antoniou, V. (eds.) *Mapping and the Citizen Sensor*. Pp. 13–35. London: Ubiquity Press. DOI: https://doi.org/10.5334/bbf.b

See, L., Mooney, P., Foody, G., Bastin, L., Comber, A., Estima, J., Fritz, S., Kerle, N., Jiang, B., Laakso, M., Liu, H.-Y., Milčinski, G., Nikšič, M., Painho, M., Pődör, A., Olteanu-Raimond, A.-M., Rutzinger, M., 2016. Crowdsourcing, citizen science or Volunteered Geographic Information? The current state of crowdsourced geographic information. *ISPRS International Journal of Geo-Information* 5, 55. DOI: https://doi.org/10.3390/ijgi5050055

Sinton, D.S., 2016. The simple map that became a global movement. *Directions Magazine*. Available at http://www.directionsmag.com/entry/osm-the-simple-map-that-became-a-global-movement/466280 [Last accessed 22March 2017].

Skopeliti, A, Antoniou, V and Bandrova, T. 2017. Visualisation and Communication of VGI Quality. In: Foody, G, See, L, Fritz, S, Mooney, P, Olteanu-Raimond, A-M, Fonte, C C and Antoniou, V. (eds.) *Mapping and the Citizen Sensor*. Pp. 197–222. London: Ubiquity Press. DOI: https://doi.org/10.5334/bbf.i.

Touya, G, Antoniou, V, Christophe, S and Skopeliti, A. 2017. Production of Topographic Maps with VGI: Quality Management and Automation. In: Foody, G, See, L, Fritz, S, Mooney, P, Olteanu-Raimond, A-M, Fonte, C C and Antoniou, V. (eds.) *Mapping and the Citizen Sensor*. Pp. 61–91. London: Ubiquity Press. DOI: https://doi.org/10.5334/bbf.d.

Zielstra, D., Hochmair, H.H., Neis, P., 2013. Assessing the effect of data imports on the completeness of OpenStreetMap - A United States case study. *Transactions in GIS* 17, 315–334. DOI: https://doi.org/10.1111/tgis.12037

CHAPTER 4

Production of Topographic Maps with VGI: Quality Management and Automation

Guillaume Touya*, Vyron Antoniou†,
Sidonie Christophe*, Andriani Skopeliti‡

*IGN, French Mapping Institute, COGIT team, Université Paris-Est, 73 avenue de Paris, 94160 Saint-Mandé, France, Guillaume.Touya@ign.fr

†Hellenic Army General Staff, Geographic Directorate, PAPAGOU Camp, Mesogeion 227-231, Cholargos, 15561, Greece

‡School of Rural and Surveying Engineering, National Technical University of Athens, 9 H. Polytechniou, Zografou, 15780, Greece

Abstract

The most common way to use geographic information is to make maps. With the ever growing amount of Volunteered Geographic Information (VGI), we have the opportunity to make many maps, but only automatic cartography (generalisation, stylisation, text placement) can handle such an amount of data with very frequent updates. This chapter reviews the recent proposals to adapt the current techniques for automatic cartography to VGI as the source data, focusing on the production of topographic base maps. The review includes methods to assess quality and the level of detail, which is necessary to handle data heterogeneity. The paper also describes automatic techniques to generalise, harmonise and render VGI.

How to cite this book chapter:

Touya, G, Antoniou, V, Christophe, S and Skopeliti, A. 2017. Production of Topographic Maps with VGI: Quality Management and Automation. In: Foody, G, See, L, Fritz, S, Mooney, P, Olteanu-Raimond, A-M, Fonte, C C and Antoniou, V. (eds.) *Mapping and the Citizen Sensor.* Pp. 61–91. London: Ubiquity Press. DOI: https://doi.org/10.5334/bbf.d.

Keywords

cartography, map generalisation, quality, OpenStreetMap, Flickr

1 Introduction

Maps are now everywhere, from the Web to smartphones, and are no longer limited to paper maps for hiking or routing. But most of the maps provided to the general public are not good maps, so they are not as effective as they could be. Whether they are static or dynamic (i.e. pan and zoom allowed), on paper or on screens of variable sizes, good maps are maps where every feature is legible, and where the user can easily understand the geography behind the map and the message of the map. Making good maps manually requires cartographic skills. However, when the amount of data is huge, for instance with the world OpenStreetMap (OSM) dataset, mapmaking has to be automated. Automating mapmaking entails two steps to obtain a legible topographic map out of a geographic database: selecting the data and the styles to be used to portray them, and refining the content in order to reach a legible map, which is complex when scale decreases, as the space in which to put the map symbols and the text reduces. These steps require the automation of three main processes: map generalisation (the simplification and abstraction of map objects when scale decreases), text placement, and cartographic symbolisation or stylisation. How to optimally automate such processes is still a research question, but, in recent years, maps have been more and more often produced through complete or partial automation. The traditional actors of automated mapmaking are the national or regional mapping agencies, the private map editors and the GIS software vendors. These actors have been used to making their maps out of traditional geographic databases, but what happens if the source data are partly or totally derived from Volunteered Geographic Information (VGI)? VGI is geographic information, and past studies on its quality (Girres and Touya, 2010; Haklay, 2010) have shown that it was satisfactory for many uses, but quite heterogeneous. Thus, the methods used for automated mapmaking should not be disrupted by the use of VGI as an input, but these methods need some adjustment to adapt to this new source of data: this adjustment is the topic of this chapter. Most of the problems presented here have been applied to the automated cartography of OSM, but we believe these problems and the proposed solutions also apply to different VGI sources, and even to cases where several VGI sources are combined into a map.

The next section of this chapter discusses the reasons why traditional automated mapping processes are not fully adapted to VGI, and is followed by a section that describes attempts to solve these problems by inferring the level of detail of VGI features. The fourth section then focuses on map generalisation, which may be the most complex of the cartographic processes. In the fifth section, the level of detailed harmonisation needed for large scale maps is discussed, while generalisation is dedicated to medium or small scale maps. The sixth part of the chapter focuses on the assessment of the quality of map features prior to applying automatic processes. Finally, in the seventh part, the issues related to advanced map stylisation with VGI are discussed.

2 Why Are Traditional Automated Mapping Processes Not Fully Adapted to VGI?

Traditional automated mapping processes have been developed to process authoritative datasets, or at least datasets with consistent and homogeneous specifications, which is clearly not the case when VGI is used as (one of) the map source(s). The first problem is that VGI datasets suffer from level of detail (LoD) heterogeneities. For instance, there is no LoD specification in OSM, which allows contributors a great deal of freedom in capturing either detailed features (e.g. the cadastral LoD buildings from Figure 1) or less detailed features (e.g. the rough built-up areas or lake outlines in Figure 1) depending partly on their skills but mostly on the data source, as precise GPS tracks allow more precision than low-resolution satellite imagery. This heterogeneity leads to LoD inconsistencies, i.e. some very detailed features and some less detailed features might coexist on a map and share spatial relations (Figure 1). Maps produced by National Mapping Agencies (NMAs), on the other hand, are based on datasets with strict specifications, where all features share the same geometrical resolution or granularity, whether they belong to the same theme or not. Thus the processes used to automate the production of such high-quality maps are not capable of handling the inconsistencies shown in Figure 1.

The main characteristic of VGI compared to traditional authoritative datasets is the heterogeneity of quality, with very-good-quality contributions and very-bad-quality ones. This is true for most types of VGI: for OSM first and

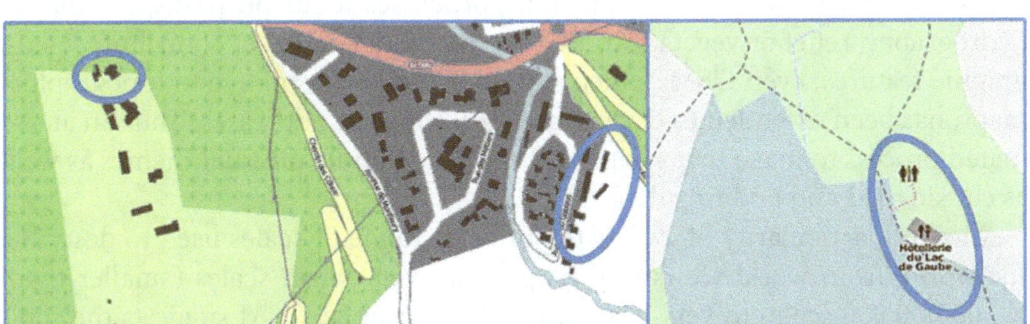

Fig. 1: Examples of LoD inconsistency in OSM. On the left, the rough built-up areas/forest limits intersect detailed buildings; on the right, detailed footpaths lie on the surface of a roughly digitised lake. ©OpenStreetMap contributors.

foremost, as shown in seminal studies by Girres and Touya (2010) and Haklay (2010), but also for photo sharing platforms such as Flickr (Zielstra and Hochmair, 2013), or even for hiking route sharing platforms (Ivanovic et al., 2015). Data quality varies from theme to theme, but also from feature to feature in the same theme (Girres and Touya, 2010). This is really different from authoritative datasets, where data quality is homogeneous, and cartography processes are developed in adaptation to this known quality. Among the quality indicators that can be heterogeneous with VGI, the most significant components are *positional accuracy*, *thematic accuracy*, *completeness* and *logical consistency*:

- *Positional accuracy* heterogeneity is, of course, a problem because it can increase the symbol overlap problems faced when creating a small scale map. Heterogeneity in positional accuracy might drive mapmakers to use incompatible features on the same scale, which can give a false picture of the reality and the relations among features.
- Heterogeneity in *thematic accuracy* is a problem because automated cartography relies on thematic information to classify the map features. The consequence of such heterogeneity is that processes should rely more on geometry and only use semantics when available.
- *Completeness* heterogeneity raises the problems of 'empty space' in the map. Empty spaces are useful to identify in automated mapmaking because they are excellent candidates to solve space conflicts during map generalisation or text placement. But, with VGI, empty might either mean really empty or just incomplete.
- *Logical consistency* heterogeneity is also a problem, because automated cartography uses, for instance, the topology of geographic networks to identify important features, and road symbolisation techniques require topologically correct networks.

Traditional NMA maps cover the classic themes of topographic maps, or road maps, and most automated mapmaking processes focus on roads, buildings, hydrography, relief or vegetation. VGI has a broader range of contributed geographic features; even OSM, which started as a free alternative to topographic maps, has been extended to cover amenities, shops or addresses. Thus an automated process to make maps with VGI needs to handle unusual themes as well as classic road and building datasets.

Another particularity of VGI is the broader range of scales used to describe the world, from world views that range from very small scales (smaller than 1: 100 000 000 scale) to very large scales. For instance, OSM suggests the capture of zebra crossings or traffic signals that can only be displayed at very large scales. Some projects even extend the OSM framework to indoor mapping (Goetz and Zipf, 2011). In contrast, traditional automated mapmaking targets a small number of fixed scales (Duchêne et al., 2014), and, even when the maps are displayed in online tools, the number of scales available is often limited by the number of scales available for paper maps (Dumont et al., 2016). In addition to the issue of the large range of scales in VGI, it should be noted that most of the automated processes were never developed for large scales that large and for small scales that small (e.g. the smallest scale produced by the French NMA only covers the whole French territory, excluding overseas territories).

Regarding symbology and stylisation, the automated processes are strongly related to the data and semantics. For instance, the choice of road symbols depends on the semantics of the road, and there has to be some consistency all along the road. When manipulating VGI data, how do we acquire these semantics? How do we handle the heterogeneities inherent to VGI?

3 Inferring LoD in VGI

3.1 LoD or Scale?

In cartography, the scale of a map is the ratio of the length of an object on the map by the length of the same object on the ground. But scale is also somehow related to map usage, and is then a proxy for map content. Maps around the scale of 1:25k are mainly used for hiking and contain information readable at this scale and useful for this purpose (e.g. footpaths, contour lines etc.); maps with a scale smaller than 1:500k are mainly used for road trips, and highlight the map themes related to roads. In contrast, it is too complex to assign a scale to VGI features, but here we consider the scale of a feature as the scale of the map at which this feature would be legible and legitimate.

LoD is a vaguer notion, which can be considered as the translation of map scale to geographic databases for which the scale is not fixed. Several factors affect the level of detail of geographic features:

- *geometric resolution*, i.e. the minimum distance between two vertices of the geometry, as an analogy with image resolution.
- *geometric precision*, i.e. the difference between the position in the database and the position in reality.
- *granularity*, i.e. the size of the smallest details in a geometry, such as the protrusions in the church in Figure 2 (left).
- *semantic resolution*, i.e. the amount of details in the semantic information attached to the geometric feature.
- *conceptual schema*, i.e. how much the ground truth information is abstracted; for instance, a wood abstracted by individual trees is more detailed than one abstracted by a polygon feature.

Thus it is difficult to infer LoD as a numerical value as one would for scale, so often categories are used, such as the LoD for 3D city models (Biljecki et al.,

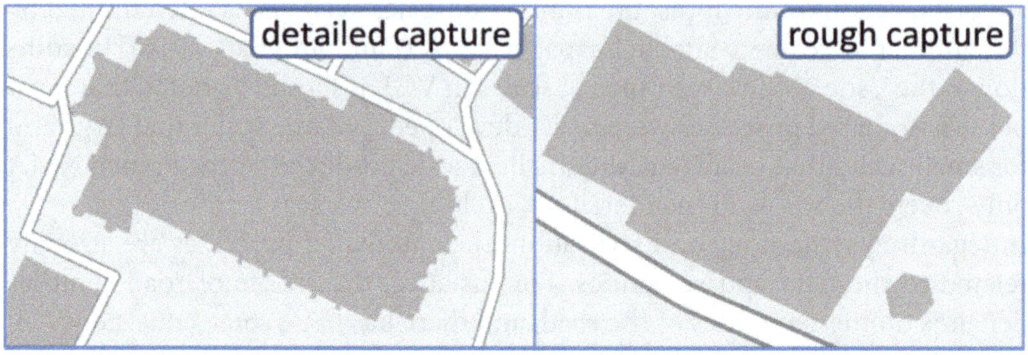

Fig. 2: Two churches with a similar granularity on the field that are captured with a different LoD: the left-hand one is captured from a scanned cadaster map and the right-hand one from Bing imagery. ©IGN, France.

2014). Touya and Brando-Escobar (2013) proposed five categories for the LoD of OSM features, from *Street* level to *Country* level. Scales can then be assigned to features if a scale range is assigned to each LoD category, e.g. the city level is assigned a scale range going from 1:15k to 1:50k (Touya and Reimer, 2015).

3.2 Reverse Engineering Scale Equivalency

Reimer et al. (2014) inferred a scale equivalency for OSM features by studying the characteristics of features in existing maps at different scales: for a given map theme, the measure that best characterises the difference in features at different scales is determined. In the example of urban areas in Reimer et al. (2014), vertex frequency (number of vertices in the polygon ring divided by the polygon perimeter) was the determining characteristic (Figure 3). Then, by inversing Töpfer's radical law (Töpfer and Pillewizer, 1966), which defines the optimal number of map features at a scale given their number at a bigger scale, and applying it to existing map features in the maps of NMAs, Reimer et al. (2014) were able to calculate the scale equivalency of any urban area in OSM.

3.3 Multiple Criteria Decision Method

We stated in Section 3.1 that LoD can be affected by a combination of five factors, all of which can be measured in a geographic dataset but are hardly comparable or can hardly be added. Multi-criteria decision methods are computational techniques that allow decision-making based on several criteria in those cases where a simple numerical value such as a mean is not a valid solution (Roy, 2005). Touya and Brando-Escobar (2013) propose a multi-criteria

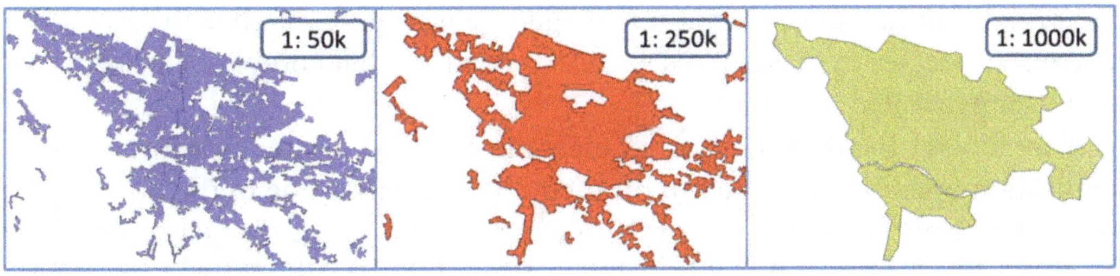

Fig. 3: Vertex frequency differences for urban areas in existing maps in France.

Fig. 4: Results of the automatic inference of LoD with the improved method from Touya and Reimer (2015) for OSM builtup areas in Tunisia (left) and OSM forest areas in France (right). ©OpenStreetMap contributors.

decision method to classify VGI features into LoD categories from street to country level. The method was improved by integrating elements from the scale equivalency in Touya and Reimer (2015). Some automatic results from the improved method are presented in Figure 4.

4 Map Generalisation of VGI

4.1 Current Generalisation in OpenStreetMap

Map generalisation is a complex process that simplifies and abstracts geographic information to produce a legible map at a given (smaller) scale. The problem of map generalisation automation has attracted research proposals for many years (see for instance Burghardt et al., 2014; Mackaness et al., 2007), and some mapping agencies are now able to use research results to produce maps with partial or total automation (Duchêne et al., 2014). One of the remaining challenges of automated generalisation research is to extend the current processes to make maps with VGI or maps that combine authoritative and user generated information.

If we look at the default maps available from OSM, there is almost no generalisation operation carried out on them. This is partly due to the philosophy of the OSM portal, which aims to show the content of the dataset rather than to display the best map possible. But it is also due to the difficulty of the generalisation process, which involves complex mechanisms that are not available in most mapping tools. However, some minimal selection operations are carried out in the default OSM map, using the semantics available to choose the zoom levels (i.e. scales) where features should be displayed. The piece of code below is extracted from the CartoCss file used to render buildings in the default OSM map. It shows that standard buildings are displayed only for zoom levels greater than 13 (zoom levels are ordered from 0 for the whole world to 19 in OSM), and with a coloured outline at zoom levels greater than 15.

```
#buildings {
        [zoom >= 13] {
                polygon-fill: @building-low-zoom;
                polygon-clip: false;
        }
        [zoom >= 15] {
                line-color: @building-line;
                polygon-fill: @building-fill;
                line-width: .75;
                line-clip: false;
        }
}
```

Besides these minimal selection operations, there are very few proposals dedicated to the issues of generalising VGI at present (Sester et al., 2014). Klammer (2013) proposed some solutions for tile-based maps such as OSM, with each tile being generalised separately, but potential problems at tile junctions are not handled: generalisation often requires an analysis of the neighbouring objects, which is not possible at the edge of the tiles. Schmid and Janetzek (2013) proposed to generalise the OSM road network at small scales on-the-fly using important placenames in the dataset. However, most of the issues remain unsolved: how can we deal with the broad range of scales in generalisation processes, with the diversity of themes or with the heterogeneities in quality and LoD?

The next two subsections address issues related to the range of scales and the diversity of themes with the generalisation of complex airports and railways from OSM. Section 4.4 addresses the generalisation of mashup maps with user generated content on top of reference datasets.

4.2 Generalisation of Complex Airports

Airports can be described in a great amount of detail in OSM, and contributors often use the OSM recommendations to capture airports as complex objects composed of runways, aprons where planes are parked, taxiways that connect aprons and runways, and terminal buildings. Figure 5 shows that such a complex structure is hard to represent legibly when the scale decreases, so generalisation algorithms dedicated to such structures must be used.

This subsection briefly describes a generalisation process presented in Touya and Girres (2014), where algorithms for the different types of features comprising airports are proposed, including, for instance, the decomposition of runways from polygons to lines. Here, we choose to focus on taxiway lines. Figure 5 shows that the junctions of taxiways are often complex, with shapes similar to slip roads. The first step in generalisation is to automatically characterise all of these complex junctions (see the coloured polygons on the right side of Figure 6) using the shapes of the lines, the angles of the connection and the number of connected taxiways. Then, each complex junction is simplified to a straight line crossing, removing all of the slip roads (Figure 6). Finally strokes are computed within the remaining taxiways. Strokes are groups of lines that follow the perceptual grouping principle of good continuity (Thomson and Richardson, 1999), like a continuous pen stroke, and have been used to simplify roads or rivers in the generalisation literature. Here, the smallest strokes are eliminated with a length threshold depending on map scale.

When algorithms for taxiways, runways, aprons and terminals (see Touya and Girres, 2014) are chained, complete airports can be generalised; the results for OSM airports with different initial complexities are presented in Figure 7, showing that the flexibility of the algorithms allows for the management of LoD heterogeneity of OSM data.

Fig. 5: The complexity of OSM airports composed of terminals, aprons, taxiways and runways, and their representation at several zoom levels. ©OpenStreetMap contributors.

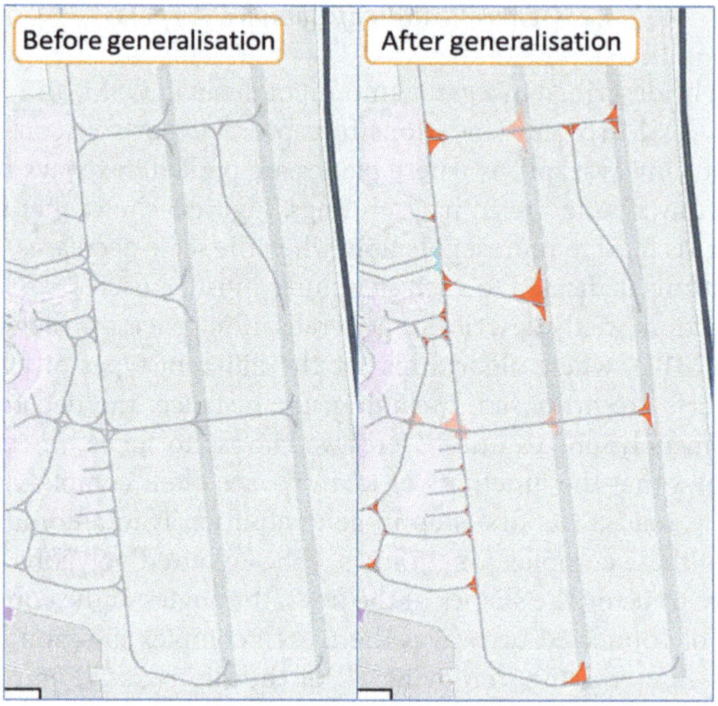

Fig. 6: Identification of different types of taxiway junctions (in red, pink and blue) and their simplification. ©OpenStreetMap contributors.

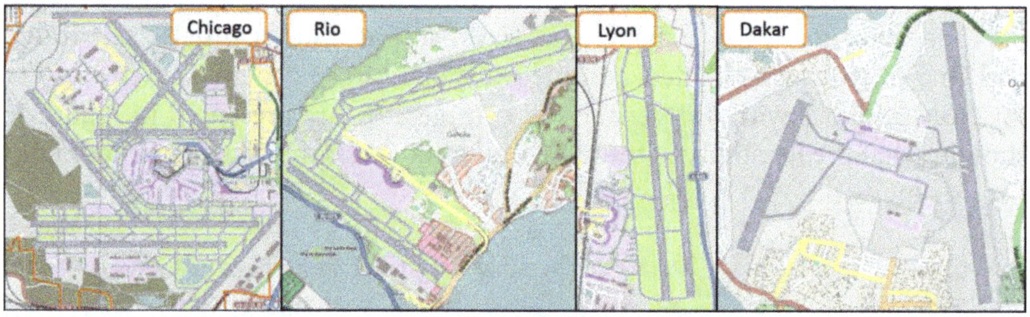

Fig. 7: 1:25k generalisation of airports of different initial complexities. ©OpenStreetMap contributors.

4.3 Generalisation of Railway Networks

Airports are not the only geographic feature that is captured with a greater complexity in OSM. The OSM specifications advise capturing each railway, even in a train station or in triage areas where a great number of lanes may exist (Figure 8). The railway lines are often very close to each other and their symbols overlap very quickly when the scale decreases. In this case, a good generalisation process is able to handle different densities of parallel railways and simplify them while preserving the connections and the patterns of the railways.

Fig. 8: A complex train station in OSM, with all of the railways captured. ©OpenStreetMap contributors.

Railway networks are composed of two very different types of patterns: the main railway lines with a small number of parallel tracks, and the train station with complex structures of tracks. The best strategy is to handle both parts of the network separately with different methods (Touya and Girres, 2014; Savino and Touya, 2015). The simplest railways to generalise are the main railway lines: the parts where several railway tracks are close and parallel have to be identified automatically and then replaced by a single track when the symbols overlap (Savino and Touya, 2015). The results of this method for railways extracted from OSM in France are presented in Figure 9.

Regarding train stations, a typification operation is required. Typification simplifies a pattern of geographic features while preserving the characteristics of the pattern more than the position of the features taken individually. Several complementary typification algorithms are proposed in Touya and Girres (2014) and Savino and Touya (2015), and Figure 10 shows a result for a 1:25k map of a small train station.

4.4 Generalisation of a Combination of Authoritative Data and VGI

When VGI is used as a thematic layer on top of a map, as in Figure 11, which is extracted from the IGN application called 'Leisure area'[1] , the issues related to generalisation are different from those related to generalisation of VGI only. The background map can be nearly generalised as a traditional topographic map, but the constraint is the preservation of the relations between the thematic layers and the background layers. If we use the example of Figure 11, the route should remain on top of the road, even if the road

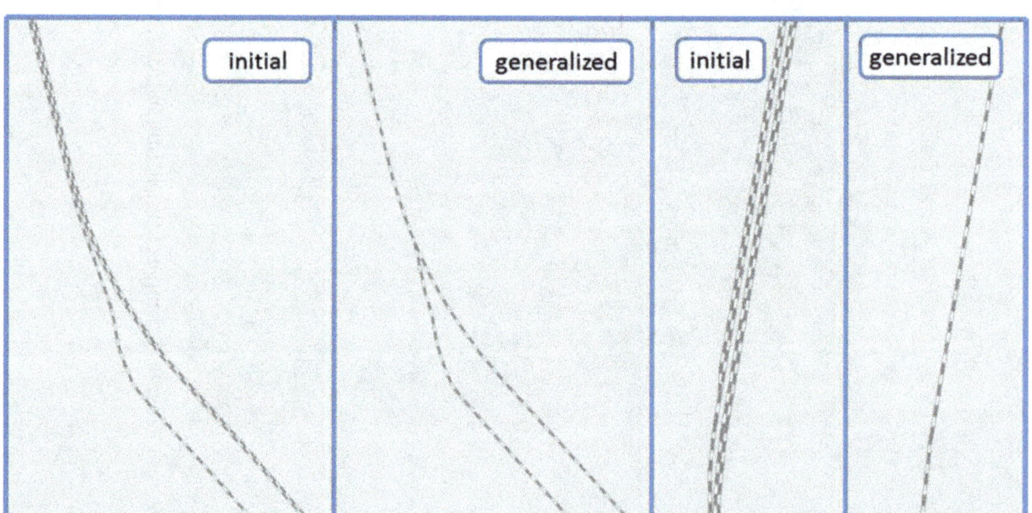

Fig. 9: Main railways with parallel lanes collapsed to single lanes (Savino and Touya, 2015). ©OpenStreetMap contributors.

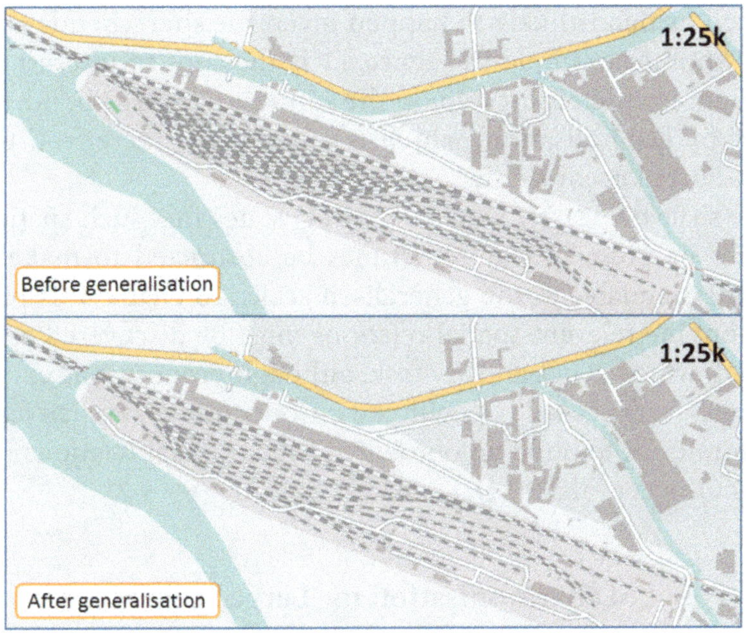

Fig. 10: 1:25k map generalisation of a small train station (Touya and Girres, 2014). ©OpenStreetMap contributors.

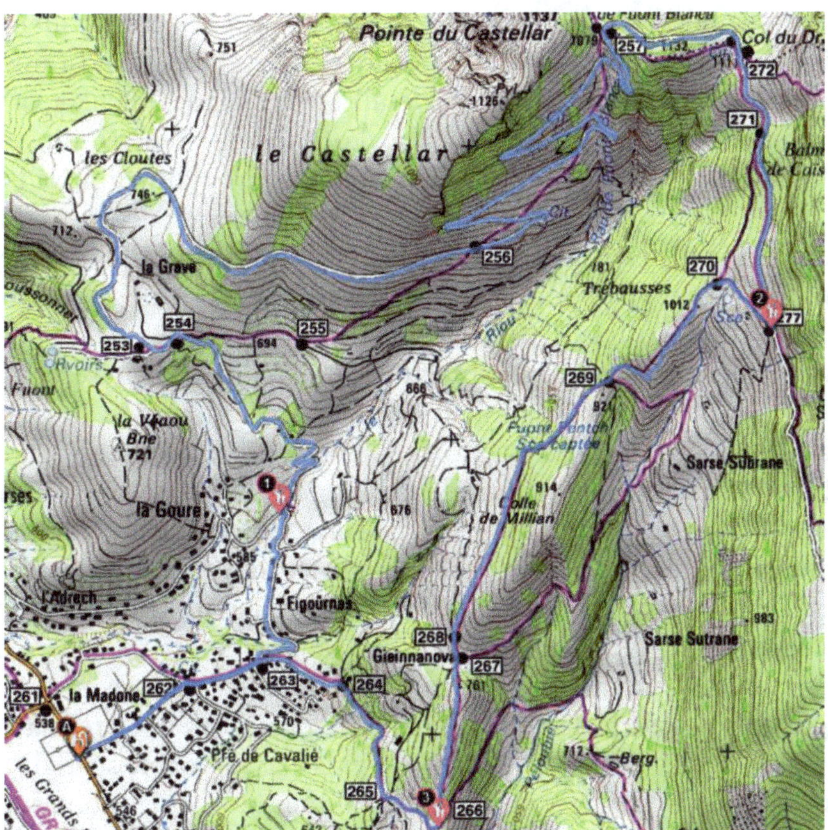

Fig. 11: Example of a crowdsourced bike route displayed on top of an IGN 1:25k topographic map, from the 'Espace loisirs IGN' application. ©IGN, France.

is generalised, which is likely to happen given the sharp bends at the top of the figure. Another example in Figure 11 is the spot of interest marked as n°2 in the figure, which is located on the summit of a large bend: if the bend is displaced by generalisation, which is a common side effect, the symbol should be adjusted accordingly.

When the scale decreases, Duchêne (2014) states that such spatial relations should either be preserved or sometimes be abstracted to make them legible and understandable at the generalised scale. To enable this preservation or abstraction, the relevant spatial relations must be discovered and properly characterised, which is not an easy task, although propositions exist to model these relations (Jaara et al., 2014) with the introduction of implicit features such as bend summits, or to build an ontology of such spatial relations relevant for cartography (Touya et al., 2014).

5 LoD harmonisation for Large Scale Maps

5.1 How can the LoD increase?

At large scales, e.g. maps at a 1:10k scale, there is no visualisation limitation for the very detailed features existing in OSM, and, as a consequence, map generalisation is not necessary. For instance, the very detailed railway networks described in Section 4 can have all of their lanes displayed without symbol overlaps at large scales. But the LoD inconsistencies illustrated in Figure 1 raise the problem of the representation of roughly digitised features at large scales. Most of the geographic meaning of maps is conveyed by relations between map features (Mackaness et al., 2014), so the solving of the problem of LoD inconsistencies should be focused on those relations that convey a specific meaning.

Following the ideas of Monmonier (1996), the idea to increase the LoD of roughly digitised features is to caricature them in order to transform the improbable relations of features into probable relations. For the examples in Figure 12, a clearing would be introduced around the group of buildings, and the bus stop would be moved to the closest road. We call this operation to artificially increase the LoD through probable spatial relations *LoD harmonisation* (Touya and Baley, in press). However, there is no clue in the data as to the real shape of the clearing required in Figure 12: we only know that there must be one. This makes harmonisation tend more towards caricature and schematic mapping than towards realistic mapping. The map does not present real and precise shapes to the reader, but rather presents very probable spatial relations. The next section briefly describes some harmonisation operations and shows some results of their implementation on OSM data, while Section 5.3 discusses the problem of automatically chaining these harmonisation operations on a complete large scale map.

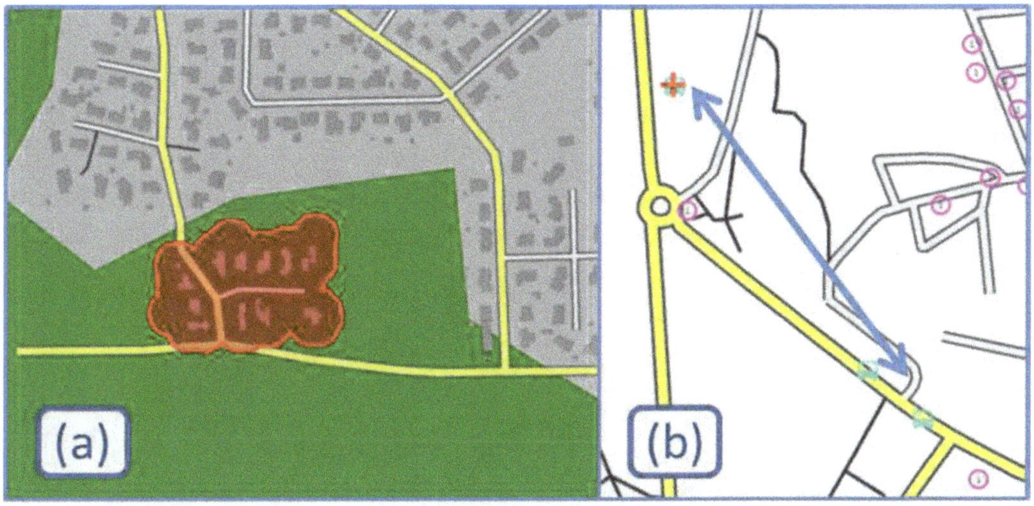

Fig. 12: (a) This automatically identified group of buildings should not be inside the forest. (b) The automatically identified bus stop (highlighted by the red cross) is too far from a road. ©OpenStreetMap contributors.

5.2 Harmonisation Operations

Different types of harmonisation operations are described by Touya and Baley (in press), and some of these are presented in this subsection. First, OSM contains some polygon features that represent functional sites such as schools, hospitals or commercial areas, which are themselves composed of other features also represented in OSM: buildings, roads, paths, parks, sports fields or helipads. For a clear understanding of what these zones mean in the map, the components should really be contained by the polygon, which is not always the case because the components are sometimes much more detailed than the zone itself. In this case, the harmonisation operation identifies the components that lie outside the zone and modifies the zone geometry so that it includes the missing components (Figure 13).

A similar problem might occur with land use/cover parcels that are often roughly digitised and some geographic features that should be inside the parcels. The most current example in OSM is the case of urban areas with buildings intersecting their limits or lying just outside. In such cases, the land use parcel geometry is extended by uniting the protruding geometries of the building just outside the area limits with the urban area geometry. The method is iterative, because new buildings can be found just outside once the geometry has been extended (see automatic results in Figure 14).

Another type of necessary harmonisation operation is disambiguation, which aims to remove spatial relations that should not exist in reality without knowing what the reality looks like. For instance, it is extremely unlikely to find a group of close buildings inside a forest without a clearing. When the forest has been roughly digitised and the buildings have a high LoD, we can infer the

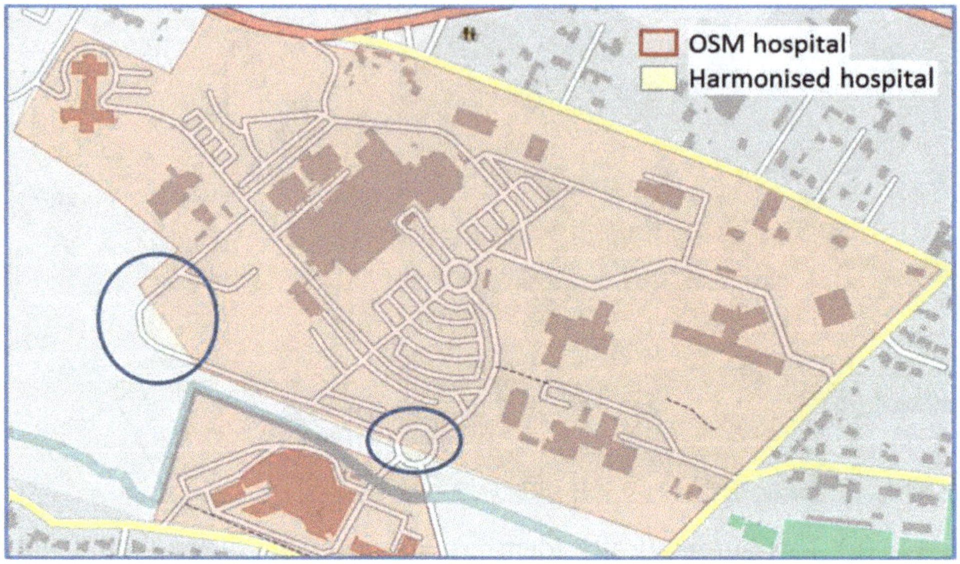

Fig. 13: The hospital zone is harmonised by extending the polygon to include all access roads. ©OpenStreetMap contributors.

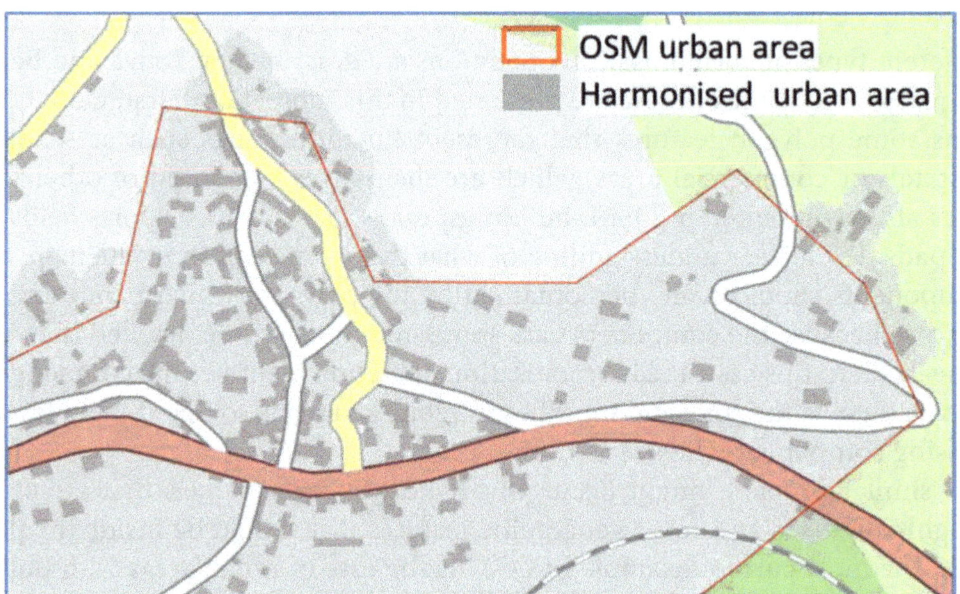

Fig. 14: The roughly digitised OSM urban area is distorted to include the buildings directly nearby. ©OpenStreetMap contributors.

presence of a clearing and try to add it in the forest. The proposed operation determines where the overlaps exist between the buildings and the forest and then crops the newly created clearing with the edges of the network elements, which are often barriers for forests (Figure 15).

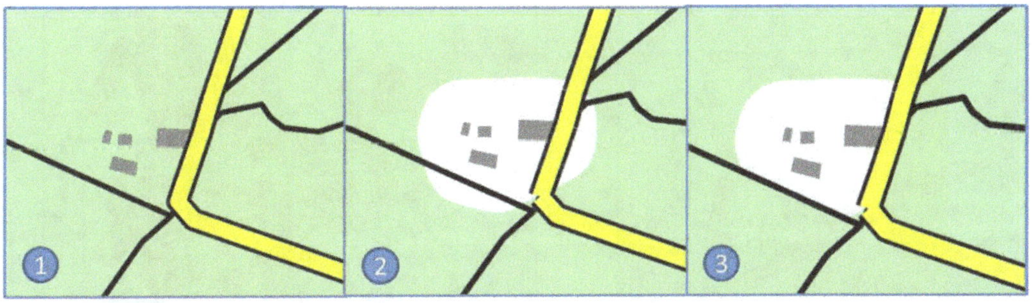

Fig. 15: The roughly digitised forest (1) contains a set overlapping buildings (2), and the newly created clearing is cropped (3) by network sections that often mark the limits of clearings/forests. ©OpenStreetMap contributors.

More examples of useful harmonisation operations can be found in Touya and Baley (in press).

5.3 How to Chain Harmonisation Operations

Harmonisation operations are the building blocks for deriving LoD harmonised large scale maps, but they are not enough, because several problems can occur:

- Harmonisation operations carried out on close parts of the map can affect each other and the last one can damage the previous harmonisations.
- Harmonisation operations that displace or distort features can cause legibility problems with other features of the map (e.g. symbol overlap).
- Harmonisation operations can be related to each other and the order of operations might have an impact; for instance, a displacement of a building that overlaps a riverbank (Figure 16) might put the building just outside the urban area, so the extension of the urban area should be implemented afterwards.

Similar problems occurred with the automation of map generalisation that first developed individual algorithms and then tried to combine them into complex processes (Harrie and Weibel, 2007; Regnauld et al., 2014). To harmonise the area shown in Figure 16, where multiple buildings overlap a riverbank, we therefore used an optimisation process inspired by map generalisation (Harrie, 1999; Sester, 2005), which combines the harmonisation of buildings that are close to each other into a least squares adjustment. Figure 16 shows that for each group of close buildings identified, all buildings have been jointly displaced, avoiding symbol overlap with the river and with other buildings.

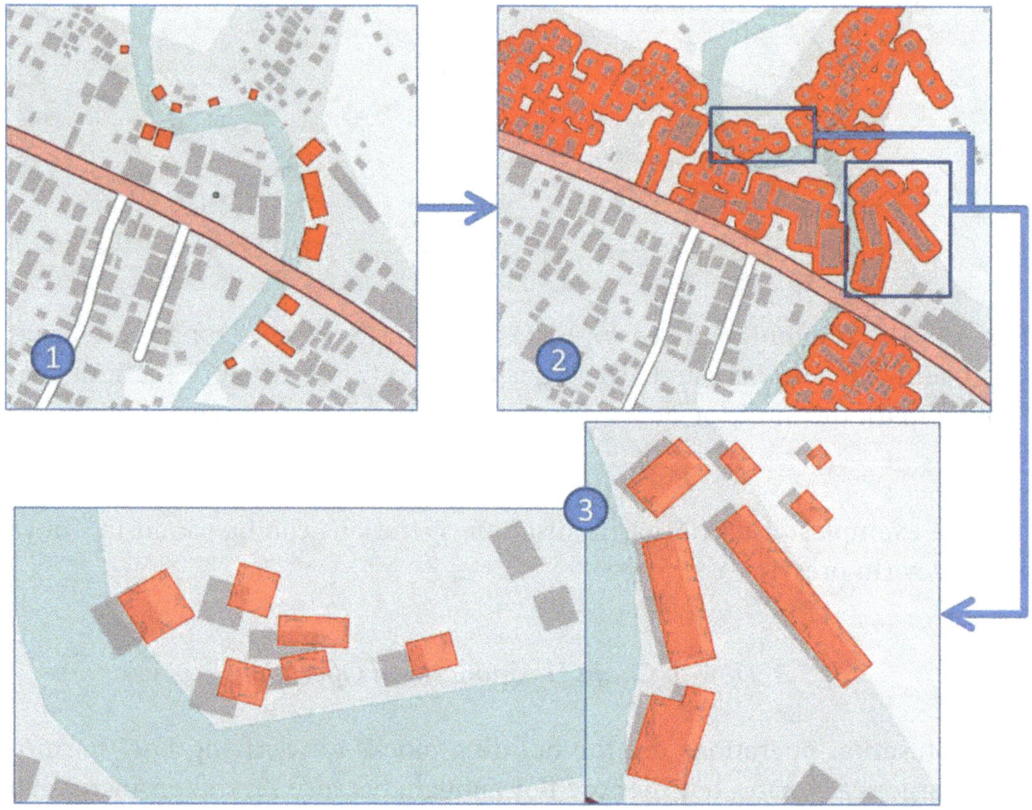

Fig. 16: 1) Detection of LoD inconsistencies (in this case a building intersecting the riverbank); 2) clusters of close buildings are created around the identified inconsistencies; 3) each cluster is harmonised as a whole to remove overlaps without creating new ones. ©OpenStreetMap contributors.

6 Quality Assessment Taking into Account Crowdsourced Ground Truth Data

As mentioned in Section 2, automatic mapmaking processes require some consistency in data quality, or some kind of assessment of this quality if consistency is not achievable, which is the case with VGI. This section describes a study to assess the quality of OSM features, using ground truth data. In many studies, OSM is usually used as a proxy for VGI data; this study is not an exception, as OSM is a prime source of vector-encoded GI that can be directly used in cartographic processes. However, any effort in mapmaking using VGI data should expand its horizons to include other sources as well. Today, VGI comes from different sources and in many flavours, such as toponyms, GPS tracks, geotagged photographs, synchronous micro-blogging, social networking content, blogs, gaming spaces, sensor measurements, etc. All of these sources can either possibly offer valuable geographic information complementary to OSM data (e.g. Geonames can provide a supplementary dataset to the OSM places) or be used as quality assessment tools (e.g. through the use of geotagged photographs from photo-sharing repositories). This latter case is the focus of this section.

Geotagged photographs are, in a sense, in-situ observations of the ground reality and thus, if properly used, can assess various quality factors of OSM data and improve the decisions in some of the cartographic processes analysed above. As explained, semantic mismatches, topological and positional errors and vague and ambiguous cases of overlaps and intersections should be expected when handling VGI. All these cases pose a challenging task when it comes to disambiguating them and can negatively affect the outcome of the cartographic processes.

When relying solely on VGI data for mapmaking, the ambiguous cases first need to be recognised and located, and then corrected or verified by the contributors themselves. Indeed, it has been documented that the positional quality of features improves as more contributors add data or modify a feature (Haklay et al., 2010). However, participation biases (Antoniou and Schlieder, 2014) and the digital divide (Graham et al., 2014) can negatively affect a widespread effort of quality improvement. Hence, we need to devise methods, by using diverse VGI data, that can more easily identify and correct such potential sources of error before they enter the cartographic chain of processes: in a sense, the mixture of diverse VGI sources might counter-balance biases and errors from individual VGI sources.

Although there is no direct link between geotagged photographs and map scales, it can be inferred that, as geotagged photographs usually capture a small ground area from a close distance in high detail, they can be of help in large scale maps. In general, cases where geotagged photographs can provide better ground truth include the efforts to:

- verify if a feature exists (i.e. assess completeness)
- verify the type of a feature (i.e. assess thematic accuracy)
- verify the topology and the relationship between features (i.e. assess logical consistency)
- verify the state of a feature for a particular time-stamp (i.e. assess temporal accuracy).

Here, as a case study, we focus on the use of other VGI sources (i.e. Flickr geotagged photographs) to evaluate the validity of OSM Points of Interest (POIs) in three different scenarios trying to i) verify the OSM points that could not have been created through image interpretation as there are objects that obscure the view (i.e. trees and wooded areas), and whose OSM updates consequently normally require the physical presence of contributors on the ground; ii) disambiguate areas of overlapping OSM land use/land cover types at a given point in time (for more, see Antoniou et al., 2016); and iii) correct problematic POIs in terms of topo-semantic consistency.

6.1 *Verify OSM POIs*

One of the comparative advantages of VGI is that it can provide timely data for areas and cases where other sources cannot be equally effective. One such case is that of the areas where satellite imagery (a prominent way of capturing authoritative data) cannot provide the needed information, e.g. under wooded areas (Figure 17). Here, local knowledge by contributors is valuable, as in-situ observations can be an important source of information. In this context, geotagged images are well placed to play a significant role.

For the verification of the OSM POIs, an online application has been developed that displays a geotagged photograph, retrieved using the Flickr API, and asks the user whether a specific POI could be recognised within approximately X meters (as computed by the location of the POI and the geotagged photograph) in the photograph. Thus, for example, the question has the form '*Do you see a* ***monument*** *about* ***2m*** *away, in the photo below?*' (for more on this, see Antoniou et al., 2016). Figure 18 shows a number of illustrative examples generated by the application.

A systematic fusion of diverse VGI sources can improve the quality of the data used for mapmaking not only in the initial phases of data gathering but also in a step-by-step implementation of cartographic processes as shown above. For example, in the case shown in Figure 15, geotagged photographs could be used to examine and verify if such openings in the forest really exist or if the constructions portrayed are hidden under the woods.

6.2 *Verify OSM Land Use / Land Cover*

The second case study for using geotagged photographs to evaluate a VGI dataset comes from the Land Use/Land Cover (LU/LC) domain. Here the challenge is to disambiguate inconsistencies regarding the actual LU/LC that arise from contradictory feature types that occur between different OSM layers, e.g. in the Landuse and the Natural OSM layers (a more thorough study can be found in

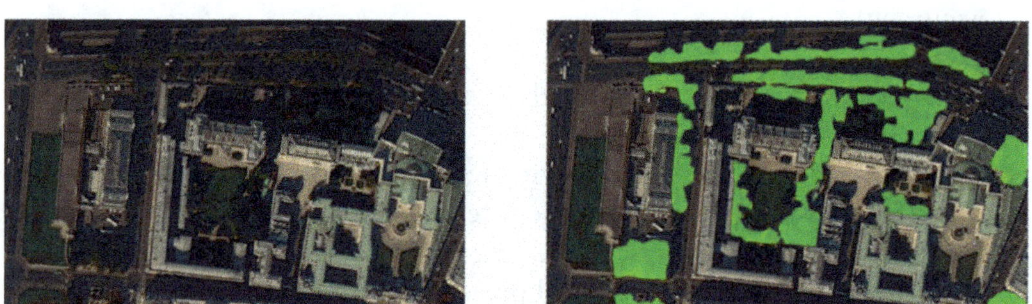

Fig. 17: A satellite image of a sample area in Paris (left) and the polygons of wooded areas (right) for the same area (©IGN, France).

Fig. 18: Illustrative screenshots of an ad-hoc application that retrieves geotagged photos for POI evaluation. Creative Commons licensed (BY-NC-ND) Flickr contributors.

Fonte et al., 2016). The LU/LC at each given point should be unambiguously retrieved: this requirement not only contributes to the overall quality of OSM and to the correct cartographic output but also enables the use of OSM data for the creation of LU/LC products. Here again, overlaps between different and contradictory LU/LC feature types create inconsistencies that could possibly be disambiguated with the use of geotagged photographs. For example, Figure 19 (left) shows the overlap of a closed construction site (purple polygon) and a residential road (green line) in OSM (green dots represent the locations of Flickr photographs). Although the VGI elements co-exist in the same VGI source (i.e. in OSM), it is obvious that it is not possible for both layers to correctly denote the actual land use of the area. The use of geotagged images could provide the necessary information to clarify the mismatch. In Figure 19 (right), a Flickr photograph taken within the polygon clearly shows that the area has been turned into a construction site. Additionally, a valuable characteristic of the VGI datasets used is the time information they contain: using the individual timestamps of features, it is possible to analyse and understand the currency of each feature, which could be valuable in updating the overlapping features that have outdated information.

Fig. 19: Mismatches between the OSM Roads and Landuse layers (left). A Flickr photograph of the area (right). ©OpenStreetMap contributors. Creative Commons licensed (BY-NC-ND) Flickr contributors.

With the two illustrations given in this and the previous section, it is shown that mixing independent VGI sources can prove a helpful way to spot possible errors, to evaluate the validity of features and to justify the implementation of various cartographic processes. In this context, the proactive disambiguation of vague cases in large scales can lead to correct decisions on the cartographic processes described above and avert the propagation of errors when moving to smaller scales.

6.3 Verifying and Correcting Topo-semantic (In)consistency

Topo-semantic consistency (Servigne et al., 2000) is a subset of logical consistency that concerns the correctness of the topological relationship between two objects according to their semantics. Topo-semantic consistency refers to the consistency of geographic objects with other geographic objects of the same theme (intra-theme consistency) or of other themes (inter-theme consistency). Inconsistency exists in VGI due to the absence of integrity constraints and, therefore, depends on the expertise of the data contributor. A map should not portray inconsistencies; thus, inconsistencies should be identified and resolved during the mapmaking process. Instead of correcting these errors in order to satisfy consistency blindly and without taking reality into account, correction can be based on ground truth provided by Flickr images, as explained earlier.

A number of tests can be applied in order to find inconsistencies in the OSM data between features from the same layer (e.g. two roads), or from different layers. Tests are based on consistency evaluation utilising topological relations that the data should satisfy, taking the data semantics captured by their attributes into account as well. In OSM, apart from the geometry capture, the existence of a plethora of tags provides a rich semantic dataset, and thus sophisticated topo-semantic relations can be explored. Here, we focus on POIs because they are more easily captured in photographs due to their dimensions. POIs that are problematic with regards to their position in comparison to other layers can be verified with Flickr images. If the Flickr images prove that the topo-semantic relation is correct, then no changes are made; otherwise the geometry (relative horizontal position) and/or the semantic information (Type tag) is updated according to the photograph. Finally, the topo-semantic relations are re-evaluated.

A case study was performed with OSM data that cover the broader Paris area (Antoniou et al., 2016). According to this study, in the area of interest there are 22,527 OSM POIs with two main attribute tags related to their identity: Name and Type. Topological relations of POIs against other thematic layers are examined based on a number of checks, and errors will be examined utilising Flickr photographs. For example, it is important to investigate the topological relationship between POIs and buildings, examining whether POIs should be situated inside or outside building polygons. Initially POIs are clipped with the convex hull of the area covered by buildings, resulting in 60136 points. A number of points (21872) are situated inside the building polygons, 2338 (4%) are situated on the building boundaries and 35926 (60%) are situated outside. It is examined whether the position of the POIs outside of the buildings is valid based on their semantics captured with the Type attribute. Based on this test, 30497 (85% of the initial estimate) can indeed be situated outside but 5429 (15% of the original estimate) should be situated inside the building polygons and need further investigation. Similarly, a number of points (24210) are situated inside the building polygons. Based on a similar test, 22047 (91%) can indeed be situated inside but 2163 (9%) should be situated outside the building polygons and need further investigation. In this study, the correct position of the points in relation to the buildings was decided according to common sense.

In another test, POIs that are semantically related to roads and railways are examined against the network geometry. Regarding POIs that are tagged as crossings (12612), 99.5% (12552) are situated on road intersections and only 60 of them (0.5%) have a different position and need checking. Regarding POIs that are tagged as traffic lights (12612), 99.2% (2292) are situated on the road intersections and only 18 of them (0.8%) have a different position that will be further checked. POIs that are tagged as 'level crossings' (209) and 'railway_crossing' (1) are situated on the rail network intersections. Points semantically related to the intersections of the rail and road network, such as level crossings, are checked in relation to the actual intersections of the road and rail network. Of the 1101 points, 949 (86%) are situated on the intersections while 152 (14%) have a different position and need further investigation. Of course map scale is also an important factor when judging distance. For example, the distance between network junctions and POIs tagged as crossings might be negligible in relation to scale.

The inspection of topo-semantic relations highlights areas where consistency is not fulfilled and should be corrected during the mapmaking process. Pre-processing based on topo-semantic relations limits the intervention of cartographers to only those cases that are problematic. Whereas an in situ visit costs time and money, the provision of ground truth through geotagged Flickr images is a welcome alternative solution emerging from the VGI universe.

7 VGI and Symbol Specification

This section discusses issues related to VGI symbolisation, and is more forward looking than the previous ones. As with the other previously described cartographic processes, the main issues regarding symbol specification with VGI are: what could be impacted by this new source of data, and what should be adapted and how? A reminder of the symbol specification process is given first. Then, we highlight aspects to be discussed and controlled to adapt this process to VGI.

7.1 The Symbol Specification Process

The symbol specification process occurs at the end of the global cartographic design process. At this stage, the input objects should be generalised for the expected map scale in order to be able to properly specify styles that are suitable at this scale. Traditional cartographic symbolisation, for instance in map series production, is based on historical knowledge of symbol specifications and cartographic practices and processes, related to a particular topographic style (Ory et al., 2015). Symbol specifications have also been considered as a user controllable problem in order to make personalised maps (Christophe, 2011). Research on style and symbol specification now focuses on processes inspired by computer graphics to mimic traditional cartographic symbolisation, or to apply artistic styles to maps (Christophe et al., 2016). The three main steps of the symbol specification process are:

- *Legend specification: themes and semantic relations between map themes.* It first requires that the legend be structured by semantic themes with semantic relationships (e.g. rivers and lakes are in the same legend theme and their symbols should be related).
- *Style specification: signs for themes.* This requires choosing and combining relevant graphic signs to enhance semantic relations on the map.
- *Map rendering.* The rendering step effectively applies the style specification to the cartographic objects on the map. It may involve complex rendering techniques, such as textures to render forest areas.

Tools such as Mapnik[2], which are used to make maps with OSM, do provide some basic rendering methods, including polygon texture fills or advanced text rendering, that could be extended to help users complete the three steps of symbol specification.

7.2 Discussion and Guidelines for Using VGI in Symbol Specification Processes

As for the other mapmaking processes, the first issue to address when using VGI in symbol specification processes is the adaptation of processes developed for consistent databases to the heterogeneity of VGI. This adaptation can be achieved by a characterisation of VGI features, i.e. its quality, semantics and LoD. But such characteristics of quality or LoD are no longer consistent on a given map theme, as each VGI feature might have its own quality or LoD. Thus a symbol specification for each map theme might not be possible with VGI. For the same map theme, for instance rivers, the symbol might be adapted to the quality, semantics and LoD of the features (e.g. darker shades of blue and wider symbols for rivers with more details/better quality).

A typical use case of maps made with VGI is the mashup map with crowd-sourced thematic data on top of existing reference data. In this case, the symbol specification for the reference background might have been designed independently from the thematic data; thus the addition of thematic VGI involves three problems:

- Management of contrasts: the thematic data should be more legible than the background and the contrast in the background should be altered to optimise the contrast with the thematic data.
- Preserving a topographic style: adding a crowdsourced thematic layer should not prevent the map reader from understanding the topographic style of the background.
- Visualising imprecision aspects: the thematic layer is both heterogeneous in terms of quality and different from the background. Thus the symbol specification should convey these differences as much as possible (see Chapter 9 by Skopeliti et al. (2017) regarding quality visualisation).

7.3 Crowdsourcing the Symbol Specification Process?

The symbol or style specification process is user-driven, as the map purpose and the map user needs are translated into a legend and rendered on the map. Additionally to the use of crowdsourced data in the map, a crowdsourced map could also include a more important interaction with the user during the mapmaking process: for example, a consensus decision among OSM contributors could be reached regarding the colour to use to render the forest areas in the standard display. Research on automated on-demand mapping tries to capture the needs of users through techniques such as ontologies and interactions (Balley et al., 2014), but allowing the users to choose the way crowdsourced data can be rendered in the legend and the map requires a step further in this direction.

8 Conclusions and Further Work

This chapter addressed the challenges of automated mapmaking using VGI as input data. VGI differs from traditional geographic databases because of heterogeneities in quality and LoD, and because of thematic diversity, so existing methods for automated mapmaking have to adapt to this situation. This chapter described a proposition to infer the LoD of VGI features to overcome heterogeneity, and then presented methods that use this inference to make maps at different scales using map generalisation or LoD harmonisation. The paper also proposed techniques to overcome the quality heterogeneity, which can alter the map legibility. Finally, the paper discussed how advanced stylisation techniques could be applied to VGI.

There is much more work to be done, as automated mapmaking itself is a large research topic. The long-term goal is to design adaptive and completely automated cartographic processes, because the amount of data is too large for manual cartography, and the content has to be adapted to different needs and display devices. Beyond continuing to improve the methods presented here, it must be noted that generalisation and harmonisation operations do not handle quality heterogeneities yet, and we should investigate how such processes can adapt to quality information that can be inferred from VGI features similarly to the handling of LoD information discussed above. For instance, a forest imported from Corine Land Cover and one captured precisely with satellite imagery do not require the same simplification algorithms. The future diffusion of web maps will be based on vector maps using vector tiling, such as the OpenScienceMap project that provides a vector mapping of OSM. Such web maps will raise several research questions, such as that of the online triggering of generalisation and harmonisation processes, when such processes are mostly designed for offline processing. The question of tiled processing is also an issue, as mapmaking processes make considerable use of the geographic neighbourhood of features to choose the best process. The development of vector web maps will also enable user customisation of stylisation, which will require addressing the research issues discussed in the last section of this chapter.

Previous publication

Section 6 was partly published in Antoniou, V., Skopeliti, A., Fonte, C., See, L., Alvanides, S. (2016). Using OSM, geo-tagged Flickr photos and authoritative data: A quality perspective, in Bandrova T., Konecny, M. (Eds.) Proceedings, 6th

International Conference on Cartography and GIS, 13–17 June 2016, Albena, Bulgaria. Available at http://cartography-gis.com/docsbca/iccgis2016/ICC-GIS2016-49.pdf [Last accessed 13 April 2017]

In section 6 the link between quality control and the topographic maps is additionally discussed as the previous paper did not focus on a particular application.

Notes

1 http://espaceloisirs.ign.fr
2 http://mapnik.org

Reference list

Antoniou, V., Schlieder, C., 2014. Participation patterns, VGI and gamification, in: Proceedings of AGILE 2014. Presented at the AGILE 2014, Castellón, Spain, pp. 3–6.

Antoniou, V., Skopeliti, A., Fonte, C., See, L., Alvanides, S., 2016. Using OSM, geo-tagged Flickr photos and authoritative data: A quality perspective, in: Bandrova T., Konecny, M. (Eds.), Proceedings, 6th International Conference on Cartography and GIS, 13–17 June 2016, Albena, Bulgaria. Available at http://cartography-gis.com/docsbca/iccgis2016/ICCGIS2016-49.pdf [Last accessed 13 April 2017]

Balley, S., Baella, B., Christophe, S., Pla, M., Regnauld, N., Stoter, J., 2014. Map specifications and user requirements, in: Burghardt, D., Duchêne, C., Mackaness, W. (Eds.), *Abstracting Geographic Information in a Data Rich World, Lecture Notes in Geoinformation and Cartography*. Springer International Publishing, pp. 17–52.

Biljecki, F., Ledoux, H., Stoter, J., Zhao, J., 2014. Formalisation of the level of detail in 3D city modelling. *Computers, Environment and Urban Systems* 48, 1–15. DOI: https://doi.org/10.1016/j.compenvurbsys.2014.05.004

Burghardt, D., Duchêne, C., Mackaness, W. (Eds.), 2014. *Abstracting Geographic Information in a Data Rich World, Lecture Notes in Geoinformation and Cartography*. Springer International Publishing, Cham.

Christophe, S., 2011. Creative Colours Specification Based on Knowledge (COLorLEGend system). *The Cartographic Journal* 48, 138–145. DOI: https://doi.org/10.1179/1743277411Y.0000000012

Christophe, S., Duménieu, B., Turbet, J., Hoarau, C., Mellado, N., Ory, J., Loi, H., Masse, A., Arbelot, B., Vergne, R., Brédif, M., Hurtut, T., Thollot, J., Vanderhaeghe, D., 2016. Map style formalization: rendering techniques extension for map design., , in: Mould, D., Bénard, P. (Eds.), *NPAR '16: Proceedings of the 14th International Symposium on Non-Photorealistic Animation and Rendering*, Lisbon, Portugal, 7–9 May 2016, pp.59–68.

Duchêne, C., 2014. Making a map from thematically multi-sourced data : the potential of making inter-layers spatial relations explicit. Proceedings of 17th ICA Workshop on Generalisation and Multiple Representation. Vienna, Austria. Available at https://kartographie.geo.tu-dresden.de/downloads/ica-gen/workshop2014/genemr2014_submission_14.pdf [Last accessed 1 May 2017].

Duchêne, C., Baella, B., Brewer, C.A., Burghardt, D., Buttenfield, B.P., Gaffuri, J., Käuferle, D., Lecordix, F., Maugeais, E., Nijhuis, R., Pla, M., Post, M., Regnauld, N., Stanislawski, L.V., Stoter, J., Tóth, K., Urbanke, S., Altena, V. van, Wiedemann, A., 2014. Generalisation in practice within National Mapping Agencies, in: Burghardt, D., Duchêne, C., Mackaness, W. (Eds.), *Abstracting Geographic Information in a Data Rich World, Lecture Notes in Geoinformation and Cartography*. Springer International Publishing, pp. 329–391.

Dumont, M., Touya, G., Duchêne, C., 2016. A comparative study of existing multi-scale maps: what content at which scale?, in: Proceedings of GIScience 2016, 9th International Conference on Geographic Information Science, Montreal, Canada.

Fonte, C., Minghini, M., Antoniou, V., See, L., Brovelli, M.A., Patriarca, J., Milčinski, G., 2016. An automated methodology for converting OSM data into a land use/cover map, in: Proceedings of the 6th International Conference on Cartography & GIS. Presented at the 6th International Conference on Cartography & GIS, Albena, Bulgaria. Available at https://cartography-gis.com/docsbca/iccgis2016/ICCGIS2016-47.pdf [Last accessed 1 May 2017].

Girres, J.-F., Touya, G., 2010. Quality assessment of the French OpenStreetMap dataset. *Transactions in GIS* 14, 435–459.

Goetz, M., Zipf, A., 2011. Extending OpenStreetMap to indoor environments: Bringing Volunteered Geographic Information to the next level, in: Rumor, M., Zlatanova, S., Ledoux, H. (Eds.), *Urban and Regional Data Management: Udms Annual 2011*. Delft, The Netherlands, pp. 47–58.

Graham, M., Hogan, B., Straumann, R.K., Medhat, A., 2014. Uneven geographies of user-generated information: Patterns of increasing informational poverty. *Annals of the Association of American Geographers* 104, 746–764.

Haklay, M., 2010. How good is volunteered geographical information? A comparative study of OpenStreetMap and Ordnance Survey datasets. *Environment and Planning B: Planning and Design* 37, 682–703. DOI: https://doi.org/10.1068/b35097

Haklay, M., Basiouka, S., Antoniou, V., Ather, A., 2010. How many volunteers does it take to map an area well? The validity of Linus' Law to volunteered geographic information. *The Cartographic Journal* 47, 315–322.

Harrie, L.E., 1999. The constraint method for solving spatial conflicts in cartographic generalization. *Cartography and Geographic Information Science* 26, 55–69. DOI: https://doi.org/10.1559/152304099782424884

Harrie, L.E., Weibel, R., 2007. Modelling the overall process of generalisation, in: Ruas, A., Sarjakoski, L.T. (Eds.), *Generalisation of Geographic Information:*

Cartographic Modelling and Applications. Elsevier Science B.V., Amsterdam, pp. 67–87.

Ivanovic, S., Olteanu-Raimond, A.-M., Mustière, S., Devogele, T., 2015. Detection of potential updates of authoritative spatial databases by fusion of volunteered geographical information from different sources, in: A. Comber, B. Bucher, S. Ivanovic (Eds.), *Proceedings of the 3rd AGILE Phd School*, Champs sur Marne, France, 17 Sep 2015. Available at: http://ceur-ws.org/Vol-1598/paper15.pdf [Last accessed 16 May 2017].

Jaara, K., Duchêne, C., Ruas, A., 2014. Preservation and modification of relations between thematic and topographic data throughout thematic data migration process, in: Buchroithner, M., Prechtel, N., Burghardt, D. (Eds.), *Cartography from Pole to Pole, Lecture Notes in Geoinformation and Cartography*. Springer Berlin Heidelberg, pp. 103–117.

Klammer, R., 2013. TileGen - an open source software for applying cartographic generalisation to tile-based mapping, in: *Proceedings of the International Cartographic Conference 2013 (ICC 2013)*. ICA, Dresden, Germany, 25–30 August 2013. Available at: http://icaci.org/files/documents/ICC_proceedings/ICC2013/_extendedAbstract/395_proceeding.pdf [Last accessed 16 May 2017].

Mackaness, W.A., Burghardt, D., Duchêne, C., 2014. Map generalisation: Fundamental to the modelling and understanding of geographic space, in: Burghardt, D., Duchêne, C., Mackaness, W. (Eds.), *Abstracting Geographic Information in a Data Rich World, Lecture Notes in Geoinformation and Cartography*. Springer International Publishing, pp. 1–15.

Mackaness, W.A., Ruas, A., Sarjakoski, L.T. (Eds.), 2007. *Generalisation of geographic information: cartographic modelling and applications*, 1st ed. Elsevier, Amsterdam.

Monmonier, M., 1996. *How to Lie with Maps*, 2nd ed. University of Chicago Press, Chicago, IL.

Ory, J., Christophe, S., Fabrikant, S.I., Bucher, B., 2015. How do map readers recognize a topographic mapping style? *The Cartographic Journal* 52, 193–203. DOI: https://doi.org/10.1080/00087041.2015.1119459

Regnauld, N., Touya, G., Gould, N., Foerster, T., 2014. Process modelling, web services and geoprocessing, in: Burghardt, D., Duchêne, C., Mackaness, W. (Eds.), *Abstracting Geographic Information in a Data Rich World, Lecture Notes in Geoinformation and Cartography*. Springer International Publishing, pp. 197–225.

Reimer, A., Kempf, C., Rylov, M., Neis, P., 2014. Scale eqivalencies for OpenStreetMap polygons, in: *Proceedings of AutoCarto 2014*. Pittsburgh, PA, USA, 5–7 October 2014. Available at: http://www.cartogis.org/docs/proceedings/2014/Reimer_Abstract_AutoCarto2014.pdf [Last accessed 16 May 2017].

Roy, B., 2005. Paradigms and challenges, in: Figueira, J., Greco, S., Ehrogott, M. (Eds.), *Multiple Criteria Decision Analysis: State of the Art Surveys, International*

Series in Operations Research & Management Science. Springer New York, New York, NY, pp. 3–24.

Savino, S., Touya, G., 2015. Automatic structure detection and generalization of railway networks, in: *Proceedings of 27th International Cartographic Conference*. ICA, Rio de Janeiro, Brazil, 23–28 August 2015. Available at: http://www.icc2015.org/trabalhos/3/920/savino_touyaicc2015.pdf [Last accessed 16 May 2017].

Schmid, F., Janetzek, H., 2013. A method for high-level road network extraction of OpenStreetMap data, in: in: *Proceedings of the International Cartographic Conference 2013 (ICC 2013)*. ICA, Dresden, Germany, 25–30 August 2013. Available at: http://icaci.org/files/documents/ICC_proceedings/ICC2013/_extendedAbstract/92_proceeding.pdf [Last accessed 16 May 2017].

Servigne, S., Ubeda, T., Puricelli, A., Laurini, R., 2000. A methodology for spatial consistency improvement of geographic databases. *Geoinformatica* 4, 7–34.

Sester, M., 2005. Optimization approaches for generalization and data abstraction. *International Journal of Geographical Information Science* 19, 871–897.

Sester, M., Jokar Arsanjani, J., Klammer, R., Burghardt, D., Haunert, J.-H., 2014. Integrating and generalising Volunteered Geographic Information, in: Burghardt, D., Duchêne, C., Mackaness, W. (Eds.), *Abstracting Geographic Information in a Data Rich World*. Springer International Publishing, Cham, pp. 119–155.

Skopeliti, A, Antoniou, V and Bandrova, T. 2017. Visualisation and Communication of VGI Quality. In: Foody, G, See, L, Fritz, S, Mooney, P, Olteanu-Raimond, A-M, Fonte, C C and Antoniou, V. (eds.) *Mapping and the Citizen Sensor*. Pp. 197–222. London: Ubiquity Press. DOI: https://doi.org/10.5334/bbf.i.

Thomson, R.C., Richardson, D., 1999. The "good continuation" principle of perceptual organization applied to the generalization of road networks, in: Proceedings of the 19th International Cartographic Conference. ICA.

Töpfer, F., Pillewizer, W., 1966. The principles of selection: a means of cartographic generalization. *The Cartographic Journal* 3, 10–16.

Touya, G., Baley, M., in press. Level of details harmonization operations in OpenStreetMap based large scale maps, in: Jokar Arsanjani, J., Leitner, M. (Eds.), *Crowdsourced Mapping*. Berlin Heidelberg: Springer.

Touya, G., Brando-Escobar, C., 2013. Detecting level-of-detail inconsistencies in volunteered geographic information data sets. *Cartographica: The International Journal for Geographic Information and Geovisualization* 48, 134–143. DOI: https://doi.org/10.3138/carto.48.2.1836

Touya, G., Bucher, B., Falquet, G., Jaara, K., Steiniger, S., 2014. Modelling geographic relationships in automated environments, in: Burghardt, D., Duchêne, C., Mackaness, W. (Eds.), *Abstracting Geographic Information in a Data Rich World, Lecture Notes in Geoinformation and Cartography*. Springer International Publishing, pp. 53–82.

Touya, G., Girres, J.-F., 2014. Generalising unusual map themes from OpenStreetMap, in: Proceedings of 17th ICA Workshop on Generalisation and Multiple Representation. ICA, Vienna, Austria.

Touya, G., Reimer, A., 2015. Inferring the scale of OpenStreetMap features, in: Arsanjani, J.J., Zipf, A., Mooney, P., Helbich, M. (Eds.), *OpenStreetMap in GIScience, Lecture Notes in Geoinformation and Cartography*. Springer International Publishing, pp. 81–99.

Zielstra, D., Hochmair, H.H., 2013. Positional accuracy analysis of Flickr and Panoramio images for selected world regions. *Journal of Spatial Science* 58, 251–273. DOI: https://doi.org/10.1080/14498596.2013.801331

CHAPTER 5

Motivating and Sustaining Participation in VGI

Steffen Fritz*, Linda See* and Maria Brovelli†

*International Institute for Applied Systems Analysis (IIASA), Schlossplatz 1, 2361 Laxenburg, Austria, fritz@iiasa.ac.at

†Department of Civil and Environmental Engineering, Politecnico di Milano, Piazza Leonardo da Vinci 32, 20133 Milano, Italy

Abstract

Volunteers are the key component in the collection of Volunteered Geographic Information (VGI), so what motivates their participation, what strategies work in recruitment and how sustainability of participation can be achieved are key questions that need to be answered to inform VGI system design and implementation. This chapter reviews studies that have examined these questions and presents the main motivational factors that drive volunteer participation, as determined from empirical research. Some best practices from broader citizen science applications are also presented that may have relevance for VGI initiatives. Finally, a set of case studies from our experiences are used to illustrate how volunteers have been motivated to collect VGI through mapping parties, gamification and working with schools.

Keywords

Motivation, recruitment, participation, incentives, retention

---

How to cite this book chapter:

Fritz, S, See, L and Brovelli, M. 2017. Motivating and Sustaining Participation in VGI. In: Foody, G, See, L, Fritz, S, Mooney, P, Olteanu-Raimond, A-M, Fonte, C C and Antoniou, V. (eds.) *Mapping and the Citizen Sensor.* Pp. 93–117. London: Ubiquity Press. DOI: https://doi.org/10.5334/bbf.e.

1 Introduction

Volunteered Geographic Information (VGI; a term originally coined by Goodchild, 2007) has two main components, i.e. the volunteer and the spatial information. Much of the literature on VGI examines either the second component, i.e. the geographic data collected, often in relation to its quality (e.g. Flanagin and Metzger, 2008; Haklay, 2010; Foody et al., 2013; Antoniou and Skopeliti, 2015), or how VGI has been used in different contexts (e.g. Zook et al., 2010; Barrington et al., 2011; Mooney and Corcoran, 2011; Connors et al., 2012). Yet it is the volunteer that is actually at the heart of VGI and the reason why there are many successful examples of it (See et al., 2016; Chapter 2 by See et al., 2017), one in particular being OpenStreetMap (OSM). Thus issues such as attracting and retaining volunteers, and understanding participant motivations and what incentives can be used to attract volunteers, are as important as the spatial information that is collected, particularly in designing new VGI applications. The importance of the volunteer has been recognised in a recent paper by Gómez-Barrón et al. (2016), where the authors consider motivational factors for VGI as a critical part of the participation planning phase in the design of any VGI system.

There are biases observed in participation that are a general characteristic of any application of user-generated content. One of these is referred to as the 1% rule (or the 90:10:1 rule), and states that 90% of the content is provided by only 1% of the users (Nielsen, 2006). Of the remaining users, 9% provide content some of the time while 90% use the content but do not contribute anything. Although these numbers may change slightly from application to application, Nielsen (2006) argues that participation inequality cannot be eliminated. Such inequalities exist even in highly successful collaborative applications such as Wikipedia; for example, He (2012) found that active users have generated around 3.5% of the content of Wikipedia and that this general pattern has not changed over time, while Wikipedia's own statistics for 2016 show that less than 0.5% of content is currently provided by active users (Wikipedia, 2016). Despite the success of OSM, there are also biases in it: Neis and Zielstra (2014) reviewed participation inequality studies for OSM and found that 10% of those registered in 2008 contributed actively while a study in 2010 showed that only 3.5% of volunteers accounted for 98% of the content (Neis et al., 2011).

Given these highly skewed figures, the aim of this chapter is to present ways in which the number of active participants can be increased in order to change the shape of the participation inequality curve (Nielsen, 2006). The starting point is to understand the nature of VGI participants and what motivates their contributions. Through a review of existing studies of VGI motivation, the factors that are relevant to the development of strategies to improve recruitment and to increase the motivation and retention of volunteers in

VGI are outlined. This is followed by a synthesis of some of the best practices from VGI and citizen science experiences. Finally, case studies of VGI are used to highlight different ways in which recruitment, motivation and retention have been tackled.

2 What Motivates Volunteers in VGI?

2.1 The Nature of Volunteers

To help understand volunteer motivations with respect to VGI and how they might differ between participants, it is useful to first understand the nature of the volunteers that take part in VGI. This is usually done by classifying volunteers into types according to factors such as their knowledge of the subject or their degree of participation. Coleman et al. (2009) offer one typology of five types that are situated along a spectrum ranging from Neophytes at one end, who include individuals that have no background in the area but have the time and interest to contribute, to Expert Authorities at the other end, who have considerable experience in mapping technologies and product specifications; in between are Interested Amateurs, Expert Amateurs and Expert Professionals. However, Coleman et al. (2009) argue that this typology is too simplistic for VGI, offering some examples of where the typology breaks down: for example, a Neophyte may have little expertise in the subject area but their local knowledge of an area might mean they can provide valuable contributions that more experienced individuals from other types cannot.

Another typology, which was developed as part of a EuroSDR Workshop, is offered by Heipke (2010). It includes:

- map lovers and experts, who would be happy to provide accurate information when, for example, maps are wrong or information is missing;
- casual mappers such as those from the biking/hiking community;
- media mappers that respond to specific campaigns in bursts of activity such as during mapping parties or post-disaster events;
- passive mappers, e.g. people who provide traffic data via their mobile phone;
- open mappers, e.g. those contributing to initiatives such as OSM;
- and mappers that would be motivated by financial incentives, e.g. through using Amazon's Mechanical Turk.

This typology already provides some insights into possible motivational factors such as interest in the subject or material gain. The open mappers were identified as being the largest group after passive mappers and one that is increasing in size over time. Although their motivations are thought to be altruistic and related to building and using open datasets as a public good (Goodchild, 2007;

Heipke, 2010), the range of motivations driving the group of open mappers is much more complex and nuanced (Budhathoki and Haythornthwaite, 2012), as outlined in the next section.

2.2 Motivational Factors for VGI Participation

Coleman et al. (2009) offer different motivations for participation in VGI that are based on empirical research from Wikipedia and the open source community. These include: altruism; professional or personal interest; intellectual stimulation; protection or enhancement of a personal investment; social reward; enhanced personal reputation; participation providing an outlet for creative and independent self-expression; and pride of place. The idea of local knowledge is captured in pride of place and is relevant to applications such as OSM where mappers more frequently map or update their local areas than areas further afield unless they are driven by mapping parties or humanitarian causes. However, other motivating factors, such as providing an outlet for creative and independent self-expression, may be less relevant to the mapping of features in OSM.

A very comprehensive identification of motivational factors for VGI has been provided by Budhathoki and Haythornthwaite (2012), who reviewed the literature on motivations from three distinct yet relevant domains: volunteerism; leisure; and the generation of knowledge online. The factors were divided into intrinsic motivations, which come directly from the individual; and extrinsic motivations, which come from the outside – such as financial incentives or gaining a positive reputation based on the quality of one's contributions or from peers. The factors are listed in Table 1 and are summarised from the original list that was provided in Budhathoki (2010). They can provide the basis for further investigation into understanding the motivations of participants in any given VGI application.

Budhathoki and Haythornthwaite (2012) used the motivational factors listed in Table 1 as the basis of a survey undertaken with OSM volunteers in order to understand which motivations were the most important for these volunteers. They also differentiated between two types of volunteers, i.e. serious mappers and casual mappers, based on the number of contributions, the length of the contributions or the frequency of contributions. The results of the survey of the 444 OSM volunteers was that two extrinsic factors, i.e. community and the project goal, and the intrinsic factors of unique ethos and altruism were the most important. However, casual mappers ranked unique ethos as more important than serious mappers. Other important factors included the importance of local knowledge (instrumentality and self-efficacy), the freedom to provide information where one wanted, trust in the system and fun. Serious mappers also positively rated learning as a motivation, and in a much stronger manner than casual mappers did. Understanding these motivations can provide strategies

Table 1: Motivational factors for VGI (adapted from Budhathoki, 2010).

Type	Factor	Relation to VGI
Intrinsic	Unique ethos	Maps should be freely available as an open public good
	Learning	Gaining new knowledge about mapping and places
	Personal enrichment	Satisfaction in contributing
	Self-actualisation	Appreciation of talents and skills in mapping and of local knowledge
	Self-expression	Ability to express skills and knowledge of mapping and local areas
	Self-image	Gaining confidence in self through contributions
	Fun	Enjoying the process of contributing and seeing contributions online
	Recreation	Mapping outdoors
	Instrumentality	Providing critical inputs to a map that would otherwise be wrong or missing information
	Self-efficacy	Feeling of being effective through contributions
	Meeting own needs	Filling gaps in spatial information needed for different applications
	Freedom of expression	Ability to choose what information to provide and how
	Altruism	Contributions to a social cause
Extrinsic	Career	Contributions become part of a CV or lead to marketable skills
	Strengthening social relations	Creating strong bonds, e.g. through mapping parties or other socially constructed events
	Project goal	Alignment between goals of the project and those of the contributor
	Community	Being part of a bigger, sustaining community
	Identity	Becoming part of a group, e.g. advancing to an expert group
	Reputation	Recognition from the system or individuals in the community
	Monetary return	Being paid for contributions or making money from the data
	Reciprocity	The idea that if you contribute, others will contribute
	System trust	Will contribute if there is trust in the system
	Networking	Contributing forms networks locally and internationally
	Socio-political	Contributing meets socio-political motivations

to turn casual mappers into more serious ones, e.g. ways that may help build confidence and emphasising the importance and strengths of local knowledge.

In a separate study by Tiwari et al. (2010), a survey of motivations was undertaken with volunteers in OSM and the GISCorps. The top motivational factors in both groups were found to be altruism, personal satisfaction and gaining new geospatial knowledge. Other factors from Table 1 were also chosen, including strengthening of social relationships and fun. Participants were also asked what incentives they would like to receive in order to increase participation. Around one quarter replied that no incentives were needed, while another quarter wanted additional geospatial training. Composto et al. (2016) considered the need to provide something back to the volunteers as a motivator: they examined two VGI initiatives, and found that the one that had more visible impact, i.e. the one that resulted in broken streetlights being reported and fixed, was the one that has had longevity and sustained participation.

3 Best Practices in Volunteer Recruitment, Motivation and Retention

To attract volunteers to contribute to a VGI initiative, there are three key issues to consider:

- What methods should be used to recruit participants?
- How will the volunteers be motivated to contribute given all the different motivational factors that have been identified through empirical research?
- How can participation be maintained in the long term?

Past initiatives have already considered many of these issues, so this section presents different approaches that have been taken in practice. In fact much of the good practice in volunteer recruitment, motivation and retention stems from citizen science initiatives, i.e. the involvement of citizens in scientific research (Bonney et al., 2009). Broader than VGI, citizen science is widespread in areas such as biodiversity monitoring (Hyvoenen et al., 2013; Clavero and Revilla, 2014) and astronomy (Clery, 2011). Although citizen science is not specifically geographic in nature, there are lessons valuable to VGI that have been learned from numerous citizen science projects, some of which are presented below.

3.1 Recruitment

The guidance document written by Tweddle et al. (2012) provides different recruitment strategies for citizen science projects, where the starting point is to determine the target audience, e.g. whether the project is targeted to the general public, to map lovers, to school children, etc. The promotion and recruitment process can then be tailored towards this group using a range of channels, including email, social media and the press. Experiences from Nature's Notebook, a citizen science project in the USA to collect phenology data (i.e. life stage data) from plants and animals, have shown the necessity to carefully identify target audiences and then to contact them with messages that are focused on explaining the personal benefits of contributing (Crimmins et al., in press). Nature's Notebook had little success when advertising its programme to the general public so instead targeted the members of another citizen science initiative with similarly rigorous protocols for data collection, and this has been a very successful method of recruitment for the project.

Holding a launch event or side event at existing conferences, workshops and festivals can be an effective way of informing potential volunteers about the aims of the project, about why their help is important and about what they will gain from the project. The project goal was ranked highly as a motivator for OSM (Budhathoki and Haythornthwaite, 2012), so communicating this aspect is clearly important for attracting volunteers.

Composto et al. (2016) examined the use of media campaigns to recruit volunteers in two VGI projects. They showed that this is a very effective way of bringing individuals to the website but that contributions decreased rapidly after the intervention, indicating that the use of the press has limited influence over time; thus other methods need to be used in combination with the media to continually stimulate recruitment.

OSM uses mapping parties as a way of recruiting new individuals and providing social contact with other OSM mappers while serving the purpose of increasing map coverage in a particular area (OSM, 2015). An interesting study by Hristova et al. (2013) showed that mapping parties did increase the amount of data collected during the event and did result in greater contributions after the event, generally for light to medium contributors in the short-term and heavy contributors in the longer-term. Mapping parties also retained more experienced users but failed to retain newcomers, possibly because it was more difficult for them to integrate socially in an already established community; thus more focus on integration of novices at these events is recommended, as well as more emphasis on easy-to-use tools and on the fun aspect. Similar events could be organised for other VGI initiatives, using the experience gained by the OSM community in running these events.

Another way of recruiting volunteers is to make explicit links to education, motivating students to take part in VGI initiatives. Some of the current partnerships between mapping agencies and schools are described by Olteanu-Raimond et al. (2017) in Chapter 13 and by Bol et al. (2016). A very successful example of citizen science linking to education is the GLOBE (Global Learning and Observations to Benefit the Environment) Program, which was initiated by Al Gore in 1995. The programme aims to increase environmental awareness by actively involving students in science, including through mapping. Similarly, integrating volunteer service directly into educational programmes is another effective way to recruit and motivate individuals. There are many examples of this in the conservation arena, such as the Master Naturalist Programs or the Conservation Stewards Programs established in different US states (Van Den Berg et al., 2009) that provide individuals with a certification and require a certain number of volunteer hours, both as part of the certification and to keep the certification once it has been gained. This type of approach could be modified to include mapping as a volunteer activity and could encourage longer term engagement.

3.2 Motivation and Retention

Nielsen (2006) provides some general advice for improving participant equality (i.e. increasing the numbers that actively contribute) in social media and online communities that also has relevance for VGI. The first recommendation is to make it as simple as possible to contribute. This is already implemented in OSM in the sense that users are free to choose what features and in what location they contribute to OSM; furthermore, this was highlighted as one of the main motivators for contributing to OSM in the study by Budhathoki and Haythornthwaite (2012). Part of this recommendation also refers to the design of the site and the ease of use, which can clearly influence participation. The Zooniverse citizen science project has put a considerable amount of effort into the design of its projects and much can be learned from its approach (Prestopnik, n.d.). Zooniverse now offers a platform to host other citizen science projects, allowing new initiatives to benefit from its design principles while also having access to a large community of citizen scientists; new VGI initiatives should consider this option of working with Zooniverse.

Another relevant recommendation from Nielsen (2006) is to make participation part of another activity so that volunteers do not find the act of contributing a burden. Passive data collection from communities such as hikers and bikers or from geotagged repositories are some examples that could be harnessed within VGI applications; alternatively, gamification, or the addition of game mechanics to applications (Deterding, 2012), can lower the burden of participation while adding an element of fun, which is another key motivator for participation in VGI (Budhathoki and Haythornthwaite, 2012; Tiwari et al., 2010). An example of gamification is the Ingress augmented reality game by Google, where players gather spatial information that is then used to update Google Maps as a side task to the main goal of the game, which is to find portals (Carney, 2012). Gamification has also been shown to help motivate participation in a citizen science application such as Project Budburst, which developed the Biotracker app for gathering phenology data: use of technology such as smartphones, coupled with competitive elements such as badges and leaderboards, was shown to appeal to the younger 'Millennial' audience (Bowser et al., 2013). A number of game apps have been built for gathering

OSM data, e.g. AddressHunter, which is a role playing game that also involves adding addresses to the OSM database, and Kort Game, for adding new features to OSM (OSM, 2013).

Motivation is also clearly linked to maintaining participation in the longer term. The use of different incentives can be a powerful way to achieve this. Reputation and confidence building measures can be effective ways to motivate volunteers. The citizen science project iNaturalist, for example, awards different levels of expertise to volunteers, from novice to expert, which recognises their knowledge and degree of contribution. Each observation is also given a stamp of quality, which can build confidence in the contributors, particularly when the observations are considered to be of research grade quality. This follows the advice of Nielsen (2006) to promote high-quality contributions. In Wikipedia, contributors can take on roles with increasing responsibilities within the community, including arbitration and administration (Bryant et al., 2005), which is also a reputation and confidence building measure.

Another incentive is related to the impact of contributions. In OSM, contributors can quickly see their changes on the map, which acts as an important form of visual feedback. Correcting areas and filling in missing information can provide a form of satisfaction that acts as a motivating factor; thus the design of VGI initiatives should include good visual displays (Budhathoki and Haythornthwaite, 2012). Experiences from Nature's Notebook with regards to retention have highlighted the need to provide frequent communication to volunteers, acknowledge the value of their contributions on a regular basis and show that their contributions are being used (Crimmins et al., in press). Nature's Notebook relies heavily on digital communication of various forms, ensuring that the content of the communication is information-rich, including summaries of publications that have used the data, which are communicated in simple language. Finally, the project provides different opportunities for volunteers to participate, which are based on problem solving approaches to keep volunteers engaged over time.

Rewarding volunteers in other ways can also be an effective approach for encouraging and supporting participation. A reward system can be implemented in several different ways; for example, Estes et al. (2016) have used Amazon's Mechanical Turk to do cropland mapping through digitisation of fields for part of South Africa using performance-based micro-payments. Maps with 91% accuracy were produced, and the authors calculated that a detailed cropland map for all of Africa could be created with 2 to 3 million USD and the crowd. Several campaigns have been run using the Geo-Wiki tool for visualisation, validation and crowdsourcing of land cover (Fritz et al., 2012; See et al., 2015), where incentives have ranged from Amazon vouchers to co-authorship on a scientific publication. However, Nielsen (2006) makes the point that participants should not be over-rewarded as this might encourage the most active volunteers to dominate and thereby disincentivise others from contributing.

4 Case Studies

This section describes a set of case studies based on our experiences to illustrate different ways in which volunteers have been motivated to contribute VGI to different applications.

4.1 Mapping Parties

As mentioned previously, mapping parties are intended to map a specific area over a short period of time while introducing newcomers to VGI. This case study describes experiences with two mapping parties that were organised as social events for delegates at the recent FOSS4G (Free and Open Source Software for Geomatics) Europe conference[1], held in July 2015 at the Politecnico di Milano, Como Campus (Figure 1). The first mapping party was a traditional OSM one, while the second focused on indoor mapping. To recruit participants, the mapping party organisers presented their ideas and calls for participation during the opening session of the conference. Information about the events was also communicated over social media, via the official conference website and via OSM in order to attract and sustain participation throughout the conference.

The OSM mapping party was designed and set up by a small number of active OSM contributors who were attending the conference (Mooney et al., 2015); their goal was to collect Points of Interest (POIs) that were missing in Como city. Around 40 participants (roughly 10% of the conference) attended

Fig. 1: Photographs from the mapping parties at the FOSS4G 2015 Europe conference.

and were taught how to collect the data using field papers, which are a specific service to print out OSM maps for annotation in the field. The POIs were then mapped in around 2.5 hours. On the second day of the conference, there was a data upload session that showed the volunteers how to insert their data into the OSM database; this session was too short, so not all data were entered into the database during the event. However, the POIs were monitored after the event and showed an increased mapping over the summer, which is attributed largely to this particular mapping party as local OSM activity in the city is not large. Thus, the mapping party motivated interested individuals by providing them with training and a social, community-based atmosphere in which to collect and upload the data. Given the increase in POIs over the summer, this may have led to some individuals continuing to contribute to OSM.

The second mapping party was focused on indoor mapping, which is something new compared to the more traditional OSM outdoor mapping parties. The main purpose of the event was to raise awareness of the scientific, technical and practical challenges associated with indoor mapping. The IndoorGML standard was used to collect the navigation pathways through rooms and in connecting spaces. The indoor mapping-party received attention from the local television and more than 30 participants took part in the event. Almost all of the mappers generated data, but only some of them contributed to the result, mainly due to technical issues and shortage of time. The overall result was a single, merged navigable graph of two floors of the University building (Figure 2).

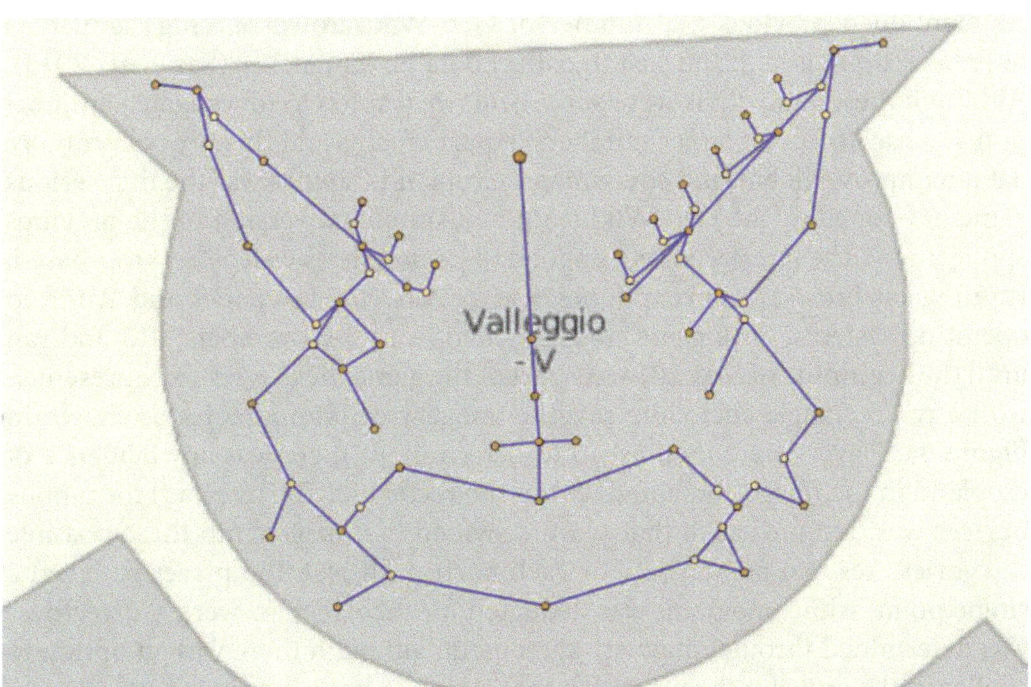

Fig. 2: Screenshot of the merged navigation graph from the participants of the Indoor Mapping Party held at the FOSS4G 2015 Europe Conference.

The indoor mapping party produced positive results as novices learned about the concepts, strategies, problems and tools for mapping indoor spaces while the researchers and developers received feedback on the techniques and tools used during the event.

Overall, the mapping parties were inclusive and friendly experiences and are recommended as side events at future FOSS4G conferences. At both parties, the incentive was the social aspect, i.e. spending time together, learning something new, making a useful social contribution and having fun. An additional incentive was offered, i.e. prizes were given to the top three contributors at the closing ceremony of each event. Thus both mapping parties appealed to a range of intrinsic and extrinsic motivations. Both events were successful in attracting participants, and the OSM mapping party may have led to the recruitment of new participants in OSM that continued to contribute to OSM beyond the actual event. The indoor mapping party was more focused on the learning element as motivator. The main disadvantage associated with both mapping parties was time, e.g. there was insufficient time to complete the uploading of POIs from the paper-based surveys, and this had to be completed by the mapping party staff after the event.

4.2 Gamification

4.2.1 Cropland Capture and Picture Pile

As mentioned previously, a number of Geo-Wiki crowdsourcing campaigns have been organised in the past to collect data on land cover (See et al., 2015). Although these campaigns were successful, we wanted to investigate gamification as a way to attract larger numbers of participants and thereby collect more data to improve global land cover maps. Cropland Capture was the first serious game developed by the Geo-Wiki team as a simplified version of the previous applications. The interface was designed to be mobile as well as desktop-based, running on browsers, smartphones and tablets (for both iOS and Android operating systems). The game was launched in mid-November 2013 and ran until the beginning of May 2014. As part of the game the players were presented with a red rectangle encircling satellite imagery or photographs, as shown in Figure 3a. Players were then asked to determine if there was any evidence of cropland in the image contained within the rectangle. The interface for mobile devices was designed such that players swiped the images into three possible categories: Yes, No or Maybe. For each correct answer, the player received a single point, while one point was deducted for incorrect answers. Correctness was determined through majority agreement, although there was an option to challenge the crowd if the player felt that they had been incorrectly penalised.

Recruitment was through the Geo-Wiki newsletter, a press release, social media and word of mouth. The game received media coverage at two different occasions during the time it was open, which resulted in a spike in participation; however, participation decreased soon afterwards, similarly to that observed by Composto et al. (2016). The game had a leader board, which was reset each week, and the top three players in terms of the total number of classifications each week were added to a prize draw that took place at the end of the game's six-month period; thus, prizes were one incentive used to motivate the players. The idea of helping science was also a strong message in the game and was meant as an additional motivating factor. In total, more than 4.5 million observations were obtained from more than 3,000 players. A survey of players was undertaken near the end of the game, which revealed that helping science, the competitive element and the beauty of the satellite images were motivating factors for participation.

Picture Pile is the direct successor to Cropland Capture, so the game mechanics are similar. However, Picture Pile was made more generic: the basic concept is that players sort or classify 'piles of pictures', where each pile represents a different task or theme including different land cover types. The idea behind having different tasks in the game is that there will be more variety for the players, which may help to retain them for longer. Another major difference between Picture Pile and Cropland Capture is the added functionality for change detection: in Picture Pile, players are presented with pairs of images from different time periods and asked to look for evidence of change over time, e.g. deforestation (see Figure 3b). Players can also view a map of their contributions and the contributions of others in real-time. Another added feature is the use of more reference data, where the images have been marked up to explain correct

(a) (b)

Fig. 3: (a) Cropland Capture and (b) Picture Pile.

answers. This is used as both feedback and training for the players, which was also intended to provide motivation to participate. Each pile has its own leader board and a chat channel, which makes it very easy for the players and the organisers to communicate with each other as the game progresses.

Recruitment strategies were similar to Cropland Capture. The game was launched in November 2015. Almost 4 million pairs of pictures were classified. Other piles will be implemented in the future.

4.2.2 FotoQuest Austria

The second game, called FotoQuest Austria, is quite different in nature from Cropland Capture and Picture Pile: instead of asking the crowd to classify imagery online, the FotoQuest Austria app is focused on getting players to go outside and document the landscape. The game is similar to geocaching except that players do not search for a physical cache. Instead, points are awarded for documenting specific locations shown on the mobile device (see Figure 4). Players are asked to take photographs in four cardinal directions and then classify the land cover and land use based on categories in a classification system developed for the EU LUCAS (Land Use and Cover Area frame Survey) survey. This EU systematic sample is collected by professional surveyors every three years in EU countries for change detection purposes, among other reasons, and therefore provides authoritative data for comparison with the crowd's results. The locations of the LUCAS points for Austria were added to the FotoQuest Austria app along with other locations to ensure sufficient numbers of points for the players to visit.

The app was specifically designed to adhere as closely as possible to the LUCAS protocol, and so only allows photographs to be taken when the user is within a certain distance of the location, the mobile device is not tilted, the compass indicates the correct direction and the horizon matches a line indicated on the app. This was to ensure that the data collected by the players would be of the highest quality possible, but also to make data collection as easy as possible. The app was launched in July 2015 and ran over a three-month period.

Recruitment was via a newsletter, social media and a more traditional media campaign, i.e. a press release was issued and interviews were held with the main television and radio stations in Austria. The app was featured as 'app of the week' in the technology section of the website of Austria's main TV channel and was featured on an afternoon programme which demonstrated how the app worked. In addition to the fun provided by the competitive elements of the game, additional motivators were interacting with the landscape and incentives such as smartphones and tablets, which were awarded at the end of the game. Overall, 2300 quests were undertaken. A second version, which was developed using feedback received from the game, will be launched in 2017.

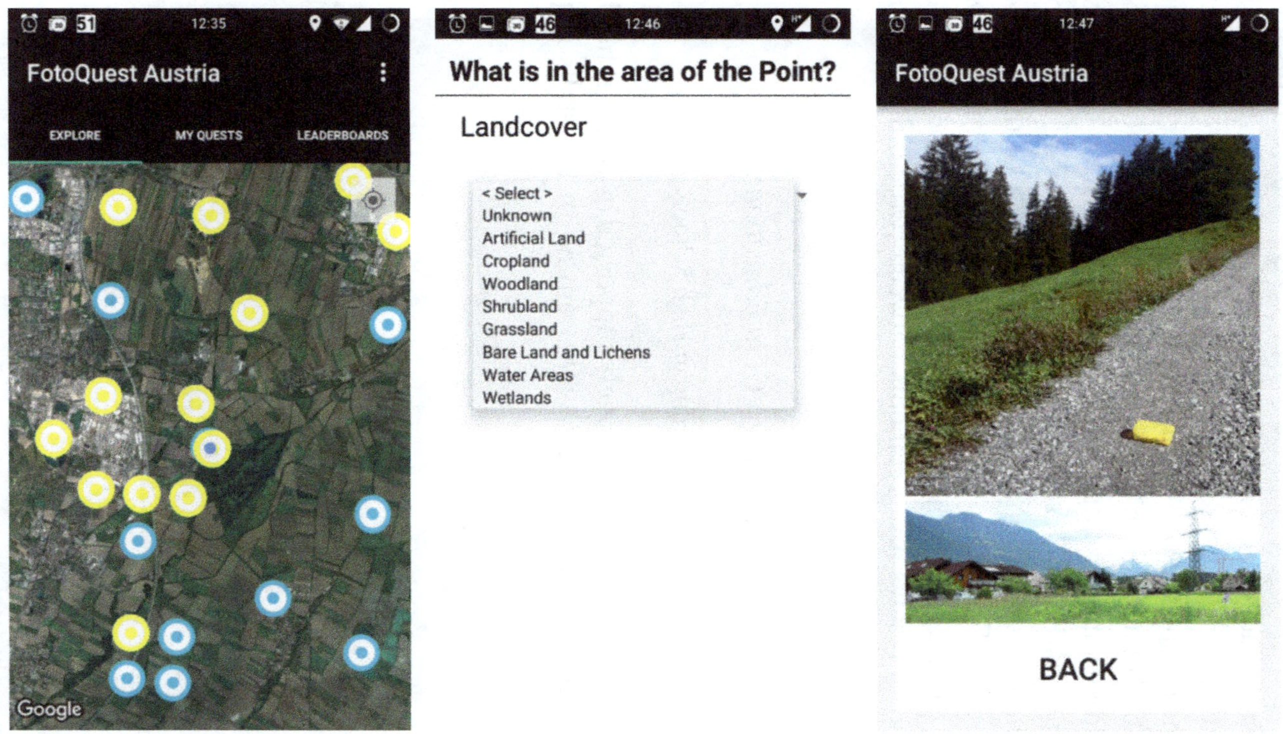

(a) (b) (c)

Fig. 4: (a) Quests in FotoQuest Austria (b) Classifying land cover (c) Geotagged photograph of the point.

4.2.3 The Land Cover Validation Game

The Land Cover Validation Game is a serious game for validating land cover (Brovelli et al., 2015). Figure 5 shows the user interface, in which players see a reference image of the land under investigation. The task is to classify the 30 m pixel shown within a blue box on the interface. Depending on the answer, the players get points, badges and a ranking on a global leaderboard. The game was introduced at the FOSS4G 2015 Europe Conference and participants played the game during the week of the conference. There were 68 participants engaged for a total of more than 20 hours of gameplay. Overall 1600 pixels were validated. A video[2] summarising the Land Cover Validation Game results was presented at the ESA Earth Observation Open Science event in October 2015. Prizes were offered as additional incentives at the end of the FOSS4G 2015 Europe Conference. The results showed that involving users in a crowdsourcing validation campaign with a gaming incentive can be an effective way to collect data and to resolve disagreements between two conflicting land cover classifications.

4.3 *Embedding VGI in Education*

4.3.1 Work Training in High Schools

Work training in schools, which is strongly supported by recent school reforms in Italy, combines classroom studies with training in the skills required to

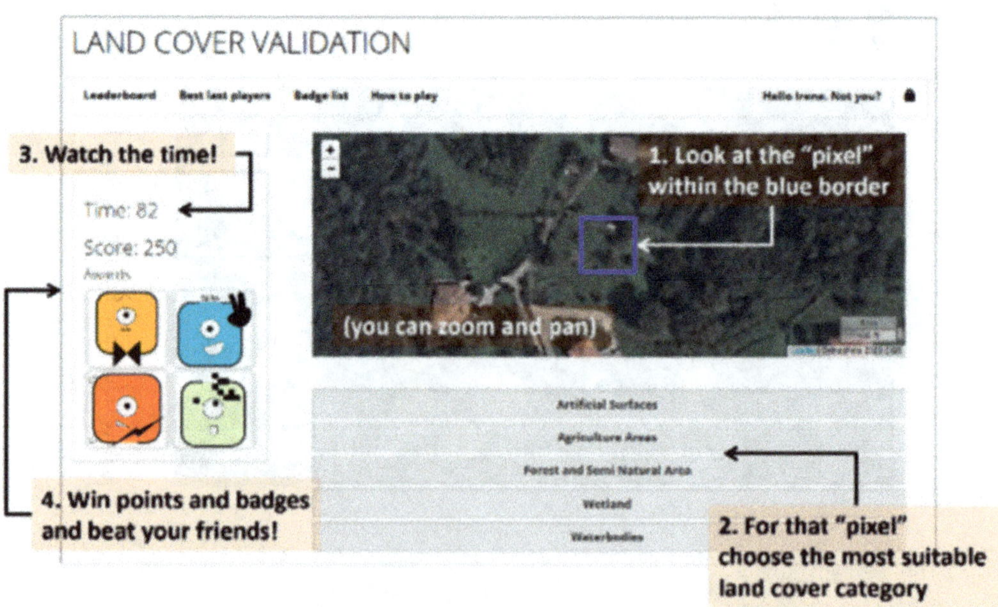

Fig. 5: Land Cover Validation Game interface, with a pixel (blue square box) to be classified (http://bit.ly/foss4game).

make a successful transition from high school to employment, and hence is aimed at students aged 15 and above. Every year since 2013, the Politecnico di Milano has organised a week-long internship for 15–20 students; the incentives for the students to participate are credits towards their course, learning new technologies and the collection of useful VGI. The collection of data is preceded by a MOOC[3] called M'appare il mondo (which is a word play in Italian, as it means 'the world appears to me', but becomes 'mapping the world' if the apostrophe is removed) and instructions on how to create a mobile app to collect the data. This latter step has been done using two applications. The first is the Open Data Kit (ODK), which is a simple, free, open tool for the Android operating system; it is very easy to implement forms in ODK for managing the collection of data, i.e. attributes, photos, videos, audio of the selected features, etc. The second was Geopaparazzi[4], which is another free, user-friendly, open source tool.

During one work training session, the students developed an app to collect data on building amenities, e.g. the presence of ramps and stairs (Figure 6). The results from the data collection exercise were then displayed on a website[5] so that the students could view their contributions online directly (Figure 7), including those features that do not conform to Italian law, simultaneously raising an issue of importance for the public. During another session, students built an app to capture local biodiversity (Figure 8).

In addition to gaining credits, the students learn how to map the world around them and collect data that are of public interest, which are displayed through a WebGIS interface. In the future there are plans to make connections between the data needs of government municipalities and of civil protection agencies and the projects undertaken by the students, which should provide additional motivation to become involved in VGI projects.

4.3.2 Humanitarian MiniMapathons in Elementary Schools

Mapathons, also known as 'armchair' mapping, are events where people come together to do mapping online. Examples are events related to natural disasters and political crises, which are supported and organised by HOT (Humanitarian OSM Team), or events devoted to mapping places that are not yet well mapped or where the most vulnerable people live, e.g. the Missing Maps project. Two MiniMapathons aimed at 10-year-old children from elementary schools were organised by the Geomatics and Earth Observation (GEO) and Hypermedia Open Center (HOC) Labs of the Politecnico di Milano with the support of HOT and Missing Maps. The first event, in which 36 children took part, was organised in Como. The second event, in Milan, saw 212 children participate. Online registration for the second event closed just a few hours after opening, having reached the maximum number of students that could be accommodated in the computer rooms of the Politecnico.

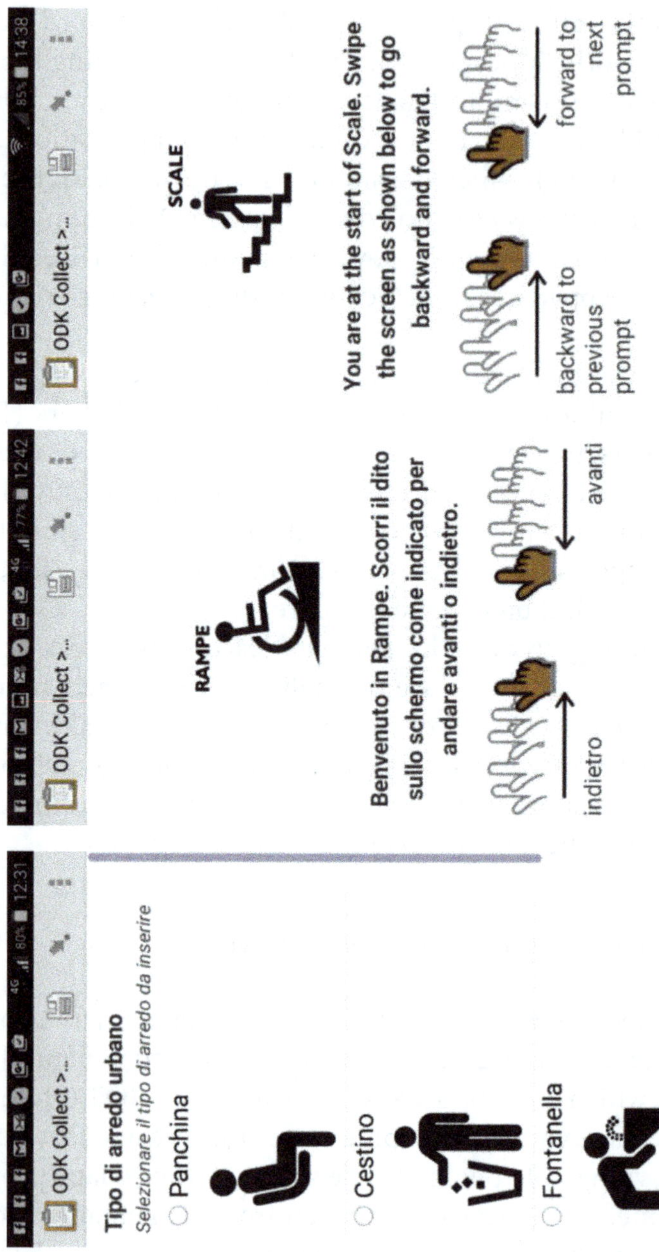

Fig. 6: Example of screenshots from ODK to collect information on building amenities.

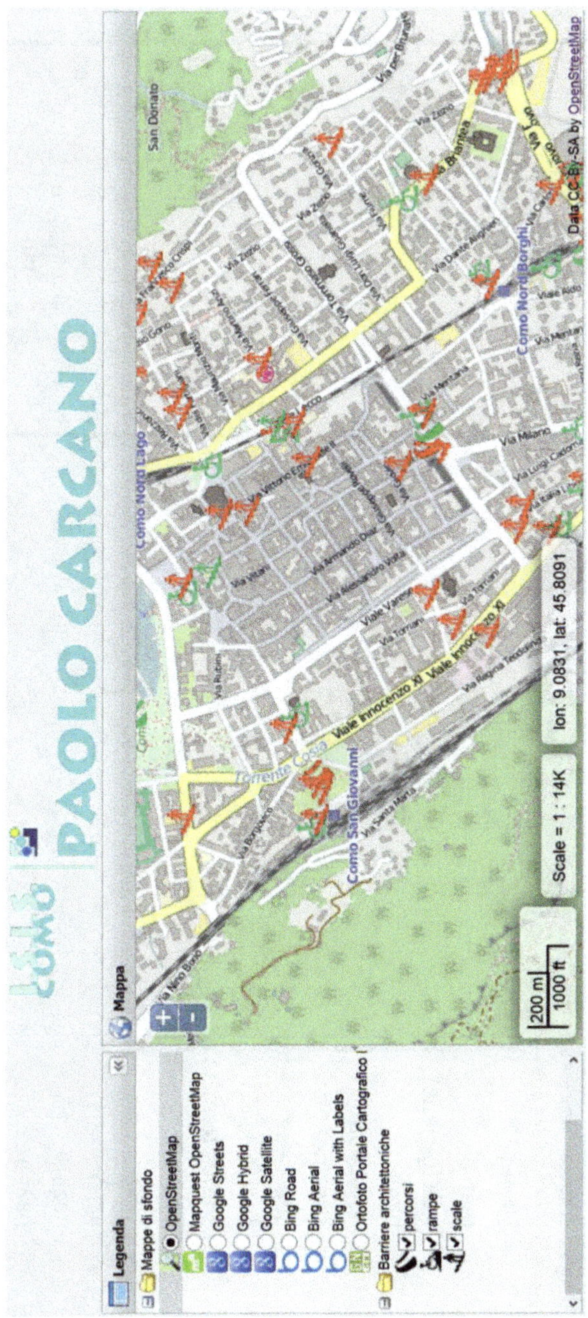

Fig. 7: The WebGIS interface showing building amenities in the city.

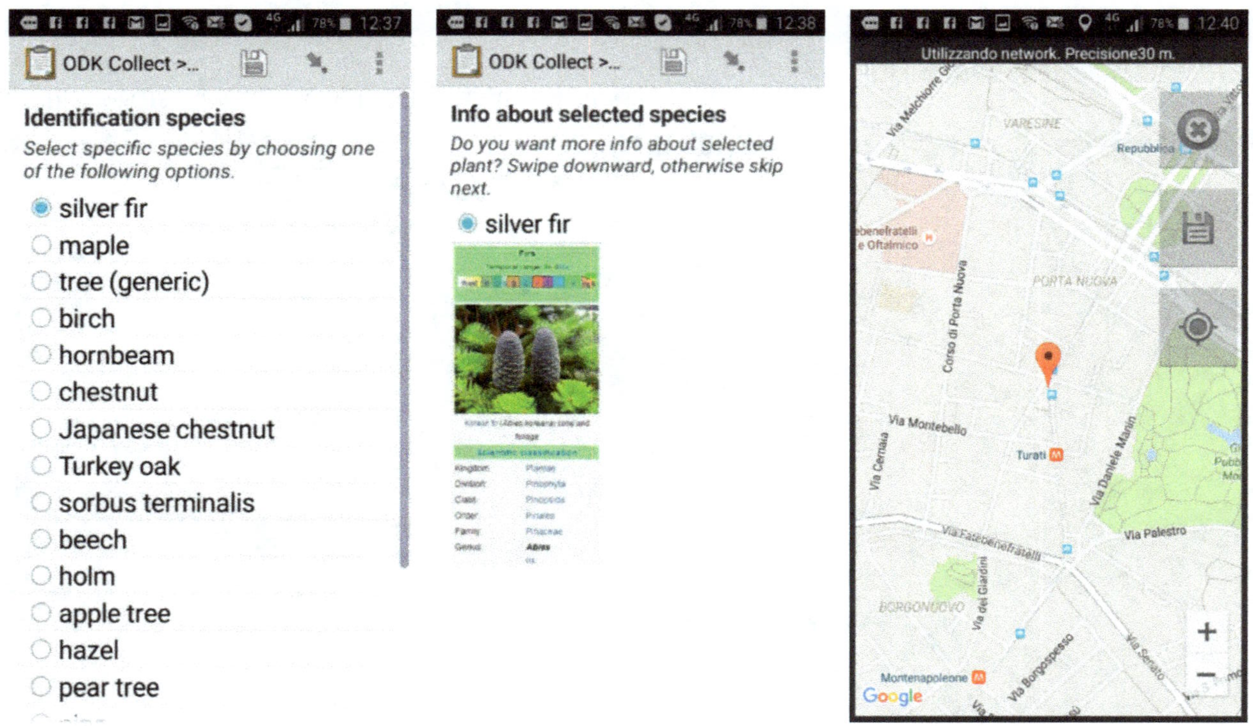

Fig. 8: Examples of screenshots from the ODK forms developed for the biodiversity app (selection of the species, information about the selected species, geolocation).

The purpose of the MiniMapathons was to map buildings in the northernmost part of Swaziland in a project related to malaria elimination. In total 5000 buildings were mapped and the quality was similar to that of adult volunteers' in terms of the shapes digitised and the ability to recognise buildings on the imagery. The teachers of the elementary schools and the children were highly motivated as they saw this as a tangible way of helping people in Swaziland, but at the same time the children acquired competencies in mapping, geometry and informatics. The second incentive for participation was a purely symbolic one, i.e. certificates of participation and baseball caps from Politecnico di Milano. The two events were highly successful and appear to be a good way to transform children into neogeographers and humanitarians and to lead them to contribute VGI for a good cause.

5 Conclusions

The success of VGI is clearly down to the participation of volunteers and of the community that supports the activities related to spatial data collection and mapping. Hence volunteer recruitment, motivation and longer-term retention are key issues when designing and implementing a VGI initiative. A number of studies have looked at typologies for characterising the nature of volunteers and the motivational factors that drive participation. These factors, which were compiled by Budhathoki and Haythornthwaite (2012), represent a comprehensive list of motivations that can be used to further investigate reasons for participation in current VGI initiatives. They can also be used in the design of new applications, drawing upon the findings of Budhathoki and Haythornthwaite (2012) for OSM volunteers. Recommendations and best practice in recruitment, motivation and retention were then provided, drawing upon experiences in the broader field of citizen science. The case studies presented here served to illustrate how recruitment and motivation are considered in a range of different VGI initiatives.

Acknowledgements

This work was supported by the EU FP7-funded ERC grant Crowdland (No. 617754).

Notes

1 http://europe.foss4g.org/2015
2 https://www.youtube.com/watch?v=Q0ru1hhDM9Q
3 https://www.youtube.com/watch?v=1nnfQgMQq4Y
4 http://geopaparazzi.github.io/geopaparazzi/
5 http://geomobile.como.polimi.it/Barriere/barriere.html

Reference list

Antoniou, V., Skopeliti, A., 2015. Measures and indicators of VGI quality: An overview, in: ISPRS Annals of the Photogrammetry, Remote Sensing and Spatial Information Sciences. Presented at the ISPRS Geospatial Week 2015, ISPRS Annals, La Grande Motte, France, pp. 345–351.

Barrington, L., Ghosh, S., Greene, M., Har-Noy, S., Berger, J., Gill, S., Lin, A.Y.-M., Huyck, C., 2011. Crowdsourcing earthquake damage assessment using remote sensing imagery. *Annals of Geophysics*. 54, 680–687. DOI: https://doi.org/10.4401/ag-5324

Bol, D., Grus, M., Laakso, M., 2016. Crowdsourcing and VGI in national map agency data collection, in: T. Bandrova and M. Konecny (Eds.), *Proceedings of the 6th International Conference on Cartography and GIS*, Albena, Bulgaria, 13–17 June 2016, pp. 493–498. Available at: https://cartography-gis.com/docsbca/iccgis2016/ICCGIS2016-50.pdf [Last access 16 May 2017].

Bonney, R., Cooper, C.B., Dickinson, J., Kelling, S., Phillips, T., Rosenberg, K.V., Shirk, J., 2009. Citizen science: A developing tool for expanding science knowledge and scientific literacy. *BioScience* 59, 977–984. DOI: https://doi.org/10.1525/bio.2009.59.11.9

Bowser, A., Hansen, D., Preece, J., 2013. Gamifying citizen science: Lessons and future directions, in: *Proceedings of CHI 2013 Conference on Human Factors in Computing*, Paris, France, 27 April to 2 May 2013. Available at: http://gamification-research.org/wp-content/uploads/2013/03/Bowser_Hansen_Preece.pdf [Last access 16 May 2017].

Brovelli, M.A., Celino, I., Molinari, M., Venkatachalam, V., 2015. Land cover validation game, in: *Geomatics Workbooks No. 12*. FOSS4G Europe Como 2015. pp. 153–157.

Bryant, S.L., Forte, A., Bruckman, A., 2005. *Becoming Wikipedian: transformation of participation in a collaborative online encyclopedia*. ACM Press, p. 1. DOI: https://doi.org/10.1145/1099203.1099205

Budhathoki, N.R., 2010. Participants' Motivations to Contribute to Geographic Information in an Online Community. Unpublished PhD Dissertation. University of Illinois at Urbana-Champaign, Urbana, Illinois, USA.

Budhathoki, N.R., Haythornthwaite, C., 2012. Motivation for open collaboration: Crowd and community models and the case of OpenStreetMap. *American Behavioral Scientist* 57, 548–575. DOI: https://doi.org/10.1177/0002764212469364

Carney, M., 2012. Google's Ingress is more than a game, its a potential data exploitation disaster. Available at https://pando.com/2012/11/19/googles-ingress-is-more-than-a-game-its-a-potential-data-exploitation-disaster/ [Last accessed on 21 August 2016].

Clavero, M., Revilla, E., 2014. Biodiversity data: Mine centuries-old citizen science. *Nature* 510, 35–35. DOI: https://doi.org/10.1038/510035c

Clery, D., 2011. Galaxy Zoo volunteers share pain and glory of research. *Science* 333, 173–175. DOI: https://doi.org/10.1126/science.333.6039.173

Coleman, D.J., Georgiadou, Y., Labonte, J., 2009. Volunteered geographic information: The nature and motivation of produsers. *International Journal of Spatial Data Infrastructures Research* 4, 332–358.

Composto, S., Ingensand, J., Nappez, M., Ertz, O., Rappo, D., Bovard, R., Widmer, I., Joost, S., 2016. How to recruit and motivate users to utilize VGI-systems? Presented at the 19th AGILE Conference on Geographic Information Science, Helsinki, Finland.

Connors, J.P., Lei, S., Kelly, M., 2012. Citizen science in the age of neogeography: Utilizing volunteered geographic information for environmental monitoring. *Annals of the Association of American Geographers* 102, 1267–1289. DOI: https://doi.org/10.1080/00045608.2011.627058

Crimmins, T.M., Barnett, L., Denny, E.G., Rosemartin, A.H., Schaffer, S., Weltzin, J.F., in press. From tiny acorns grow mighty oaks: What we've learned from nurturing Nature's Notebook, in: Lepczyk, C.A. (Ed.), *Handbook of Citizen Science in Ecology and Conservation.*

Deterding, S., 2012. Gamification: designing for motivation. *Interactions* 19, 14. DOI: https://doi.org/10.1145/2212877.2212883

Estes, L.D., McRitchie, D., Choi, J., Debats, S., Evans, T., Guthe, W., Luo, D., Ragazzo, G., Zempleni, R., Caylor, K.K., 2016. A platform for crowdsourcing the creation of representative, accurate landcover maps. *Environmental Modelling & Software* 80, 41–53. DOI: https://doi.org/10.1016/j.envsoft.2016.01.011

Flanagin, A., Metzger, M., 2008. The credibility of volunteered geographic information. *GeoJournal* 72, 137–148.

Foody, G., See, L., Fritz, S., Van der Velde, M., Perger, C., Schill, C., Boyd, D.S., 2013. Assessing the accuracy of volunteered geographic information arising from multiple contributors to an internet based collaborative project. *Transactions in GIS* 17, 847–860. DOI: https://doi.org/10.1111/tgis.12033

Fritz, S., McCallum, I., Schill, C., Perger, C., See, L., Schepaschenko, D., van der Velde, M., Kraxner, F., Obersteiner, M., 2012. Geo-Wiki: An online platform for improving global land cover. *Environmental Modelling & Software* 31, 110–123. DOI: https://doi.org/10.1016/j.envsoft.2011.11.015

Gómez-Barrón, J.-P., Manso-Callejo, M.-Á., Alcarria, R., Iturrioz, T., 2016. Volunteered Geographic Information system design: Project and participation guidelines. *ISPRS International Journal of Geo-Information* 5, 108. DOI: https://doi.org/10.3390/ijgi5070108

Goodchild, M.F., 2007. Citizens as sensors: the world of volunteered geography. *GeoJournal* 69, 211–221. DOI: https://doi.org/10.1007/s10708-007-9111-y

Haklay, M., 2010. How good is volunteered geographical information? A comparative study of OpenStreetMap and Ordnance Survey datasets. *Environment and Planning B: Planning and Design* 37, 682–703. DOI: https://doi.org/10.1068/b35097

He, Z., 2012. *Digital by-product data in Web 2.0: exploring mass collaboration of Wikipedia*. Cambridge Scholars, Newcastle.

Heipke, C., 2010. Crowdsourcing geospatial data. *ISPRS Journal of Photogrammetry and Remote Sensing* 65, 550–557. DOI: https://doi.org/10.1016/j.isprsjprs.2010.06.005

Hristova, D., Quattrone, G., Mashhadi, A., Capra, L., 2013. The life of the party: Impact of social mapping in OpenStreetMap, in: Proceedings of the 7th International AAAI Conference on Weblogs and Social Media, Boston, MA, USA, 8–10 July 2013, pp. 234–243. Available at: https://www.aaai.org/ocs/index.php/ICWSM/ICWSM13/paper/view/6098 [Last access 16 May 2017]

Hyvoenen, E., Alonen, M., Koho, M., Tuominen, J., 2013. BirdWatch: Supporting citizen scientists for better linked data quality for biodiversity management. Presented at the First International Workshop on Semantics for Biodiversity (S4BioDiv 2013), University of Montpellier, Montpellier, France.

Mooney, P., Corcoran, P., 2011. Can Volunteered Geographic Information be a participant in eEnvironment and SDI?, in: Hřebíček, J., Schimak, G., Denzer, R. (Eds.), *Environmental Software Systems. Frameworks of eEnvironment, IFIP Advances in Information and Communication Technology*. Springer Berlin Heidelberg, pp. 115–122.

Mooney, P., Minghini, M., Stanley-Jones, F., 2015. Observations on an OpenStreetMap mapping party organised as a social event during an open source GIS conference. *International Journal of Spatial Data Infrastructures Research* 10, 138–150. DOI: https://doi.org/10.2902/ijsdir.v10i0.395

Neis, P., Zielstra, D., 2014. Recent developments and future trends in volunteered geographic information research: The case of OpenStreetMap. *Future Internet* 6, 76–106. DOI: https://doi.org/10.3390/fi6010076

Neis, P., Zielstra, D., Zipf, A., 2011. The street network evolution of crowdsourced maps: OpenStreetMap in Germany 2007–2011. *Future Internet* 4, 1–21. DOI: https://doi.org/10.3390/fi4010001

Nielsen, J., 2006. The 90-9-1 Rule for Participation Inequality in Social Media and Online Communities. Nielsen Norman Group. Available at https://www.nngroup.com/articles/participation-inequality/ [Last accessed on 21 August 2016].

Olteanu-Raimond, A-M, Laakso, M, Antoniou, V, Fonte, C C, Fonseca, A, Grus, M, Harding, J, Kellenberger, T, Minghini, M, Skopeliti, A. 2017. VGI in National Mapping Agencies: Experiences and Recommendations. In: Foody, G, See, L, Fritz, S, Mooney, P, Olteanu-Raimond, A-M, Fonte, C C and Antoniou, V. (eds.) *Mapping and the Citizen Sensor*. Pp. 299–326. London: Ubiquity Press. DOI: https://doi.org/10.5334/bbf.m.

OpenStreetMap (OSM), 2015. Mapping Weekend Howto. Available at http://wiki.openstreetmap.org/wiki/Mapping_Weekend_Howto [Last accessed on 21 August 2016].

OpenStreetMap (OSM), 2013. Gamification. Available at http://wiki.openstreetmap.org/wiki/Gamification [Last accessed on 21 August 2016].

Prestopnik, N.R., n.d. Citizen Science Case Study: Galaxy Zoo / Zooniverse. Available at http://www.imperialsolutions.com/research/galaxyzoo.pdf [Last accessed on 21 August 2016].

See, L, Estima, J, Pőd.r, A, Arsanjani, J J, Bayas, J-C L and Vatseva, R. 2017. Sources of VGI for Mapping. In: Foody, G, See, L, Fritz, S, Mooney, P, Olteanu-Raimond, A-M, Fonte, C C and Antoniou, V. (eds.) *Mapping and the Citizen Sensor*. Pp. 13–35. London: Ubiquity Press. DOI: https://doi.org/10.5334/bbf.b.

See, L., Fritz, S., Perger, C., Schill, C., McCallum, I., Schepaschenko, D., Duerauer, M., Sturn, T., Karner, M., Kraxner, F., Obersteiner, M., 2015. Harnessing the power of volunteers, the internet and Google Earth to collect and validate global spatial information using Geo-Wiki. *Technological Forecasting and Social Change* 98, 324–335. DOI: https://doi.org/10.1016/j.techfore.2015.03.002

See, L., Mooney, P., Foody, G., Bastin, L., Comber, A., Estima, J., Fritz, S., Kerle, N., Jiang, B., Laakso, M., Liu, H.-Y., Milčinski, G., Nikšič, M., Painho, M., Pődör, A., Olteanu-Raimond, A.-M., Rutzinger, M., 2016. Crowdsourcing, citizen science or Volunteered Geographic Information? The current state of crowdsourced geographic information. *ISPRS International Journal of Geo-Information* 5, 55. DOI: https://doi.org/10.3390/ijgi5050055

Tiwari, R., Agrawal, A., Shekhar, S., 2010. Contributions of volunteered geographic world: Motivation behind contribution. Presented at the GIScience 2010: The Role of Volunteered Geographic Information in Advancing Science, Oak Ridge National Laboratory Distributed Active Archive Center, Zurich, Switzerland.

Tweddle, J.C., Robinson, L.D., Pocock, M.J.O., Roy, H., 2012. *Guide to citizen science: developing, implementing and evaluating citizen science to study biodiversity and the environment in the UK*. Natural History Museum and NERC Centre for Ecology & Hydrology.

Van Den Berg, H.A., Dann, S.L., Dirkx, J.M., 2009. Motivations of adults for non-formal conservation education and volunteerism: Implications for programming. *Applied Environmental Education & Communication* 8, 6–17. DOI: https://doi.org/10.1080/15330150902847328

Wikipedia, 2016. Wikipedia:Statistics. Available at https://en.wikipedia.org/wiki/Wikipedia:Statistics [Last accessed on 21 August 2016].

Zook, M., Graham, M., Shelton, T., Gorman, S., 2010. Volunteered Geographic Information and crowdsourcing disaster relief: A case study of the Haitian earthquake. *World Medical & Health Policy* 2, 7–33. DOI: https://doi.org/10.2202/1948-4682.1069

CHAPTER 6

Considerations of Privacy, Ethics and Legal Issues in Volunteered Geographic Information

Peter Mooney*, Ana-Maria Olteanu-Raimond[†], Guillaume Touya[†], Niels Juul[‡], Seraphim Alvanides[§] and Norman Kerle[¶]

*Department of Computer Science, Maynooth University, Co. Kildare, Ireland, peter.mooney@nuim.ie

[†]IGN, French Mapping Institute, COGIT Laboratory, Université Paris-Est, 73 avenue de Paris, 94160 Saint-Mandé, France

[‡]Roskilde University, Roskilde, Denmark

[§]Department of Architecture and the Built Environment, Northumbria University, Newcastle, UK

[¶]Faculty of Geo-Information Science and Earth Observation, University of Twente, Enschede, The Netherlands

Abstract

Today almost any kind of User Generated Content (UGC) can be situated within a geographic context. Volunteered Geographic Information (VGI) can include many types of UGC, such as georeferenced photographs, social media and text, geographic data themselves, etc. There are legal, privacy and ethical issues raised by VGI, and at present these are not very well studied or understood despite the rise in popularity of VGI. This chapter will discuss, investigate and define some

How to cite this book chapter:

Mooney, P, Olteanu-Raimond, A-M, Touya, G, Juul, N, Alvanides, S and Kerle, N. 2017. Considerations of Privacy, Ethics and Legal Issues in Volunteered Geographic Information. In: Foody, G, See, L, Fritz, S, Mooney, P, Olteanu-Raimond, A-M, Fonte, C C and Antoniou, V. (eds.) *Mapping and the Citizen Sensor.* Pp. 119–135. London: Ubiquity Press. DOI: https://doi.org/10.5334/bbf.f.

of the most prominent issues related to the legal, privacy and ethics topic within VGI. The chapter argues that these issues are not well understood by all of the actors in VGI, and in particular by the producers of this information as well as the users or consumers of this new data source. Creating a better understanding of these issues will be very important in the future development and evolution of VGI in society.

Keywords

Data privacy, ethics, legal issues, Volunteered Geographic Information

1 Introduction

The public collection and exchange of geospatial data and information as Volunteered Geographic Information (VGI) involve many privacy, legal and ethical issues (Blatt, 2015). These issues are exacerbated with the further distribution and dissemination of these data by third parties such as libraries, online data services, etc. In many examples of VGI, the collection of geographic data involves the use of location-based devices that record the identities, positions and movements of the contributors of the information. Other examples of VGI, such as social media, can embed geographic position into imagery, video, sound, text, message data, etc. These data and information objects can then be accessed by other citizens, systems and services. As crowdsourced geographic information becomes more prevalent in society today, more detailed spatial data are constantly being collected from citizens, particularly through the proliferation of spatially aware devices such as smartphones, smart devices and sensors. The major issue developing here is that these sources of spatial data can be combined or linked to other databases and data sources and can potentially expose sensitive private information, such as the personal data, living habits and health conditions of the citizen contributor themselves (Shen et al., 2016). The further usage, storage and integration of these data are often the subject of complex legal and ethical considerations.

1.1 The role of the citizen within privacy, legal and ethical issues in VGI

In this chapter we consider the position of the citizen and the VGI that they can generate, and we discuss the privacy, legal and ethical issues relating to the production of this VGI and its further usage. In VGI projects and activities the citizen is at the very core of almost all aspects of VGI data production, management, dissemination and usage. Yet we argue in this paper that there is still a large gap in our understanding of the privacy, legal and ethical issues connected to these activities. VGI is still a relatively new field of research; subsequently there is not a great deal of published knowledge or guidelines available on these issues in VGI.

Although VGI tends to be associated with the collection and supply of explicitly geographic material, such as OSM (see Chapters 3 and 4 – Mooney and Minghini, 2017; Touya et al., 2017) or citizen science projects (see Chapters 1 and 2 – Foody et al., 2017; See et al., 2017), it is certainly not limited to this type of materials. As means of a short motivating example, we consider geotagged photographs. Geotagged photographs are not associated explicitly with VGI, in the sense that geotagging has become so implicit with the use of smartphones that most citizens may not be aware of this feature, i.e. that our holiday photographs, for example, are being geotagged when we take them and upload them to various social media sites. In this case, this information is volunteered passively (Fast and Rinner, 2014), without realizing that it is actually geographic information nor that it can be reused and integrated with other geographic information. Indeed many citizens are not aware that when, for example, we contribute geotagged photographs to a citizen science project, one cannot always predict what the downstream future usages of those photographs will be given the myriad of mashup tools and technologies available. Overall this means that although crowdsourced geographic information can be both volunteered, as in VGI, or harvested in a passive or ambient way (Stefanidis et al., 2013), for the most part citizens are not fully aware of the additional intelligence that can be elicited by the powerful combinations of software, cloud computing and data processing technologies available today. Dienlin and Trepte (2015) emphasise that even though citizens today have substantial concerns with regard to their online privacy, they are often engaged in self-disclosing behaviours that do not adequately reflect their concerns. It is therefore necessary to attempt to highlight the types of privacy, ethical and legal issues that can be faced knowingly or unknowingly by citizens involved in VGI today.

The remainder of this chapter is organised as follows. In Section 2 we provide a brief discussion of the current understanding of the issues of privacy, ethical and legal frameworks in VGI today by considering simple actor/use case scenarios. In the three sections that follow it, we discuss privacy (Section 3), ethics (Section 4) and legal issues (Section 5). In Section 6 we summarise the paper with some concluding remarks while highlighting future directions for this work.

2 Positioning the Issues of Privacy, Ethics and Legality in VGI

At the time of writing, the issues of privacy, ethics and legality in VGI have not received widespread or in-depth treatment by the research community. The exact nature of the VGI or data used and which use case it is applied to may help to determine which legal, ethical and privacy issues are most prominent. When information about individual citizens is transferred and presented within a geographic context, the resulting profile information could be both 'highly revelatory and involuntary' (Scassa, 2013:5), and this can raise important privacy and ethical issues. The ability for VGI data and information to be mashed up or integrated with other VGI datasets, proprietary datasets or other information sources means that new sources of data are created. The privacy, ethics and legal issues that existed for the original VGI dataset may not have completely changed due to this transformative change. In this section, we provide a simple table (Table 1) that situates privacy, ethics and legal issues for the principal actors involved in the collection, production and dissemination of VGI, namely citizens, national mapping agencies (NMAs), commercial companies, researchers and other entities such as small and medium-sized enterprises (SMEs). While this table is not a fully comprehensive overview of all of the possible actor interactions with privacy, ethics and legal issues, it will allow us to situate our discussions in the subsequent sections of this chapter. Each cell in the table provides a simple example of considerations that are made by the corresponding actor when producing, collecting, managing, using or disseminating VGI.

As we can see, there is some overlap in the table. All of the actors will confront and deal with many of the same privacy, ethics and legal issues but they will respond to these issues differently. For example, how an NMA deals with the liability and legal aspects of VGI will be different to how an academic researcher deals with the same problem. With these examples in mind we will now look at privacy (Section 3), ethics (Section 4) and legal issues (Section 5) in the next three sections.

3 Privacy Issues

Privacy is probably the most well known aspect of the three issues considered in this chapter; protecting it is very important, and this is no different when considering VGI. Privacy of user data and information should be considered in the initial design of VGI systems, as adding privacy protection to existing systems can be very cumbersome, and this is no different for VGI systems and projects.

3.1 Understanding Privacy within the VGI context

Private data in the VGI context are any geographic data or information that can be linked to an individual contributor who created, collected or edited those data. Thus, to prevent VGI data being used to violate the privacy of individuals, we need to look at the character of the data and investigate the entire process from the collection of data to the submission of the VGI to data repositories, and then onwards to the usage of the data. The most efficient measure is not to

Table 1: Privacy, ethics and legal issues for actors involved in the collection, production and dissemination of VGI.

Who am I? I am …	An example of a privacy issue	An example of an ethical consideration	An example of the legal issues involved
A citizen contributing my data to a VGI or Crowdsourcing Project	Ensuring that personal information is not linked to any contribution. Nor will the contribution provide the personal or private information of another individual.	A contributor will report observations honestly and truthfully, and, to the best of their ability, they will contribute accurate information.	The citizen will obtain consent or permission to survey, record or measure a specific area or geographic feature.
A National Mapping Agency, Environmental Ministry, Geological Survey, etc.	Ensuring that the organisation gives careful consideration to the scale at which it provides geographic data and to the contents of the metadata attached to those data.	The organisation will not knowingly provide false or inaccurate data or information nor report with bias on specific geographic themes.	The organisation is bound by many legal requirements to produce mapping products. The organisation is legally bound to the quality of data produced and could be liable to consequences of the use of these data.
A commercial mapping company	Privacy can mean keeping the information about the sources of the VGI hidden from public view.	Acting responsibly, ensuring the privacy of citizens is maintained and that the VGI is distributed using an appropriate licence.	The focus here is on the terms and conditions of the type of licence applied to the VGI, whether it be for the commercial usage of the data or the integration of the data with other data products.

Continued.

Table 1: (Continued).

Who am I? I am …	An example of a privacy issue	An example of an ethical consideration	An example of the legal issues involved
An academic or researcher looking to use VGI data	During the dissemination of research outputs, care must be taken not to expose the identities of or other private information related to the citizens who contributed to the VGI project. Patterns and inferences made about the contributors of the data must be carefully considered so as not to breach the privacy of those citizens.	If carrying out an analysis/ survey of citizens involved in VGI, the consent process ensures that individuals are voluntarily participating in the research with full knowledge of relevant risks and benefits.	Many geography-based research projects can mix and integrate multiple datasets for investigation. It is important that the licences are compatible to allow this, so that future research results can be disseminated legally.
An SME wishing to use VGI for an application or service that it is looking to develop and bring to market	VGI as a product must be used in such a way that the producer's privacy is protected from any potential future commercial exploitation.	Using the data 'as is' without any embellishment or corrections to the VGI data that are untrue or incorrect.	Ensuring that the licences and terms of use of the VGI data are compatible with commercial development. In some open licences it might be necessary that changes made to the VGI dataset/ database by the SME also be made available under the same licence.

collect private data at all or at least not to collect data that are linkable to individuals. If linkable private data are collected, it then becomes necessary to set up protection mechanisms to ensure that the data are only used according to the original purpose defined before the collection of the VGI started. As VGI data collections are considered a resource for new and maybe unforeseen usages and research, it becomes all the more important that these data do not provide linkable private data about individuals. The question that must be asked is whether location information in itself is private data or can be linked to individuals: the answer depends on the location accuracy. Many location data are accurate enough to be bound to one individual or to a small group of individuals, e.g. an office or home, and are sometimes even combined with precise time and date. There is no one-size-fits-all solution here; the collection of point-based geographic data for a specific purpose may need to have high geographic accuracy. With this requirement for accuracy comes a possibility that the geographic features close to the collected points could be used to infer other information.

3.2 Approaches to Privacy Preservation in VGI

The guiding principle of privacy protection is to collect as little private data as possible. Cho (2014) argues that there must be privacy and legal protection for volunteers in VGI data collection and projects, otherwise 'the ensuing litigation may destroy the VGI model before it reaches its full potential'. Calderoni et al. (2015) remark that we, as citizens, are only starting to grasp the privacy risks associated with the constant tracking of our whereabouts by the very devices that we carry around with us. In order to continue using location-based services in the future without compromising personal privacy and security, there is an urgent need for privacy-friendly applications and protocols.

There exists some literature related to privacy concerns and possible solutions related to VGI. There are a number of prevalent technological approaches, including perhaps the popular approach of blurring or fuzzing information from its original data (Luther et al., 2009). Anonymising data and selectively revealing information according to volunteer preference is another approach (Kim et al., 2013). In the Geographic Privacy-Aware Knowledge Discovery and Delivery (GeoPKDD) project, Giannotti and Pedreschi (2008) investigated various scientific and technological issues of mobility data, open problems and roadmaps. They found that privacy issues related to Information and Communications Technology (ICT) can only be addressed through an alliance of technology, legal regulations and social norms. In the meanwhile, increasingly sophisticated privacy-preserving data mining techniques are being studied and need to be further developed. These approaches aim to achieve appropriate levels of anonymity by means of controlled transformation of data and/or patterns with limited distortion, to avoid the undesired side effects on privacy while preserving the possibility of discovering useful patterns and trends.

The most common question asked about privacy in VGI is whether data collection services and systems can be enhanced so that the spatial data collected or generated by a contributor cannot be traced back to that individual contributor. The contributor should not be identifiable through their contributions to a VGI project; more precisely, the contributor should be identifiable within the VGI project (such as through a pseudonym username in a project) but their contribution should not be linkable to the personal and private data and information for their actual person. There is a need to consider the sensitivity of the privacy issues within contributions to VGI: are there situations where a contributor would prefer *not* to be linked to a set of contributions or a single contribution? In the capture of aerial imagery, geotagged photographs and street-level photography, people can also potentially be identifiable as subjects. There are thus many privacy issues, and these issues have not been adequately addressed as of yet.

3.3 Privacy for non-human subjects in VGI

Privacy can also be related to non-human subjects in VGI. Suppose there is a crowdsourcing or VGI campaign in the area of biodiversity and a very rare or precious plant species is found and geolocated. To protect this species (and potentially its habitat), this information needs to be kept private. But other species identified by the campaign may not need privacy. This example could also extend to similar scenarios for a geological survey. Suppose a contributor identifies the potential location of a precious metal; there might be very good reasons related to why this location and find must be kept private. The discussions above for both human privacy and the privacy of non-human subjects raises the question of the need to have manual checking of contributions for these privacy issues: is it necessary to moderate contributions for their privacy characteristics and not just their data quality aspects? The moderation question in VGI already raises many obstacles to its implementation (Neis and Zielstra, 2014). It might not be possible to automate this process to include the consideration of privacy aspects.

While the focus above has been on the individual VGI contributor, it is often the case that contributors to VGI projects are institutions and organisations that provide datasets for VGI; institutions or organisations must also be aware of and familiar with the licence terms within which they provide content.

4 Ethics Issues

As far back as the work of Mitchell and Draper (1983), the issue of ethics has been subject to research conversation in geography. In their work, they indicate that geographers have not always been sensitive to ethical issues, and that, as geography researchers, one has to balance the obligations of understanding and knowledge with those of respecting the dignity and integrity of research subjects.

4.1 Key Ethical Issues in VGI

In VGI, the citizens who collect, manage and work with the data are very often the subject of research. Little work has been carried out specifically on ethics in VGI. Many studies on contributors have been performed and published in the literature in the last few years (Granell and Ostermann, 2016). Hartter et al. (2013) outline that ethical standards in science require that research with human subjects respect individuals, commit to nondisclosure of participants' identities, minimise potential harm and ensure that the benefits and burdens of research be fairly distributed, and that subjects be informed of the full nature of the research so they can decide against participation if they wish. Ethical standards and plans now usually require ethics approval funding review boards and research authorities. Luppicini (2010) introduces the term technoethics to refer to an interdisciplinary study of technological impacts on the morals and ethics in a society. Ethical conduct and social responsibility are important factors within contemporary society to maintain respect and harmony. Lingel and Bishop (2014) consider the 'labour ethics' surrounding VGI in terms not only of what is technically possible, but of what is also ethically responsible. The authors argue that the introduction of ethical considerations should not discourage the production of VGI within volunteer communities; rather, those involved in instigating this VGI or managing it must give careful consideration to how these communities are managed.

Ethical considerations can be performed by both the data producer (the volunteers) and the users (VGI project coordinator/platform operator). As before, the volunteers have to consider and adopt an ethical approach to their reporting of information and data. For example, in a disaster or crisis situation, this involves not engaging in the false reporting of damage, casualties, fatalities, etc. Indeed, ethical considerations must be given by volunteers to information and data that they provide that can lead to the action of authorities such as emergency services (Haworth and Bruce, 2015). Volunteers wilfully contributing false or misleading data or information not only undermine the VGI project in which they are involved, but also causes a further lack of trust and suspicion from users about the quality and usability of VGI in general. From the coordinator side, the volunteer must be made aware of the purpose of the project that they are volunteering for; voluntary submissions must not be used for commercial purposes, or shared with other entities for different purposes without the consent of the volunteers. At this point, it is clear that the consideration of ethics combines the issues of data privacy and the legal aspects of VGI – these issues are not easily disengaged from each other.

4.2 Summary of Ethical Issues

As communicated by Sula (2016), the key ways to respect ethics in data-based research include involving participants throughout the research process, avoiding collecting information that should remain private, notifying participants of their inclusion and providing them with options to correct or delete personal information, and using public channels to disseminate research, such as Open Data. Ethical research has the least possible impact on subjects, asking or collecting only as much as is needed to answer its questions. In the case of VGI research, the researchers involved may not know exactly what knowledge they are trying to extract or patterns they are trying to uncover; the data are being used in an exploratory way. In these circumstances, it seems nearly impossible to inform participants of all anticipated harms and benefits in advance.

Today, datasets collected through VGI and crowdsourced means have a potentially very long lifespan. Given the longevity of these datasets and their potential interoperability and integration with other datasets, researchers and scientists must, in general and where possible, avoid data with personally identifiable information or information that could later be used to identify participants in connection with other datasets, e.g. screennames, usernames, etc. The potential for unintended consequences are high, but entirely mitigated when no personally identifiable information is collected in the first place (Sula, 2016). The integration of many datasets with each other creates a brand new dataset that is essentially an unknown quantity in terms of its ethical characteristics. In this situation the creators of these new datasets must be conscious of how the new dataset will be used, distributed, analysed and even itself potentially integrated with other datasets in the future.

5 Legal Issues

In Olteanu-Raimond et al. (2017), one of the six obstacles described for NMAs in using VGI is the legal issue. The most relevant of these legal issues in using VGI are intellectual property and liability. With the new trend of open data, more and more public bodies have adopted a policy of open data. Generally there are two concepts of open data: one concept means that 'data and content can be freely used, modified, and shared by anyone for any purpose' and the other involves open source licensing applied on software. Intellectual property concerns both data producers and users. From the producers' point of view, it defines ownership rights of the data, licences, and how data can be used and under which conditions. From the users' point of view, it defines rules to enrich and disseminate the data.

5.1 Liability as a Legal Issue in VGI

Concerning liability, the main question is that of who is liable and under what circumstances if harm is caused, economic loss happens or incorrect decisions are taken. This issue is linked closely to the concerns with data quality, i.e. precision and accuracy. Liability can be different from country to country and from product to product. When crowdsourced data are used by a legally mandated organisation such as an NMA, what are the implications for that organisation? Does the NMA take all of the legal responsibility? Is there any citizen responsibility? Should there be? Indeed, Cho (2014:10) argues that there must be legal protection for volunteers in VGI data collection and projects, otherwise 'the ensuing litigation may destroy the VGI model before it reaches its full potential'. Rak et al. (2012) studied the integration of VGI into Canadian authoritative datasets from the liability point of view by proposing four primary risk management techniques to manage risks resulting such an incorporation. One of the most important and difficult of these risk management techniques sees the information provider being required to show that steps were taken to ensure the accuracy of VGI that has been integrated into their data.

5.2 Legal Issues Surrounding Data Licence Types

The type of licence applied to VGI data for their subsequent dissemination has an important influence on their usage. There are three main types of open data licences:

- Share alike licences, which require the derived datasets to be released with the same licence as the original one(s); the most famous such licence in the area of geographic information is the Open Database License (ODbL) used by OpenStreetMap (OSM).
- Open licences, which allow any type of use provided the citation of the data provider is given; it allows, for instance, commercial use of derived datasets. An example of such a licence is the French 'Licence ouverte', which is used to release governmental open data in France.
- Limited use open licences, which limit the use of the dataset to personal use, or non-commercial use. For instance, the IGN (the French mapping agency) releases its datasets openly for research and education purposes.

The choice of a licence conveys a political or commercial strategy, and the strategies of these licences might not be compatible. So what happens when projects with different strategies plan to merge their datasets? And what happens when one or more of these datasets are from VGI? It is useful at this point to provide a real-world example. The most typical case regarding geographic information is the following: how is it possible to integrate non-ODbL open data into OSM? The case of the French national address dataset is interesting to study, as it plans to integrate data from the IGN, which is a governmental administration, the French Post Office company, which is a public limited company, and OSM (Figure 1). All three already have address datasets updated by crowdsourcing communities. They also have different licensing strategies. OSM uses the ODbL while the French Post Office would prefer a licence that allows commercial use of derived datasets. Figure 1 shows a possible integration scenario for the architecture of the project and the licensing strategy. Two new datasets are created in this scenario: a common and central address dataset, and a copy of this dataset using the OSM technologies (in RDF format). The OSM-like copy is under the ODbL licence, which allows OSM contributions regarding addresses to be directly included, and the other way around. The common address dataset is under two licences: a limited open licence that only allows personal and non-commercial use of the data, and a charged licence for other uses. The OSM-like dataset is only a partial copy, as the French Post Office does not want to release all the information of its dataset (e.g. the standardised spelling of addresses). A quality control step is included in the common dataset

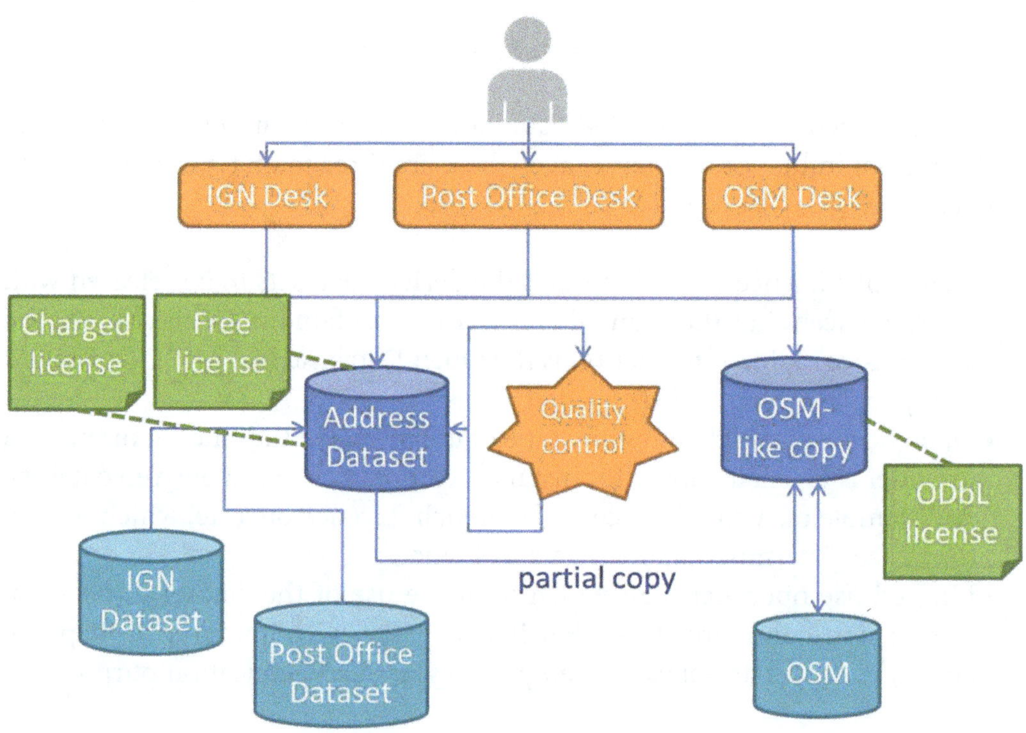

Fig. 1: Possible architecture to mix licences and dissemination strategies between OSM, the IGN and private companies.

to improve contributions through both field survey (by mail carriers and IGN surveyors) and automatic tools.

In this scenario, different access desks are proposed for citizens, derived from existing tools. The IGN desk, which fills the common address dataset, is dedicated to community-sourcing (from city administrations, firefighters, police officers, etc.); the Post Office desk, which also fills the common address dataset, is dedicated to citizens and administrations that report updates on addresses; and the OSM desk is based on OSM software, such as iD[1], and could fill both the common dataset and the OSM-like dataset. The tricky part of the integration scenario is that the contributions go to both datasets at the same time, so it is not 'infected' by ODbL. This architecture seeks to attract OSM contributors to this project, but the contributors should accept that their contribution will fill both address datasets, which have different licences.

5.3 *Summary of Legal Issues in VGI*

In summary, the legal issues in VGI must be considered from the side of both the data producers or collectors (i.e. the volunteers or citizens) and the users or facilitators (i.e. VGI project management, VGI data portal operators) of the data. From the position of the volunteer, their legal role and their contribution may not always be clearly defined and this can lead to potentially exposing them to legal problems. On the other hand, if a data provider or data portal only facilitates the transfer or access to VGI data, then who carries the legal responsibilities related to consequences of future use of these data? For example, submissions from volunteers to a VGI project may indicate natural hazards in a particular location or the vulnerabilities of a property. This (potentially false) information could be used by an insurance company to raise insurance premiums. Then, from the VGI project coordinators' side, to what extent must a portal/project coordinator provide a disclaimer about legal aspects? Under what circumstances can a portal be held liable for omissions (e.g. damaged areas not mapped during a disaster), or mistakes (e.g. infrastructure shown to be intact that is actually broken, leading to inaccessibility) be challenged? In reality, there are no clear cut answers to these questions at this point in time. Christin et al. (2011) indicate that the research community should provide open datasets that can serve as a baseline for performance, security and legal evaluation in order to begin addressing these critical issues.

6 Conclusions and Future Directions

In this chapter we have provided a brief overview and discussion of privacy, ethics and legal issues in the production, collection, storage, dissemination and integration of VGI. These are complex issues. As VGI continues to grow rapidly in terms of popularity amongst contributors and as an alternative or complementary source of spatial data for researchers, authoritative agencies, commercial companies, etc., these issues will become more prevalent and urgent. In their study of privacy concerns in the use of location-based services such as social media, Fodor and Brem (2015) found that privacy concerns do influence citizen adoption of these services but that the answer is more complex and multi-faceted than just a simple case of trusting such services. Even now, with VGI, new technologies are emerging all of the time, offering citizens new and exciting ways to generate and collect spatial data. Luppicini and So (2016) argue that in technologies such as the use of drones for collecting data and information, a lack of understanding of the factors of ethics and privacy often causes the prohibition of the use of these technologies. A lack of understanding does not often really mitigate the issues, but can hinder the development of devices and technologies that can be used in many positive ways.

When VGI is collected and subsequently disseminated, it can be reused, displayed, integrated and transformed in a myriad of ways. The model for understanding what happens with data once they are released by the individual, or what this means on an aggregate scale, is thus fluid and uncertain (Hallinan et al., 2012). In reality, citizens often have a poor basis on which to form a picture of the data relationships, the consequences and the issues in VGI. Citizens often struggle to comprehend how these issues add to the importance of these data flows in relation to other social structures or issues. Hallinan et al. (2012:271) go on to argue that due to the complexity of the issues of privacy, ethics and legality, 'it appears that the public are being forced to act in an environment they have little template for approaching'. The concepts of VGI and Open Data are still relatively new. Consequently, it will take time for citizens to become deeply familiar with the issues discussed above.

Christin et al. (2011) argue that at the moment, privacy research usually operates on either private or synthetic datasets. These datasets do not allow new mechanisms for privacy, ethical and legal considerations to be harmonised or benchmarked against. In any case, Torra and Navarro-Arribas (2014:277) indicate after their wide scale review of the issues of data privacy online that the development of methods to protect citizens 'has to take into account the specificities of the data involved'. No two VGI datasets are the same; indeed, it can be the case that within a VGI dataset different objects might be collected by different citizens in different circumstances. VGI is an exciting and powerful source of geospatial data that is likely to continue growing. Understanding how to protect the citizen while enhancing their role in the production of VGI is a big research challenge for the next few years. Indeed this research issue has not really been tackled at all by the research community at this point in time. Protection of the citizen's privacy and ethical rights under suitable legal conditions is very important. However, the frameworks or structures developed to implement these protections must not place insurmountable barriers to citizen participation in VGI. The act of being involved in VGI as citizens should continue to be a leisure activity pursued by those motivated to volunteer. There is a fine balance between, on the one hand, encouraging and fostering participation in VGI activities and, on the other hand, ensuring that the complex issues of privacy, ethics and legality are understood and adhered to by a potentially large cohort of individuals (Rak et al., 2012; Torra and Navarro-Arribas, 2014). Finding this balance will have a major influence on the future trajectory of VGI.

Notes

[1] http://ideditor.com/

Reference list

Blatt, A.J., 2015. The Benefits and Risks of Volunteered Geographic Information. *Journal of Map & Geography Libraries* 11(1), 99–104. DOI: http://dx.doi.org/10.1080/15420353.2015.1009609

Calderoni, L., Palmieri, P., Maio, D., 2015. Location privacy without mutual trust: The spatial Bloom filter. *Computer Communications* 68, 4–16. DOI: https://doi.org/10.1016/j.comcom.2015.06.011

Cho, G., 2014. Some legal concerns with the use of crowd-sourced Geospatial Information. *IOP Conference Series: Earth and Environmental Science* 20(1), p.012040. Available at http://iopscience.iop.org/1755-1315/20/1/012040 [Last accessed 1 May 2017]

Christin, D., Reinhardt, A., Kanhere, S.S., Hollick, M., 2011. A survey on privacy in mobile participatory sensing applications. *Journal of Systems and Software* 84(11), 1928–1946. DOI: https://doi.org/10.1016/j.jss.2011.06.073

Dienlin, T., Trepte, S., 2015. Is the privacy paradox a relic of the past? An in-depth analysis of privacy attitudes and privacy behaviors. *European Journal of Social Psychology* 45(3), 285–297. DOI: https://doi.org/10.1002/ejsp.2049

Fast, V., Rinner, C., 2014. A Systems Perspective on Volunteered Geographic Information. *ISPRS International Journal of Geo-Information* 3(4), 1278–1292. DOI: https://doi.org/10.3390/ijgi3041278

Fodor, M., Brem, A., 2015. Do privacy concerns matter for Millennials? Results from an empirical analysis of Location-Based Services adoption in Germany. *Computers in Human Behavior* 53, 344–353. DOI: https://doi.org/10.1016/j.chb.2015.06.048

Foody, G, Fritz, S, Fonte, C C, Bastin, L, Olteanu-Raimond, A-M, Mooney, P, See, L, Antoniou, V, Liu, H-Y, Minghini, M and Vatseva, R. 2017. *Mapping and the Citizen Sensor.* In: Foody, G, See, L, Fritz, S, Mooney, P, Olteanu-Raimond, A-M, Fonte, C C and Antoniou, V. (eds.) *Mapping and the Citizen Sensor.* Pp. 1–12. London: Ubiquity Press. DOI: https://doi.org/10.5334/bbf.a.

Giannotti, F., Pedreschi, D. (Eds.), 2008. *Mobility, Data Mining and Privacy*. Berlin, Heidelberg: Springer. DOI: https://doi.org/10.1007/978-3-540-75177-9

Granell, C., Ostermann, F.O., 2016. Beyond data collection: Objectives and methods of research using VGI and geo-social media for disaster management. *Computers, Environment and Urban Systems* 59, 231–243. DOI: https://doi.org/10.1016/j.compenvurbsys.2016.01.006

Hallinan, D., Friedewald, M., McCarthy, P., 2012. Citizens' perceptions of data protection and privacy in Europe. *Computer Law & Security Review* 28(3), 263–272. DOI: https://doi.org/10.1016/j.clsr.2012.03.005

Hartter, J., Ryan, S.J., MacKenzie, A.C., Parker, J.N., Strasser, C.A., 2013. Spatially Explicit Data: Stewardship and Ethical Challenges in Science. *PLoS Biology* 11(9). DOI: https://doi.org/10.1371/journal.pbio.1001634

Haworth, B., Bruce, E., 2015. A Review of Volunteered Geographic Information for Disaster Management. *Geography Compass* 9, 237–250. DOI: https://doi.org/10.1111/gec3.12213

Kim, S., Mankoff, J., Paulos, E., 2013. Sensr: Evaluating a Flexible Framework for Authoring Mobile Data-collection Tools for Citizen Science, in: Proceedings of the 2013 Conference on Computer Supported Cooperative Work. CSCW '13. New York, NY, USA: ACM, pp. 1453–1462. DOI: https://doi.org/10.1145/2441776.2441940

Lingel, J., Bishop, B.W., 2014. The GeoWeb and everyday life: An analysis of spatial tactics and volunteered geographic information. *First Monday* 19(7). DOI: http://dx.doi.org/10.5210/fm.v19i7.5316

Luppicini, R., 2010. *Technoethics and the Evolving Knowledge Society: Ethical Issues in Technological Design, Research, Development, and Innovation*. IGI Global. DOI: https://doi.org/10.4018/978-1-60566-952-6

Luppicini, R., So, A., 2016. A technoethical review of commercial drone use in the context of governance, ethics, and privacy. *Technology in Society* 46,109–119. DOI: https://doi.org/10.1016/j.techsoc.2016.03.003

Luther, K., Counts, S., Stecher, K.B., Hoff, A., Johns, P., 2009. Pathfinder: An Online Collaboration Environment for Citizen Scientists, in: Proceedings of the SIGCHI Conference on Human Factors in Computing Systems. CHI '09. New York, NY, USA: ACM, pp. 239–248. DOI: http://doi.acm.org/10.1145/1518701.1518741

Mitchell, B., Draper, D., 1983. Ethics in Geographical Research. *The Professional Geographer* 35(1), 9–17. DOI: https://doi.org/10.1111/j.0033-0124.1983.00009.x

Mooney, P and Minghini, M. 2017. A Review of OpenStreetMap Data. In: Foody, G, See, L, Fritz, S, Mooney, P, Olteanu-Raimond, A-M, Fonte, C C and Antoniou, V. (eds.) *Mapping and the Citizen Sensor*. Pp. 37–59. London: Ubiquity Press. DOI: https://doi.org/10.5334/bbf.c.

Neis, P., Zielstra, D., 2014. Recent Developments and Future Trends in Volunteered Geographic Information Research: The Case of OpenStreetMap. *Future Internet* 6(1), 76–106. DOI: https://doi.org/10.3390/fi6010076

Olteanu-Raimond, A.-M., Hart, G., Foody, G., Touya, G., Kellenberger, T., Demetriou, D., 2017. The scale of VGI in map production: A perspective of European National Mapping Agencies. *Transactions in GIS* 21, 74–90. DOI: https://doi.org/10.1111/tgis.12189

Rak, A., Coleman, D.,, Nichols, S., 2012. Legal liability concerns surrounding volunteered geographic information applicable to Canada, in: Rajabifard, A., Coleman, D. (Eds.), *Spatially enabling government, industry and citizens*. USA: GSDI Association Press, pp.125–143.

Scassa, T., 2013. Legal issues with volunteered geographic information. *The Canadian Geographer / Le Géographe canadien* 57(1), 1–10. DOI: https://doi.org/10.1111/j.1541-0064.2012.00444.x

See, L, Estima, J, Pőd.r, A, Arsanjani, J J, Bayas, J-C L and Vatseva, R. 2017. Sources of VGI for Mapping. In: Foody, G, See, L, Fritz, S, Mooney, P, Olteanu-Raimond, A-M, Fonte, C C and Antoniou, V. (eds.) *Mapping and the Citizen Sensor*. Pp. 13–35. London: Ubiquity Press. DOI: https://doi.org/10.5334/bbf.b.

Shen, N., Yang, J., Yuan, K., Fu, C., Jia, C., 2016. An efficient and privacy-preserving location sharing mechanism. *Computer Standards & Interfaces* 44, 102–109. DOI: https://doi.org/10.1016/j.csi.2015.06.001

Stefanidis, A., Crooks, A., Radzikowski, J., 2013. Harvesting ambient geospatial information from social media feeds. *GeoJournal* 78(2), 319–338. DOI: https://doi.org/10.1007/s10708-011-9438-2

Sula, C.A., 2016. Research Ethics in an Age of Big Data. *Bulletin of the Association for Information Science and Technology* 42(2), 17–21. DOI: https://doi.org/10.1002/bul2.2016.1720420207

Torra, V., Navarro-Arribas, G., 2014. Data privacy. *Wiley Interdisciplinary Reviews: Data Mining and Knowledge Discovery* 4(4), 269–280. DOI: https://doi.org/10.1002/widm.1129

Touya, G, Antoniou, V, Christophe, S and Skopeliti, A. 2017. Production of Topographic Maps with VGI: Quality Management and Automation. In: Foody, G, See, L, Fritz, S, Mooney, P, Olteanu-Raimond, A-M, Fonte, C C and Antoniou, V. (eds.) *Mapping and the Citizen Sensor*. Pp. 61–91. London: Ubiquity Press. DOI: https://doi.org/10.5334/bbf.d.

CHAPTER 7

Assessing VGI Data Quality

Cidália Costa Fonte*, Vyron Antoniou†, Lucy Bastin‡, Jacinto Estima§, Jamal Jokar Arsanjani¶, Juan-Carlos Laso Bayas‖, Linda See‖ and Rumiana Vatseva**

*Department of Mathematics, University of Coimbra, 3001-501 Coimbra, Portugal / INESC Coimbra, Rua Sílvio Lima, Pólo II, 3030-290 Coimbra, Portugal

†Hellenic Army General Staff, Geographic Directorate, PAPAGOU Camp, Mesogeion 227-231, Cholargos, 15561, Greece

§NOVA IMS, Universidade Nova de Lisboa, 1070-312, Lisbon, Portugal

¶University of Aalborg, Denmark, Copenhagen

‖International Institute for Applied Systems Analysis (IIASA), Laxenburg, Austria

**National Institute of Geophysics, Geodesy and Geography, Bulgarian Academy of Sciences, Bulgaria.

Abstract

Uncertainty over the data quality of Volunteered Geographic Information (VGI) is the largest barrier to the use of this data source by National Mapping Agencies (NMAs) and other government bodies. A considerable body of literature exists that has examined the quality of VGI as well as proposed methods for quality assessment. The purpose of this chapter is to review current data quality indicators for geographic information as part of the ISO 19157 (2013) standard and how these have been used to evaluate the data quality of VGI in the past. These indicators include positional, thematic and temporal accuracy, completeness, logical consistency and usability. Additional indicators that have

How to cite this book chapter:

Fonte, C C, Antoniou, V, Bastin, L, Estima, J, Arsanjani, J J, Bayas, J-C L, See, L and Vatseva, R. 2017. Assessing VGI Data Quality. In: Foody, G, See, L, Fritz, S, Mooney, P, Olteanu-Raimond, A-M, Fonte, C C and Antoniou, V. (eds.) *Mapping and the Citizen Sensor.* Pp. 137–163. London: Ubiquity Press. DOI: https://doi.org/10.5334/bbf.g.

been proposed for VGI are then presented and discussed. In the final section of the chapter, the idea of integrated indicators and workflows of quality assurance that combine many assessment methods into a filtering system is highlighted as one way forward to improve confidence in VGI.

Keywords

Spatial data quality, ISO 19157, positional accuracy, thematic accuracy, usability

1 Introduction and Background

Quality is a key component of any dataset. Decisions on using a spatial dataset for a certain purpose are heavily based on quality measures such as positional accuracy, thematic quality, completeness and usability. This also applies to Volunteered Geographic Information (VGI), a new and growing source of data, contributed by citizens, that can take many different forms, e.g. geotagged photographs through sites such as Panoramio and Flickr, online maps such as OpenStreetMap (OSM) and Wikimapia, and 3D VGI such as OSM-3D and OSM2World. For a more detailed overview of the diverse range of current VGI data sources, see Chapter 2 (See et al., 2017).

A set of elements is specified in the ISO 19157 standard for spatial data quality (ISO, 2013). This framework adequately serves communities such as National Mapping Agencies (NMAs), which have professional staff following rigorous protocols and multiple quality control processes so as to produce high-quality products of a minimum acceptable specification. However, these spatial data quality guidelines have not been developed with any consideration of the nature of VGI. The data quality of VGI brings new challenges into the quality assessment field, and therefore it is possible to consider VGI data quality using this standard and then recommend additional measures that take the specific nature of VGI into account.

One characteristic of VGI is its heterogeneous nature, e.g. there is often a spatial bias in the information, with more data collected in urban than in rural areas (Estima et al., 2014; Neis and Zielstra, 2014; Ma et al., 2015) or a bias towards specific types of features, influenced by the interests of the volunteers (Bégin et al., 2013). Moreover, even inside the urban fabric, the more popular and touristic areas are getting more attention, and thus more data with higher detail, than obscure and fairly unknown urban areas (Antoniou and Schlieder, 2014; Estima et al., 2014). These biases can be further influenced by access to, and knowledge of, digital resources, the language of the VGI application, cultural differences and how much time users have to participate (Holloway et al., 2007; Zook and Graham, 2007).

Another issue with VGI is the lack of rigorous data specifications of the kind that accompany more authoritative Geographic Information (GI), an issue which can lead to heterogeneous data quality (Hochmair and Zielstra, 2012). While collaborative mapping can improve data quality to a certain extent (Haklay et al., 2010), frequent changes to the same features can deteriorate the overall quality and usability of the data; examples of this phenomenon can be found in location-based services (Mooney and Corcoran, 2012) and gazetteers (Antoniou et al., 2016b). Moreover, the fact that there is no standard way in which the data are collected, as well as data specifications that vary between and also within initiatives, means that quality will vary over space and time; see e.g. OSM, where free tagging of features is possible.

For some types of VGI applications, such as OSM or Instagram, the volunteers may contribute information in any location. However, some VGI campaigns have been promoted with a more specific objective in mind and consequently have employed a statistical sampling system to make sure that the data are collected where they are needed, that a more global coverage is obtained or that more accurate results are achieved. These campaigns have been promoted to citizen scientists, eliciting their help with specific goals, e.g. quantifying human impact (See et al., 2013) or assessing cropland and other land use area estimates (Waldner et al., 2015), or even collecting photographs around the world, such as for the Degree Confluence Project[1]. Some of the statistical sampling systems used include systematic allocation of points in a grid; and random or stratified random samples, whether these are points, polygons or pixels. One of the key advantages of using statistical samples includes having a stricter control on what data the users can contribute and where, allowing for more straight-forward measures of quality, e.g. through estimation of statistical uncertainties and determination of possible sample augmentation to reduce these uncertainties. Additionally, and depending on the design of these systems, comparisons between users are easier to do, since the location is fixed and shared between the contributors. A key disadvantage of predetermined sampling systems, however, might be precisely their strictness, e.g. bounding the users to a pre-defined set of geographic locations, with usually little possibility of reporting local and sometimes more relevant characteristics from the surroundings that might contribute to a better understanding and achievement of a given objective; this, in itself, could be detrimental to the quality of the information by providing information that is very precise but off-target.

VGI quality has been the subject of a considerable amount of research, particularly with regard to the quality of OSM. For example, a number of studies have tried to assess VGI quality based on comparisons with authoritative data provided by NMAs or commercial companies (e.g. Girres and Touya, 2010; Haklay, 2010; Zielstra and Zipf, 2010; Antoniou, 2011; Estima and Painho, 2013; Fan et al., 2014). These comparisons are based on the belief that authoritative data are always of a minimum, acceptable quality and created according to high standards and that it is thus reasonable to assume that authoritative data can play the role of reference datasets during a quality evaluation process of VGI datasets. In these studies, a number of methods are used, e.g. data matching, generalisation evaluation, etc., that consider different elements of data quality such as positional or thematic accuracy. However, the application of these methods is not always possible, because of limited data availability, licence restrictions or the lack of access to costly authoritative datasets. Moreover, as VGI datasets are often richer than their authoritative counterparts, and will only continue to increase in richness, the use of authoritative data as a reference dataset for quality evaluation may no longer be the most valid choice. In some parts of the world, VGI is more complete and more accurate than authoritative datasets (Neis et al., 2011; Vandecasteele and Devillers, 2015), which poses challenges to the assessment of VGI data quality.

This chapter provides a review of data quality indicators for geographic information that are part of the ISO 19157 (2013) standard, of how these have been used to evaluate the data quality of VGI in the past and of other approaches that could be used. Additional indicators that have been proposed for VGI in particular are also presented, as well as initiatives to develop quality assessment frameworks combining several quality measures and indicators.

2 Measures and Indicators to Assess VGI Quality

ISO 19157 is the latest release (2013) of a data quality standard among the internationally known standards for describing spatial data quality, e.g. the International Cartographic Association (ICA), Federal Geographic Data Committee (FGDC) and Committee on Standardization (CEN) standards. It attempts to define a set of measures for evaluating and reporting data quality. The conceptual model for geodata quality as specified in ISO 19157 represents data quality by a series of data quality elements, e.g. positional accuracy. Each data quality element is then further described by measures that allow the data quality to be evaluated, and the results of the evaluation can be documented and reported to any interested party. The ISO 19157 standard does not attempt to define any minimum acceptable levels of quality for spatial data, and it considers only conventional datasets without proposing any data quality elements or measures specific to VGI. The next subsection outlines the different spatial data quality elements that are part of ISO 19157 and how they can be used to measure VGI quality, drawing upon examples from the literature and VGI practices.

2.1 ISO Quality Measures Applicable to VGI

The first five spatial data quality elements of ISO 19157 (Sections 2.1.1 to 2.1.5) are focused on the quality of the product from a producer's point of view, or on what is termed the 'internal quality' of a dataset (Devillers and Jeansoulin, 2006). The sixth spatial data quality element (Section 2.1.6) is focused on the user needs and requirements and is referred to as the 'external quality' of a dataset (Devillers and Jeansoulin, 2006). Thus there may be situations where the internal quality is high (i.e. it is produced according to a set of specifications) but the external quality poor (i.e. it does not fulfil a particular purpose from a user's perspective). The same will apply to VGI, so the fact that a VGI dataset is created according to some initial specifications does not necessarily mean that it can be used to cover all or any requirements stated by potential end users. This is of particular importance when we consider that in many implicit VGI sources, the existing specifications might have no direct relation to spatial or geomatics aims. Some additional quality elements have been proposed for crowdsourced data that fall in between internal and external quality (Meek et al., 2014), corresponding to what the authors called the stakeholder model; these additional quality elements have also been referred to as quality indicators (Antoniou and Skopeliti, 2015) and are discussed in more detail in Section 2.2.

2.1.1 Positional Accuracy

Positional accuracy refers to the accuracy of the position of features (i.e. points, lines or areas) within a spatial reference system, and is usually assessed by comparing the position of features with their counterparts in reference data, which are considered to represent the 'true' position. This assessment, however, requires the existence of reference data with similar characteristics and a valid time frame to make the comparison.

The use of portable data collection technologies, such as Global Navigation Satellite Systems (GNSS) receivers embedded in smartphones, is one of the most common methods to collect the geographic position associated with crowdsourced data. Previously, these technologies were capable of delivering a spatial precision exceeding ±10m (Coleman, 2010). However, the precision is continuously improving, and accuracies of 2–3 m or even higher can now be achieved, depending on the receivers used, the observation method or the observation conditions (Pesyna et al., 2015). When combined with the increasing availability of Web-based maps and imagery (in some cases with very high spatial resolution) that can be used, for example, as digitising backdrops, it is not surprising that the positional accuracy of VGI has increased, and is now appropriate for a wide range of applications.

Several studies have been conducted to assess the positional accuracy of VGI data. An analysis of positional accuracy of OSM in relation to Google Maps and Bing Maps was undertaken by Ciepłuch et al. (2010) for sites in Ireland, and concluded that in some locations there were differences of up to 10m (for Google Maps) between these sources, although only for some types of features, which seemed to result from digitisation over low-resolution images. For a set of OSM road features compared to the UK's Ordnance Survey data, the average errors identified were 5.8m (Haklay, 2010) – a distance unlikely to be seriously problematic for most land cover maps, but one which could cause small or narrow features (ponds, hedges, riparian habitats, etc.) to be missed or misplaced. Canavosio-Zuzelski et al. (2013) performed a positional accuracy assessment of OSM as part of a vector adjustment correction. However, in this case, rather than accepting official survey data as truth, both official data and OSM data were assessed against independent stereo imagery, which means the technique can be applied to other national agency and topographic datasets and has the potential to identify areas where the VGI surpasses the accepted dataset. Thus the authors were able to assess OSM against USGS (United States Geological Survey) and TIGER (Topologically Integrated Geographic Encoding and Referencing) road data on a more-or-less equal footing – albeit for a very small area for which the aerial imagery was available. In general, the availability of such accurate benchmarking data is restricted, and this (or a requirement for very current information) may be the very reason why VGI is being elicited. The most successful examples of such quality control analyses are where feedback is given to the volunteers to enable them to improve their contributions, e.g. in OSM.

The positional accuracy of points representing geotagged photographs may also be considered and analysed, once the specifications are available regarding what feature should be positioned. In Hochmair and Zielstra (2012), the location associated with the Flickr and Panoramio photographs was compared to the location of the photograph as determined by the authors analysing what was represented in the photograph. Several aspects were identified that may influence positional quality; for example, the position assigned to some photographs was the location from which the photograph was taken, while for others it was the position of what was represented in the photograph (potentially some distance away), without any additional indication of what the position represented. Another aspect identified that influenced the positional accuracy was the confusion between similar features that are present in the region (such as different bridges over a river close to each other), which became apparent when the location of the photographs was viewed on a satellite image or digital map.

The assessment of the positional accuracy or the extent mapping of patchy vegetation, highly-textured land use types and ecotones presents much more of a challenge. For land cover mapping, it is often the case that categorical labels (or degrees of similarity to those labels) are being elicited from contributors for attachment to user-supplied location points or to predefined polygon features. Absolute positional accuracy is still important, but more often relates to boundaries between mapped areas or to the location of single survey points, and the predominant source of inaccuracy is thematic misclassification (to which, of course, these positional inaccuracies can contribute).

Other approaches may, however, be considered for assessing or increasing positional accuracy of VGI, due to the amount of data available and their dynamic characteristics (Section 2.2). To correct and quantify positional errors, conflation approaches that use a set of reference features are common for discrete data that fit an existing taxonomy (Coleman, 2010; Girres and Touya, 2010; Haklay, 2010).

2.1.2 Thematic Accuracy

Thematic accuracy refers to the accuracy of classes or thematic tags associated with specific locations or objects placed in geographic space, such as classes assigned to pixels in a land cover map or tags assigned to a vector-encoded entity, e.g. a highway, river, building or green area. The assessment of thematic accuracy in VGI may be performed using a traditional approach, where the information is compared to reference data, e.g. satellite imagery or authoritative data, by experts. For instance, Estima and Painho (2013; 2015) and Jokar Arsanjani et al. (2015b) investigated the thematic accuracy of the classification of OSM features using the Corine Land Cover database and the pan-European GMESUA dataset as authoritative reference data, respectively. However, the assessment of the thematic accuracy of VGI raises new challenges, due to the lack of strict specifications, the characteristics of the contributors and contributions, and the type of thematic information at stake. Therefore, additional quality indicators may be used, which are further explained in Section 2.2. The assignment of thematic information in VGI has many similarities to the extensive tagging and relevance assessment of documents by volunteers or paid contractors working via systems such as Amazon's Mechanical Turk. Many land cover mapping challenges are effectively labelling problems, where predefined pixels or spatial features must be assigned to particular classes; therefore, some of the work developed in these areas of application to assure data quality may be applied to VGI.

Currently, the majority of VGI is contributed for free, by volunteers, but there is an increasing interest in contracting out classification tasks such as land cover labelling to paid workers in the cloud. In such contexts, spam and errors are common, whether these stem from a lack of skill or from deliberate attempts to mislead (including attempts to cheat the system in a way that cannot be easily detected). A number of strategies have been proposed and evaluated for getting the best value out of contracted labellers, and in particular for trading off the value of new information about unlabelled entities against the value of reinforcing or correcting information about entities that have been labelled repeatedly (Ipeirotis et al., 2014). This corresponds to the use of additional quality indicators, which are further addressed in Section 2.2. One consideration when deciding between accuracy improvement and new data acquisition must be the possible impact of errors when a dataset is used in the real world – a balancing act similar to the calculation of ROC (Receiver Operating Characteristic) curves or sensitivity/specificity calculations for classifiers and prediction algorithms. The problem of risk and liability, when considered in the VGI world, is usually sidestepped through the use of disclaimers, but if VGI begins to seriously underpin Spatial Data Infrastructures (SDIs) – see Chapter 12 (Demetriou et al., 2017) – and commercial products, the issue will become more pressing.

Many of the non-VGI labelling tasks described have marked parallels to VGI problems: for example, data points are often being collected, like 'ground truth', in order to carry out a supervised classification, and in many cases the labelling is not simply binary or categorical. In such cases, when redundant observations exist for each particular item, the variation between labellers is not simply noise; often, the uncertainty and disagreement, if recorded and analysed, can yield important information about the real world. In the case of VGI, this could include conditions on the ground such as vegetation succession, change of ownership or mixing of land covers. Many papers in the field also note the importance of training for labellers as well as for models (e.g. Clark and Aide, 2011; Fritz et al., 2012), and show the sorts of learning curves that are possible with varying quantities and qualities of reference data.

Of course, even well trained users vary in their accuracy, and differences between experts and non-experts are also likely to exist. A comparison of the quality results of expert and non-expert volunteers for tag assignment was done by See et al. (2013). The results showed that in some types of tags (in this particular case, 'human impact'), non-expert volunteers produced results as good as the experts, probably because the concept was new to both non-experts and experts alike so both had the same learning curves. However, for some land cover classes, the experts (some of whom had considerable experience in image classification) performed better, but the non-experts showed improvements over time, especially when feedback on the quality of their results was provided to them.

2.1.3 Completeness

Completeness refers to the presence or absence of features, of their attributes and of relationships compared to the product's specification; it is divided into a) commission, which explains excess data presence in a dataset, and b) omission, which explains data absence from a dataset. Completeness is of major concern/importance in VGI, since many volunteered datasets are demonstrably biased towards particular spatial regions (see e.g. Haklay, 2010), but also towards certain features that are easier to measure or towards themes or 'pet features' (Bégin et al., 2013) that are of particular interest to the contributing individual, or even motivated by accessibility or digital inclusion (Zielstra and Zipf, 2010). This reliance on the motivation of individual volunteers will determine the resolution, homogeneity, representativity and domain consistency of the resulting data. Where a principled sampling strategy can be imposed on volunteers, e.g. a probabilistic schema or the systematic, even grid of the Degree Confluence Project, the volunteered data have the potential to be more broadly applicable, but the value of the data will depend on the coverage by volunteers, meaning that many platforms must actively direct users to the desired locations, trading off potentially rich information elsewhere against an even placement of observations.

The lack of specifications and the nature of VGI makes, in some cases, the assessment of completeness a complex process, which cannot rely only on direct unit-based comparisons, and instead requires the development of new approaches. Moreover, in many areas, the number of digitised VGI features may exceed that found in an authoritative dataset (Neis et al., 2011), making a simple comparison of feature counts inappropriate, and requiring a subtler consideration of commission and omission (Jackson et al., 2013). Koukoletsos et al. (2012) present a method that holds promise for such contexts, combining geometric and attribute constraints to match road segments in OSM with those found in an authoritative dataset, and to achieve a tile-by-tile completeness assessment. In another study, Hecht et al. (2013) proposed an object-based approach to assess the completeness of building footprints. Haklay (2010) identified a bias in UK OSM data coverage towards more affluent areas, and relates this to the fact that socially marginal (and less-mapped) areas may be the very locations where charities and agencies requiring free data are operating. Brovelli et al. (2017) developed a web application to compare OSM road data with authoritative road data, enabling the assessment of completeness and positional accuracy of OSM data. Ciepłuch et al. (2010) also compared the spatial coverage of OSM to that of Google Maps and Bing Maps, and identified regions with different levels of coverage in the three datasets. Globally, this bias is being somewhat redressed by the volunteers' own efforts to improve coverage, and by focused initiatives such as KompetisiOSM in Indonesia[2], but it remains the case that coverage is extremely heterogeneous in VGI, both spatially and thematically, and that the absence of information in an area makes it difficult to draw robust conclusions about trends. Brunsdon and Comber (2012) specifically addressed the lack of experimental design in a volunteered dataset recording the first flowering date of lilacs in the USA by applying random coefficient modelling and bootstrapping approaches to tease out more reliable information on phenological trends.

2.1.4 Temporal Quality

Temporal quality refers to the quality of the temporal attributes, such as date of collection, date of publication, update frequency, last update or temporal validity (also referred to as currency), and also to relationships between the temporal validity of features. Currency is one aspect of traditional data quality where VGI can be expected to surpass authoritative data, especially in dynamically changing environments, given the large numbers of citizens who are acting as sensors at any one time. However, there is often a trade-off between currency and other facets of data quality. The issue of representativeness becomes even more vexed when the spatial domain is extended to the spatio-temporal domain, and, unless a temporal sampling scheme is also imposed upon contributors, the density and coverage of a VGI dataset over a small time range can be very limited. For citizen sensor networks, which are largely made up of automated instruments, such as the Weather Underground, the observation pattern across time is fairly consistent. However, in other contexts (e.g. presence-only species observations and the mapping of urban infrastructure), a user will need to carefully consider the ranges of data that are appropriate for their purpose, and whether cumulative observations are valuable. In making this decision, they will probably require metadata on the individual features, e.g. date stamps and data on feature updates. An important consideration here is that the date stamp should reflect the time at which the measurement or observation was made, rather than the time at which it was uploaded or digitised, depending on the application to which the data are applied (see e.g. Antoniou et al., 2016a).

Even though the potential of VGI to provide updated information is large, it is relevant to notice that a large heterogeneity is likely to occur over space and for different types of phenomena or features to be mapped, since VGI is dependent on the availability of interested volunteers to collect each particular type of data at the required locations.

2.1.5 Logical Consistency

Logical consistency refers to the degree of adherence to logical rules of data structure, attribution and relationships as described in a product's specifications. Logical consistency of an observation makes little sense in isolation: it must usually be assessed with reference to other data from the same source, or from independent (and sometimes authoritative) data, and lends itself to automated quality assessment – for example, to the use of rules such as 'forest fires are highly unlikely in dense urban areas'. Hashemi and Ali Abbaspour (2015) used the concept of spatial similarity in a multi-representation data combination to build a framework to determine the probable inconsistencies in OSM, aiming to help in evaluating the logical consistency of VGI data. Bonter and Cooper (2012) discuss the use of a smart filter system in the context of species identification in Project FeederWatch: when participants enter counts of species that are too high or species that do not normally appear on standard lists, the filter is activated and users are informed of unusual observations, thereby correcting potential errors in real-time. Similar smart filters could be devised and put into place in other types of VGI projects, thereby addressing some aspects of logical consistency.

2.1.6 Usability

As mentioned above, usability (or fitness-for-use) refers to the external quality of a dataset and is focused on the needs of the user. The five aforementioned data quality elements may be aggregated in order to describe the overall usability of a specific dataset for a particular use, i.e., fitness-for-purpose. In other words, usability acts as a complementary element by linking both user requirements and data quality measures to check whether the data for a specific application can be used (Guptill and Morrison, 1995; Devillers et al., 2007).

Table 1 summarises the requirements and specific aspects regarding the application of ISO quality measures to VGI. In Section 3, establishing workflows and combining quality indices to assess VGI quality in order to assess usability is further developed.

2.2 Quality Measures Specific to VGI

When considering VGI, other data quality indicators are required to supplement those proposed in the ISO framework. This occurs not only because in many situations comparison with authoritative datasets is not possible, but also because the characteristics and nature of VGI enable the use of indicators that do not usually make sense when applied to data created by professionals. These indicators may provide valuable information even though in most situations they do not assess accuracy but instead assess data reliability or credibility (which are considered as synonyms in this chapter). As these indicators may

Table 1: ISO quality elements, their requirements and issues related to their use with VGI.

<table>
<tr><th colspan="2">ISO quality elements</th><th>Requirements</th><th>Issues for the application to VGI</th></tr>
<tr><td rowspan="5">Internal quality</td><td>Positional accuracy</td><td rowspan="4">• Data specification
• Existence of reference data with similar characteristics and valid time frame</td><td rowspan="4">• Lack of specifications
• Dynamic nature of VGI
• Inexistence of comparable reference data
• Spatial and thematic heterogeneity</td></tr>
<tr><td>Thematic accuracy</td></tr>
<tr><td>Completeness</td></tr>
<tr><td>Temporal Quality</td></tr>
<tr><td>Logical Consistency</td><td>• Other data of the same source or independent data</td><td>• Applicable to VGI
• May enable automatic validation checks</td></tr>
<tr><td>External quality</td><td>Usability</td><td>• Specification of user needs</td><td>• May be assessed by combining quality measures and indicators</td></tr>
</table>

provide data that allow quality estimation in real-time or near real-time, they enable the development of automated approaches that may be used to improve the process of data collection, requiring, for example, confirmation and/or additional checks by the contributors.

Different suggestions have been put forth regarding what these indicators might look like (Table 2). For example, Goodchild and Li (2012) provide three broad categories of measures to ensure VGI data quality: i) crowdsourcing revision, where data quality can be ensured by multiple contributors; ii) social measures, which focus on the assessment of contributors themselves as a proxy measure for the quality of their contributions; and iii) geographic consistency, through an analysis of the consistency of contributed entities. Meek et al. (2014) provide three models of data quality, where the stakeholder model sits in between the more traditional internal (producer) and external (consumer) quality indicators, and they suggest a number of different quality elements, including vagueness, ambiguity, judgement, reliability, validity and trust. Bordogna et al. (2014) also provide a set of quality indicators for VGI that are arranged into internal and external quality, where the internal quality measures are grouped by type of VGI, i.e. measurements or text-based VGI, and the external quality measures are grouped by reliability of the individual and reputation of the organisation. Senaratne et al. (2016) review VGI quality assessment methods and separate them into measures and indicators of quality, where the former correspond to the traditional accuracy assessment measures described in the previous section, and the latter are referred to as qualitative and more abstract quality indicators, such as local knowledge, experience and reputation. They also suggest that an additional approach to ensure data quality, referred to as 'data mining', should be added to the ones proposed by Goodchild and Li (2012). Antoniou and Skopeliti (2015) propose the aggregation of the quality indicators into three broad categories: i) data indicators; ii) demographic and other socio-economic indicators; and iii) indicators about the contributors. These may be considered to integrate the types of indicators mentioned in the above different frameworks and are developed further in this chapter.

Table 2: Categories of quality measures proposed for VGI.

Goodchild and Li (2012)	Meek et al. (2014)	Bordogna et al. (2014)	Antoniou and Skopeliti (2015)	Senaratne et al. (2016)
• Crowdsourcing revision • Social measures • Geographic consistency	• Internal quality indicators • Stakeholder model • External quality indicators	• Internal quality • External quality	• Data indicators • Demographic and socio-economic indicators • Contributor indicators	• Measures of quality • Indicators of quality • Data mining

2.2.1 Data-based Indicators

One important group of quality indicators of VGI are those that involve comparison with other sources of crowdsourced data (Table 3). One possibility is to measure the 'agreement' to the corresponding data, which we define here as the coherence of the data with other sources of crowdsourced data. Agreement can be measured between datasets using a Boolean measure or a continuous variable with traditional measures such as distance between corresponding elements, attribute comparisons, etc., and may be considered an indicator of data reliability. Logical consistency of data available in different data sources can also be used to estimate data reliability, identifying if, according to the types of features present in all available data sources, a particular contribution is likely to be correct or not. As stressed by Sui et al. (2013), approaches that compare data based on their geographic location have not yet been developed enough. Note, however, that all these indicators may be used to measure data reliability, but not to assess data accuracy if none of the data under comparison can be considered as reference data.

Another set of indicators can also be calculated that could reveal VGI quality by solely examining the VGI dataset itself and the associated metadata (Table 3). The work in this area has focused primarily on assessing OSM data quality. Such indicators could include the total length of features and the point density in a square-based grid, as calculated by Ciepłuch et al. (2010), or the number of versions, the stability against changes and the corrections and rollbacks of features, as examined by Keßler and de Groot (2013). The provenance of features contributed to OSM (i.e. whether the data were captured using a GPS, were manually digitised or resulted from a bulk import) has been the

Table 3: Data-based quality indicators proposed for VGI.

Indicators Category	Indicators	Description / Examples
Data-based indicators (assess data reliability)	Coherence with other sources of corresponding data (not considered as reference)	Compare, for example, geometric attributes such as distance between corresponding elements or overlaps
	External logical consistency	Logical consistency of VGI with non-corresponding data available in other data sources
	Internal logical consistency	Logical consistency of the VGI dataset itself
	VGI metadata	Number of versions, features corrections, stability against changes, observation methods, used equipment, date of observation

focus of the quality-related work of Van Exel et al. (2010). Finally, Barron et al. (2014) have developed iOSMAnalyzer, which uses more than 25 methods and indicators to assess OSM data quality based solely on data history. Although some of these indicators are related to the aforementioned quality component of completeness (Section 2.1.3), completeness in authoritative GI would not be measured in this way. Hence there is a need to find completeness and other data indicators that are customised to the nature of VGI.

Some of the facets of traditional metadata are of particular interest in assessing and using VGI. For example, the lineage of a record or dataset may include its edit history and information on how it was measured, and can be especially important in the automated assessment of VGI fitness-for-use. Examples of metadata potentially useful for VGI are equipment used in measurements; data about the volunteer (contributor indicator); date and time of data collection; or atmospheric conditions at the time a particular observation was taken. Individual metadata about heterogeneous observations can be extremely useful in identifying bias and likely trustworthiness, as seen, for example, in the context of amateur weather monitoring (Bell et al., 2013) and digitised trails (Esmaili et al., 2013). However, metadata are often not available for VGI, which limits, to some extent, the use of these approaches. To overcome this difficulty, methodologies have already been proposed to create metadata for VGI (Kalantari et al., 2014).

2.2.2 Demographic and Socio-economic Indicators

Empirical studies have revealed that there is a correlation between the demographics of an area and the completeness and positional accuracy of the data (Mullen et al., 2015). It has also been shown that areas with lower population density (i.e. rural areas) can have a negative effect on the completeness of VGI data (Zielstra and Zipf, 2010). At the same time, population density correlates positively with the number of contributions, thus affecting data completeness

Table 4: Demographic and Socio-economic quality indicators proposed for VGI.

Indicators Category	Indicators	Relevance
Demographic and Socio-economic indicators of the region (indicators of data quality)	Demographics	Show correlation with data quality parameters
	Population density	
	Social deprivation	
	Socio-economic reality	
	Income	
	Population age	

or positional accuracy (see e.g. Zielstra and Zipf, 2010; Haklay, 2010; Haklay et al., 2010; Jokar Arsanjani and Bakillah, 2015) .

Closely related to demographics are other socio-economic factors, which may also influence the overall quality (Tulloch, 2008; Elwood et al., 2013). For example, it has been shown that social deprivation and the underlying socio-economic reality of an area can have a considerable effect on completeness and positional accuracy of OSM data (Haklay et al., 2010; Antoniou, 2011). Similarly, other factors such as high income and low population age can result in a higher number of contributions and therefore higher VGI quality in terms of positional accuracy and completeness (Girres and Touya, 2010; Jokar Arsanjani and Bakillah, 2015).

Thus, if census or social survey data are available for an area, they might be used to make inferences about the quality of VGI data over geographic space. Table 4 summarises the above mentioned indicators.

2.2.3 Contributor Indicators

Quality indicators can include the history of contributions, the profiling of contributors or the experience, recognition and local knowledge of the individual (van Exel et al., 2010; Table 5). Moreover, the number of contributors in certain areas or features has been examined, and has been positively correlated with data completeness and positional accuracy (Keßler and de Groot, 2013). Methods for the automatic computation of contributor reliability regarding

Table 5: Contributor quality indicators proposed for VGI.

Indicators Category	Indicators	Description	Relevance
Contributor indicators (assess contributor reliability)	Contributors' interests	Infer contributor bias to particular features	Expected correlation with data reliability
	Contributors' history of contributions	Infer contributor trustworthiness	
	Contributors' recognition by other contributors	Infer contributor reliability	
	Contributors' location	Infer contributor local knowledge	
	Contributors' behaviour	Infer contributor difficulty in contributing	
	Contributors' education	Infer contributor expertise	
	Profiling of contributors	Created by aggregating several contributor indicators	

thematic information in VGI have been proposed by several authors. Haklay et al. (2010) and Tang and Lease (2011) stress the need for multiple observations and observers to enable consensus-based data quality assessments. Foody and Boyd (2012) and Foody et al. (2013) proposed a method for using these repeated observations to concretely assess the quality of VGI contributors using a latent class analysis of VGI in relation to land cover.

Differences between volunteers are always likely to exist, and, therefore, in the examples of 'social' quality assessment described above, known individuals could be identified and given a more trusted status, and these individuals could then be actively responsible for reviewing the work of others. However, when considering thematic quality, the issue of contributor reliability can be more complicated than a single ranking. Some contributors excel at labelling particular types of objects or habitats, but perform poorly elsewhere in the problem domain. Knowledge of the strengths and weaknesses of the volunteers allows a more nuanced consideration of the trustworthiness of their contributions, but often requires independent reference data to be computed. For example, Comber et al. (2013) calculated the consistency and skill of each volunteer in relation to each land cover class, using a number of control points for which the land cover had been independently determined by experts, and demonstrated that at least some concerns about the quality of VGI can be addressed through careful data collection, the use of control points to evaluate volunteer performance and spatially explicit analyses.

In the context of labelling for commercial gain, the workers do not see the submissions of others, and it is necessary to automate the process of identifying trustworthy experts against whom the work of others can be benchmarked (Raykar and Yu, 2012). Vuurens and de Vries (2012) tackle this issue by deriving patterns from the behaviour of different worker types, and attempt to diagnose the nature, and thus the likely error rate, of particular workers. For example, they note that 'diligent' workers are less likely to differ in their votes by more than one step on an ordinal scale of labels, and they exploit this fact to interpret the difference between contributors' judgements to identify their trustworthiness. However, there are many contexts where no natural ordering is present in the labels from which a contributor can choose.

Some of the facets of metadata regarding the volunteer, such as age, address, level of education or interests, are of interest in assessing VGI reliability. It is also possible to construct metadata based on the past behaviour of a user or the number of times their contributions have been identified as erroneous by other volunteers, which requires the storing of all alterations and changes made to the system. This may enable, through the definition of a set of rules, the automatic extraction of quality information, which may be used as an initial indicator of credibility, enabling the exclusion of some VGI from an analysis based on the likelihood that it might be less trustworthy. An example of these procedures is the approach proposed by Lenders et al. (2008), where the contributor's reliability is assessed using the information about the volunteer's location and the time of the contribution. These types of approaches may be particularly useful for NMAs (see Chapter 13 by Olteanu-Raimond et al., 2017), for example, to identify which contributions are more reliable and therefore worthy of allocations of resources for their validation, as all crowdsourced data used by NMAs need to be validated by professionals (Fonte et al., 2015a).

It is also possible to measure the 'vagueness' of contributions, defined by Meek et al. (2014) as the inability of a contributor to make a clear-cut decision. For example, when volunteers are asked to interpret satellite imagery in Geo-Wiki, they attach a confidence rating to their choice, which ranges from highly uncertain to full confidence in their answer (Fritz et al., 2012). These vagueness measures can be used as filters on the data or to apply weights to those answers with higher vagueness.

3 Developing Quality Assurance Workflows and Combining Indicators

Although many different quality indicators and measures for VGI have been emerging over the last decade, combining these indicators into an integrated quality assessment is an ongoing area of VGI data quality research. For example, Bishr and Mantelas (2008) have proposed a 'trust and reputation model', where these two concepts together are proxies for data quality (Figure 1). Users rate each other's contributions on a score range of 1 to 10, which makes up the reputation component. Users are also linked to one another through a social network, which can be used to measure the strength of the relationship between two individuals. These two components are combined and then divided by the logarithm of the distance between a contributor's location and the observation to calculate a trust rating. This trust model therefore takes both spatial context and reputation, through user ratings and the relationships between contributors, into account. The model remains theoretical and was not applied in the paper cited above, but an example of data collection for an urban growth scenario was outlined. The inclusion of relationships via social networking could give greater weight to the ratings of certain individuals.

Jokar Arsanjani et al. (2015a) have for their part proposed a multivariate indicator, referred to as the contribution index (CI), that combines diverse classic quality indicators, as well as user perspectives of data, including the number of volunteers involved in mapping a particular feature along with the frequency of contributions (Figure 2).

However, the main problem with the assessment of VGI based on fitness-for-use is that many methods and measures are designed to assess a specific VGI dataset or a single use case, and are not generalisable or transferable to other VGI datasets or purposes. However, some papers have appeared in which quality assurance workflows have been proposed. For example, Bordogna et al. (2015) propose a flexible system that allows users to specify minimum acceptable quality levels based on their requirements (Figure 3). The system contains a series of quality indicators, including both standard

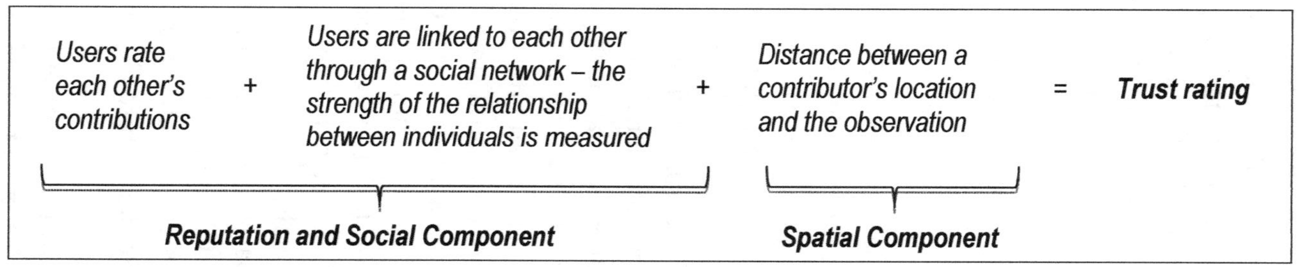

Fig. 1: Scheme of the trust and reputation model proposed by Bishr and Mantelas (2008).

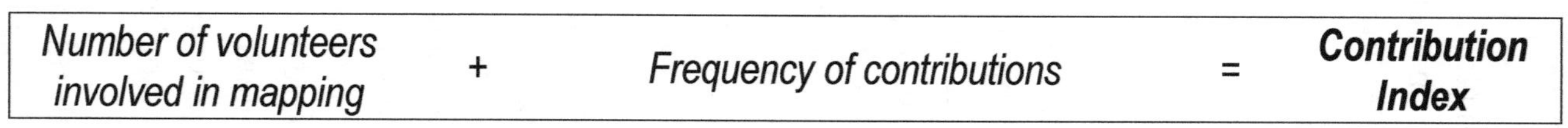

Fig. 2: Scheme of the determination of the contribution index proposed by Jokar Arsanjani et al. (2015a).

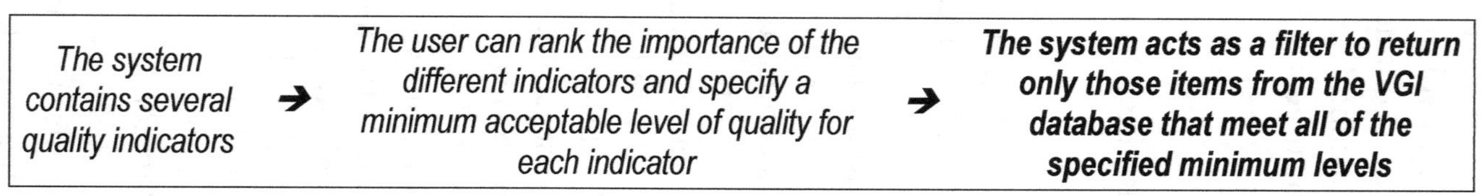

Fig. 3: Scheme of the system proposed by Bordogna et al. (2015) that allows users to specify minimum acceptable quality levels based on their requirements.

internal quality measures such as positional accuracy and ones specifically geared towards VGI (see Section 2.2). The user can rank the importance of the different indicators and specify a minimum acceptable level of quality for each indicator, and then the system acts as a filter to return only those items from the VGI database that meet all of these minimum levels; the authors perform a demonstration of the system on a VGI dataset of glaciological observations.

The creation of workflows that allow for the assessment of different aspects of quality has also been proposed. The framework proposed by COBWEB includes a quality assessment workflow that uses some automatic validation procedures to obtain data quality indicators to insert in the information metadata (Meek et al., 2016), while Ballatore and Zipf (2015) have proposed a multidimensional framework to assess conceptual quality.

The need to assess fitness-for-use has been present even without considering VGI, and methodologies to make this assessment have already been proposed in other contexts. For example, Lush (2015) proposed the creation of a GEO label that aims to be a mechanism to assist users to determine the fitness-for-use of datasets: a visual tool was developed that aggregates information about the producer, data lineage, compliance with standards, existence of quality information, user's feedback, expert reviews and citation information. These types of tools may be adapted to the characteristics of VGI and generate user friendly tools that can assist the user in identifying which data are appropriate for each application, according to their needs.

This is an area of research that we anticipate will continue to grow in the future.

4 Conclusions

This chapter considered the quality of VGI from the perspective of ISO 19157 and then presented additional quality measures designed to handle the specific nature of VGI, e.g. data-specific indicators, demographic and socio-economic indicators, and indicators related to the contributors. Authoritative data and VGI have similarities, i.e. both are examples of spatial data that can be assessed using the measures set out in ISO 19157. However, there are also some differences between these two data sources that require new ways of quality assessment, since the specific nature of VGI presents some problematic issues as well as new challenges. These issues and challenges include the heterogeneity of the data and contributors, spatial bias, lack of specifications, the dynamic nature in which the data are updated, the patchiness of the contributions and the lack of authoritative data, all of which have driven the development of new assessment methods for VGI. For example, the lack of reference data (as well as the static nature of reference data) has led to studies that have moved away from the need to use authoritative data to assess the quality of VGI; this has resulted in the creation of new data indicators, e.g. consistency related to multiple contributions at the same place or agreement of multiple contributions of the same set of features. At the same time, the social element of VGI has led to research into socio-economic and demographic indicators, while the pivotal role of the contributor in VGI has stimulated research around a diverse set of indicators related to quantifying them.

Another area of more recent VGI quality-related research has been in combining indicators, either as a way to visualise the quality using graphical approaches, such as through a GEO label (Lush, 2015), or to create workflows that allow for the assessment of different aspects of quality. However, few attempts have yet been implemented that use automated processes to assess VGI quality in addition to the use of the crowd self-correction or of selected volunteers for data validation (Fonte et al., 2015b). Nevertheless, these combinations are particularly desirable due to the dynamic characteristic of VGI, which makes the use of traditional approaches, which take time and require expert intervention, less suitable.

Although VGI has many similarities to authoritative GI, one of the main difference is the much more relaxed nature of the data collection protocols. The need for more VGI protocols, including the need for a framework that considers quality as one element, is addressed in Chapter 10 (Minghini et al., 2017). Chapter 10 also considers how quality assurance can be influenced by technological solutions that can help to seamlessly enforce protocols and thereby increase data quality, while recognising the trade-offs between the complexity of the protocol and participant motivation and retention.

The quality of VGI will continue to be one of the most important barriers to the integration of VGI to authoritative data, and developing generic and flexible solutions such as the system proposed by Bordogna et al. (2015) represents one tangible step forward; thus, we envisage that workflow developments will be a key area of research in the future. Standards agencies also need to recognise that there are new sources of spatial data and that existing standards must be adapted to include these sources or new standards must be developed. A first step in this direction has been made by the W3C with a document (currently in a draft form; Tandy et al., 2016) on best practices that should be taken into consideration when publishing and using spatial data on the Web. The document highlights another aspect, and, in a sense, extends the notion of usability, by drawing attention to the discoverability and accessibility of the spatial data published.

Notes

[1] http://confluence.org/

[2] https://www.hotosm.org/projects/indonesia-0

Reference list

Antoniou, V., 2011. User Generated Spatial Content: An Analysis of the Phenomenon and its Challenges for Mapping Agencies. Unpublished PhD Thesis. University College London (UCL), London, UK.

Antoniou, V., Fonte, C., See, L., Estima, J., Arsanjani, J., Lupia, F., Minghini, M., Foody, G., Fritz, S., 2016a. Investigating the feasibility of geo-tagged photographs as sources of land cover input data. *ISPRS International Journal of Geo-Information* 5, 64. DOI: https://doi.org/10.3390/ijgi5050064

Antoniou, V., Schlieder, C., 2014. Participation patterns, VGI and gamification, in: *Proceedings of AGILE 2014*, Castellón, Spain, 3–6 June 2014, pp. 3–6. Available at: http://www.geogames-team.org/agile2014/submissions/Antoniou_Schlieder_2014_Participation_Pattern_VGI_and_Gamification.pdf [Last accessed 16 May 2017].

Antoniou, V., Skopeliti, A., 2015. Measures and indicators of VGI quality: An overview, in: Proceedings of the ISPRS Geospatial Week 2015, La Grande Motte, France, 28 Sep – 03 Oct 2015, pp. 345–351.Available at: http://www.isprs-ann-photogramm-remote-sens-spatial-inf-sci.net/II-3-W5/345/2015/isprsannals-II-3-W5-345-2015.pdf [Last accessed 16 May 2017].

Antoniou, V., Touya, G., Olteanu-Raimond, A.-M., 2016b. Quality analysis of the Parisian OSM toponyms evolution, in: Capineri, C., Haklay, M., Huang, H., Antoniou, V., Kettunen, J., Ostermann, F., Purves, R. (Eds.), *European Handbook of Crowdsourced Geographic Information*. Ubiquity Press, London, UK, pp. 97–112. DOI: https://doi.org/10.5334/bax

Ballatore, A., Zipf, A., 2015. A conceptual quality framework for Volunteered Geographic Information, in: Fabrikant, S.I., Raubal, M., Bertolotto, M., Davies, C., Freundschuh, S., Bell, S. (Eds.), *Spatial Information Theory*. Springer International Publishing, Cham, pp. 89–107.

Barron, C., Neis, P., Zipf, A., 2014. A comprehensive framework for intrinsic OpenStreetMap quality analysis. *Transactions in GIS* 18, 877–895. DOI: https://doi.org/10.1111/tgis.12073

Bégin, D., Devillers, R., Roche, S., 2013. Assessing volunteered geographical information (VGI) quality based on contributors mapping behaviours, in: International Archives of the Photogrammetry, Remote Sensing and Spatial Information Sciences. Presented at the 8th International Symposium on Spatial Data Quality, Hong Kong, China, pp. 149–154.

Bell, S., Cornford, D., Bastin, L., 2013. The state of automated amateur weather observations. *Weather* 68, 36–41. DOI: https://doi.org/10.1002/wea.1980

Bishr, M., Mantelas, L., 2008. A trust and reputation model for filtering and classifying knowledge about urban growth. *GeoJournal* 72, 229–237. DOI: https://doi.org/10.1007/s10708-008-9182-4

Bonter, D.N., Cooper, C.B., 2012. Data validation in citizen science: a case study from Project FeederWatch. *Frontiers in Ecology and the Environment* 10, 305–307. DOI: https://doi.org/10.1890/110273

Bordogna, G., Carrara, P., Criscuolo, L., Pepe, M., Rampini, A., 2015. A user-driven selection of VGI based on minimum acceptable quality levels. *ISPRS Annals of Photogrammetry, Remote Sensing and Spatial Information Sciences II-3/W5*, 277–284. DOI: https://doi.org/10.5194/isprsannals-II-3-W5-277-2015

Bordogna, G., Carrara, P., Criscuolo, L., Pepe, M., Rampini, A., 2014. On predicting and improving the quality of Volunteer Geographic Information projects. *International Journal of Digital Earth* 0, 1–22. DOI: https://doi.org/10.1080/17538947.2014.976774

Brovelli, M.A., Minghini, M., Molinari, M., Mooney, P., 2017. Towards an automated comparison of OpenStreetMap with authoritative road datasets. *Transactions in GIS* 21, 191–206. DOI: https://doi.org/10.1111/tgis.12182

Brunsdon, C., Comber, L., 2012. Assessing the changing flowering date of the common lilac in North America: a random coefficient model approach. *GeoInformatica* 16, 675–690. DOI: https://doi.org/10.1007/s10707-012-0159-6

Canavosio-Zuzelski, R., Agouris, P., Doucette, P., 2013. A photogrammetric approach for assessing positional accuracy of OpenStreetMap© roads. *ISPRS International Journal of Geo-Information* 2, 276–301. DOI: https://doi.org/10.3390/ijgi2020276

Ciepłuch, B., Jacob, R., Winstanley, A., Mooney, P., 2010. Comparison of the accuracy of OpenStreetMap for Ireland with Google Maps and Bing Maps, in: Proceedings of the Accuracy 2010 Symposium, Leicester, UK, 20–23 July. Available at: http://www.spatial-accuracy.org/system/files/img-X07133419_0.pdf [Last accessed 16 May 2017].

Clark, M.L., Aide, T.M., 2011. Virtual Interpretation of Earth Web-Interface Tool (VIEW-IT) for collecting land-use/land-cover reference data. *Remote Sensing* 3, 601–620. DOI: https://doi.org/10.3390/rs3030601

Coleman, D., 2010. Volunteered geographic information in spatial data infrastructure: An early look at opportunities and constraints, in: Rajabifard, A., Crompvoets, J., Kanantari, M., Kok, B. (Eds.), *Spatially Enabling Society: Research, Emerging Trends and Critical Assessment*. Leuven University Press, Leuven, Belgium, pp. 1–18.

Comber, A., See, L., Fritz, S., Van der Velde, M., Perger, C., Foody, G., 2013. Using control data to determine the reliability of volunteered geographic information about land cover. *International Journal of Applied Earth Observation and Geoinformation* 23, 37–48. DOI: https://doi.org/10.1016/j.jag.2012.11.002

Demetriou, D, Campagna, M, Racetin, I, Konecny, M. 2017. Integrating Spatial Data Infrastructures (SDIs) with Volunteered Geographic Information (VGI) for creating a Global GIS platform. In: Foody, G, See, L, Fritz, S, Mooney, P, Olteanu-Raimond, A-M, Fonte, C C and Antoniou, V. (eds.) *Mapping and the Citizen Sensor*. Pp. 273–297. London: Ubiquity Press. DOI: https://doi.org/10.5334/bbf.l.

Devillers, R., Bédard, Y., Jeansoulin, R., Moulin, B., 2007. Towards spatial data quality information analysis tools for experts assessing the fitness for use of spatial data. *International Journal of Geographical Information Science* 21, 261–282. DOI: https://doi.org/10.1080/13658810600911879

Devillers, R., Jeansoulin, R., 2006. *Fundamentals of Spatial Data Quality*. John Wiley & Sons, New York, USA.

Elwood, S., Goodchild, M.F., Sui, D., 2013. Prospects for VGI research and the emerging fourth paradigm, in: Sui, D., Elwood, S., Goodchild, M. (Eds.), *Crowdsourcing Geographic Knowledge*. Springer Netherlands, Dordrecht, pp. 361–375.

Esmaili, R., Naseri, F., Esmaili, A., 2013. Quality assessment of volunteered geographic information. *American Journal of Geographic Information System* 2, 19–26. DOI: https://doi.org/10.5923/j.ajgis.20130202.01

Estima, J., Fonte, C.C., Painho, M., 2014. Comparative study of Land Use/Cover classification using Flickr photos, satellite imagery and Corine Land Cover database, in: Huerta, J., Schade, S., Granell, G. (Eds.), *Proceedings of the 17th AGILE International Conference on Geographic Information Science: Connecting a Digital Europe through Location and Place*, Castellón, Spain, 3–6 June 2014, Available at: https://agile-online.org/conference_paper/cds/agile_2014/agile2014_141.pdf [Last accessed 16 May 2017].

Estima, J., Painho, M., 2015. Investigating the potential of OpenStreetMap for land use/land cover production: A case study for continental Portugal, in: Jokar Arsanjani, J., Zipf, A., Mooney, P., Helbich, M. (Eds.), *OpenStreetMap in GIScience, Lecture Notes in Geoinformation and Cartography*. Springer International Publishing, Cham, pp. 273–293.

Estima, J., Painho, M., 2013. Exploratory analysis of OpenStreetMap for land use classification, in: Proceedings of the Second ACM SIGSPATIAL International Workshop on Crowdsourced and Volunteered Geographic Information, GEOCROWD '13. ACM, New York, NY, USA, pp. 39–46. DOI: https://doi.org/10.1145/2534732.2534734

Fan, H., Zipf, A., Fu, Q., Neis, P., 2014. Quality assessment for building footprints data on OpenStreetMap. *International Journal of Geographical Information Science* 28, 700–719. DOI: https://doi.org/10.1080/13658816.2013.867495

Fonte, C.C., Bastin, L., Foody, G., Kellenberger, T., Kerle, N., Mooney, P., Olteanu-Raimond, A.-M., See, L., 2015a. VGI quality control. *ISPRS Annals of Photogrammetry, Remote Sensing and Spatial Information Sciences II-3/W5*, 317–324. DOI: https://doi.org/10.5194/isprsannals-II-3-W5-317-2015

Fonte, C.C., Bastin, L., See, L., Foody, G., Lupia, F., 2015b. Usability of VGI for validation of land cover maps. *International Journal of Geographical Information Science* 29(7), 1269–1291. DOI: https://doi.org/10.1080/13658816.2015.1018266

Foody, G., Boyd, D.S., 2012. Using volunteered data in land cover map validation: Mapping tropical forests across West Africa, in: Geoscience and

Remote Sensing Symposium (IGARSS), 2012 IEEE International. Presented at the Geoscience and Remote Sensing Symposium (IGARSS), 2012 IEEE International, pp. 6207–6208. DOI: https://doi.org/10.1109/IGARSS.2012.6352675

Foody, G., See, L., Fritz, S., Van der Velde, M., Perger, C., Schill, C., Boyd, D.S., 2013. Assessing the accuracy of volunteered geographic information arising from multiple contributors to an internet based collaborative project. *Transactions in GIS* 17, 847–860. DOI: https://doi.org/10.1111/tgis.12033

Fritz, S., McCallum, I., Schill, C., Perger, C., See, L., Schepaschenko, D., van der Velde, M., Kraxner, F., Obersteiner, M., 2012. Geo-Wiki: An online platform for improving global land cover. *Environmental Modelling & Software* 31, 110–123. DOI: https://doi.org/10.1016/j.envsoft.2011.11.015

Girres, J.-F., Touya, G., 2010. Quality assessment of the French OpenStreetMap dataset. *Transactions in GIS* 14, 435–459.

Goodchild, M.F., Li, L., 2012. Assuring the quality of volunteered geographic information. *Spatial Statistics* 1, 110–120. DOI: https://doi.org/10.1016/j.spasta.2012.03.002

Guptill, S.C., Morrison, J.L., 1995. *Elements of Spatial Data Quality*. Elsevier Science Limited.

Haklay, M., 2010. How good is volunteered geographical information? A comparative study of OpenStreetMap and Ordnance Survey datasets. *Environment and Planning B: Planning and Design* 37, 682–703. DOI: https://doi.org/10.1068/b35097

Haklay, M., Basiouka, S., Antoniou, V., Ather, A., 2010. How many volunteers does it take to map an area well? The validity of Linus' Law to volunteered geographic information. *The Cartographic Journal* 47, 315–322.

Hashemi, P., Ali Abbaspour, R., 2015. Assessment of logical consistency in OpenStreetMap based on the spatial similarity concept, in: Jokar Arsanjani, J., Zipf, A., Mooney, P., Helbich, M. (Eds.), *OpenStreetMap in GIScience*. Springer International Publishing, Cham, pp. 19–36.

Hecht, R., Kunze, C., Hahmann, S., 2013. Measuring completeness of building footprints in OpenStreetMap over space and time. *ISPRS International Journal of Geo-Information* 2, 1066–1091. DOI: https://doi.org/10.3390/ijgi2041066

Hochmair, H.H., Zielstra, D., 2012. Positional accuracy of Flickr and Panoramio images in Europe, in: Car, A., Griesebner, G., Strobl, J. (Eds.), Proceedings of the Geoinformatics Forum. Presented at the Geospatial Crossroads @ GI Forum '12, Wichman, Heidelberg, Germany, pp. 14–23.

Holloway, T., Bozicevic, M., Börner, K., 2007. Analyzing and visualizing the semantic coverage of Wikipedia and its authors. *Complexity* 12, 30–40. DOI: https://doi.org/10.1002/cplx.20164

Ipeirotis, P.G., Provost, F., Sheng, V.S., Wang, J., 2014. Repeated labeling using multiple noisy labelers. *Data Mining and Knowledge Discovery* 28, 402–441. DOI: https://doi.org/10.1007/s10618-013-0306-1

ISO, 2013. *ISO 19157: 2013 Geographic Information – Data quality.*

Jackson, S., Mullen, W., Agouris, P., Crooks, A., Croitoru, A., Stefanidis, A., 2013. Assessing completeness and spatial error of features in Volunteered Geographic Information. *ISPRS International Journal of Geo-Information* 2, 507–530. DOI: https://doi.org/10.3390/ijgi2020507

Jokar Arsanjani, J., Bakillah, M., 2015. Understanding the potential relationship between the socio-economic variables and contributions to OpenStreetMap. *International Journal of Digital Earth* 8, 861–876. DOI: https://doi.org/10.1080/17538947.2014.951081

Jokar Arsanjani, J., Mooney, P., Helbich, M., Zipf, A., 2015a. An exploration of future patterns of the contributions to OpenStreetMap and development of a Contribution Index: Future Patterns of the Contributions to OpenStreetMap. *Transactions in GIS* 19, 896–914. DOI: https://doi.org/10.1111/tgis.12139

Jokar Arsanjani, J., Mooney, P., Zipf, A., Schauss, A., 2015b. Quality assessment of the contributed land use information from OpenStreetMap versus authoritative datasets, in: Jokar Arsanjani, J., Zipf, A., Mooney, P., Helbich, M. (Eds.), *OpenStreetMap in GIScience, Lecture Notes in Geoinformation and Cartography*. Springer International Publishing, Cham, pp. 37–58.

Kalantari, M., Rajabifard, A., Olfat, H., Williamson, I., 2014. Geospatial Metadata 2.0 – An approach for Volunteered Geographic Information. *Computers, Environment and Urban Systems* 48, 35–48. DOI: https://doi.org/10.1016/j.compenvurbsys.2014.06.005

Keßler, C., de Groot, R.T.A., 2013. Trust as a proxy measure for the quality of Volunteered Geographic Information in the case of OpenStreetMap, in: Vandenbroucke, D., Bucher, B., Crompvoets, J. (Eds.), *Geographic Information Science at the Heart of Europe*. Springer International Publishing, Cham, pp. 21–37.

Koukoletsos, T., Haklay, M., Ellul, C., 2012. Assessing data completeness of VGI through an automated matching procedure for linear data. *Transactions in GIS* 16, 477–498. DOI: https://doi.org/10.1111/j.1467-9671.2012.01304.x

Lenders, V., Koukoumidis, E., Zhang, P., Martonosi, M., 2008. Location-based trust for mobile user-generated content: Applications, challenges and implementations, in: Proceedings of the 9th Workshop on Mobile Computing Systems and Applications, HotMobile '08. ACM, New York, NY, USA, pp. 60–64. DOI: https://doi.org/10.1145/1411759.1411775

Lush, V., 2015. Visualisation of Quality Information for Geospatial and Remote Sensing Data - Providing the GIS Community with the Decision Support Tools for Geospatial Dataset Quality Evaluation. Unpublished PhD Thesis. Aston University, Birmingham, UK.

Ma, D., Sandberg, M., Jiang, B., 2015. Characterizing the heterogeneity of the OpenStreetMap data and community. *ISPRS International Journal of Geo-Information* 4, 535–550. DOI: https://doi.org/10.3390/ijgi4020535

Meek, S., Jackson, M.J., Leibovici, D.G., 2016. A BPMN solution for chaining OGC services to quality assure location-based crowdsourced data. *Computers & Geosciences* 87, 76–83. DOI: https://doi.org/10.1016/j.cageo.2015.12.003

Meek, S., Jackson, M.J., Leibovici, D.G., 2014. A flexible framework for assessing the quality of crowdsourced data, in: Huerta, J., Schade, S., Granell, G. (Eds.), *Proceedings of the 17th AGILE International Conference on Geographic Information Science: Connecting a Digital Europe through Location and Place*, Castellón, Spain, 3–6 June 2014, Available at: https://agile-online.org/conference_paper/cds/agile_2014/agile2014_112.pdf [Last accessed 16 May 2017].

Minghini, M, Antoniou, V, Fonte, C C, Estima, J, Olteanu-Raimond, A-M, See, L, Laakso, M, Skopeliti, A, Mooney, P, Arsanjani, J J, Lupia, F. 2017. The Relevance of Protocols for VGI Collection. In: Foody, G, See, L, Fritz, S, Mooney, P, Olteanu-Raimond, A-M, Fonte, C C and Antoniou, V. (eds.) *Mapping and the Citizen Sensor*. Pp. 223–247. London: Ubiquity Press. DOI: https://doi.org/10.5334/bbf.j.

Mooney, P., Corcoran, P., 2012. The annotation process in OpenStreetMap. *Transactions in GIS* 16, 561–579. DOI: https://doi.org/10.1111/j.1467-9671.2012.01306.x

Mullen, W.F., Jackson, S.P., Croitoru, A., Crooks, A., Stefanidis, A., Agouris, P., 2015. Assessing the impact of demographic characteristics on spatial error in volunteered geographic information features. *GeoJournal* 80, 587–605. DOI: https://doi.org/10.1007/s10708-014-9564-8

Neis, P., Zielstra, D., 2014. Recent developments and future trends in volunteered geographic information research: The case of OpenStreetMap. *Future Internet* 6, 76–106. DOI: https://doi.org/10.3390/fi6010076

Neis, P., Zielstra, D., Zipf, A., 2011. The street network evolution of crowdsourced maps: OpenStreetMap in Germany 2007–2011. *Future Internet* 4, 1–21. DOI: https://doi.org/10.3390/fi4010001

Olteanu-Raimond, A-M, Laakso, M, Antoniou, V, Fonte, C C, Fonseca, A, Grus, M, Harding, J, Kellenberger, T, Minghini, M, Skopeliti, A. 2017. VGI in National Mapping Agencies: Experiences and Recommendations. In: Foody, G, See, L, Fritz, S, Mooney, P, Olteanu-Raimond, A-M, Fonte, C C and Antoniou, V. (eds.) *Mapping and the Citizen Sensor*. Pp. 299–326. London: Ubiquity Press. DOI: https://doi.org/10.5334/bbf.m.

Pesyna, K.M., Heath, R.W., Humphreys, T.E., 2015. Accuracy in the palm of your hand. Centimeter Positioning with a Smartphone-Quality GNSS Antenna. *GPSWorld* Available at http://gpsworld.com/accuracy-in-the-palm-of-your-hand/ [Last accessed 1 May 2017].

Raykar, V.C., Yu, S., 2012. Eliminating spammers and ranking annotators for crowdsourced labeling tasks. *Journal of Machine Learning Research* 13, 491–518.

See, L., Comber, A., Salk, C., Fritz, S., van der Velde, M., Perger, C., Schill, C., McCallum, I., Kraxner, F., Obersteiner, M., 2013. Comparing the quality of crowdsourced data contributed by expert and non-experts. *PLoS ONE* 8, e69958. DOI: https://doi.org/10.1371/journal.pone.0069958

See, L, Estima, J, Pőd.r, A, Arsanjani, J J, Bayas, J-C L and Vatseva, R. 2017. Sources of VGI for Mapping. In: Foody, G, See, L, Fritz, S, Mooney, P,

Olteanu-Raimond, A-M, Fonte, C C and Antoniou, V. (eds.) *Mapping and the Citizen Sensor*. Pp. 13–35. London: Ubiquity Press. DOI: https://doi.org/10.5334/bbf.b.

Senaratne, H., Mobasheri, A., Ali, A.L., Capineri, C., Haklay, M. (Muki), 2016. A review of volunteered geographic information quality assessment methods. *International Journal of Geographical Information Science* 1–29. DOI: https://doi.org/10.1080/13658816.2016.1189556

Sui, D.Z., Goodchild, M.F., Elwood, S., 2013. Volunteered Geographic Information, the exaflood, and the growing digital divide, in: Sui, D.Z., Elwood, S., Goodchild, M. (Eds.), *Crowdsourcing Geographic Knowledge*. Springer Netherlands, pp. 1–12.

Tandy, J., Barnaghi, P., Van den Brink, L., 2016. W3C. Spatial Data on the Web Best Practices. W3C Working Draft. *W3C OGC* Available at http://w3c.github.io/sdw/bp/ [Last accessed 1 May 2017].

Tang, W., Lease, M., 2011. Semi-supervised consensus labeling for crowdsourcing. Presented at the SIGIR 2011 Workshop on Crowdsourcing for Information Retri eval, Beijing, China, pp. 36–41.

Tulloch, D.L., 2008. Is VGI participation? From vernal pools to video games. *GeoJournal* 72, 161–171. DOI: https://doi.org/10.1007/s10708-008-9185-1

van Exel, M., Dias, E., Fruijtier, S., 2010. The impact of crowdsourcing on spatial data quality indicators, in: Wallgrün, J.O., Lautenschütz, A.K. (eds.), *Proceedings of the GIScience 2010 Doctoral Colloquium*, Zurich, Switzerland, 14–17 September 2010, Akademische Verlagsgesellschaft Aka GmbH (IOS Press), Heidelberg, Germany. Available at: http://www.giscience2010.org/pdfs/paper_213.pdf [Last accessed 16 May 2017].

Vandecasteele, A., Devillers, R., 2015. Improving volunteered geographic information quality using a tag recommender system: The case of OpenStreetMap, in: Jokar Arsanjani, J., Zipf, A., Mooney, P., Helbich, M. (Eds.), *OpenStreetMap in GIScience, Lecture Notes in Geoinformation and Cartography*. Springer International Publishing, Cham, Switzerland, pp. 59–80.

Vuurens, J.B.P., de Vries, A.P., 2012. Obtaining high-quality relevance judgments using crowdsourcing. *IEEE Internet Computing* 16, 20–27. DOI: https://doi.org/10.1109/MIC.2012.71

Waldner, F., Fritz, S., Di Gregorio, A., Defourny, P., 2015. Mapping Priorities to Focus Cropland Mapping Activities: Fitness Assessment of Existing Global, Regional and National Cropland Maps. *Remote Sensing* 7, 7959–7986. DOI: https://doi.org/10.3390/rs70607959

Zielstra, D., Zipf, A., 2010. A comparative study of proprietary geodata and volunteered geographic information for Germany. Presented at the 13th AGILE International Conference on Geographic Information Science 2010, Guimarães, Portugal, pp. 1–15.

Zook, M.A., Graham, M., 2007. The creative reconstruction of the Internet: Google and the privatization of cyberspace and DigiPlace. *Geoforum* 38, 1322–1343. DOI: https://doi.org/10.1016/j.geoforum.2007.05.004

CHAPTER 8

The Impact of the Contribution Micro-environment on Data Quality: The Case of OSM

Vyron Antoniou* and Andriani Skopeliti†

*Hellenic Army General Staff, Geographic Directorate, PAPAGOU Camp, Mesogeion 227-231, Cholargos, 15561, Greece, v.antoniou@ucl.ac.uk

†School of Rural and Surveying Engineering, National Technical University of Athens, 9 H. Polytechniou, Zografou, 15780, Greece

Abstract

OpenStreetMap (OSM) is the most successful example of Volunteered Geographic Information (VGI). It is also the most frequently used case study in research that focuses on VGI quality, as it is usually considered a proxy for other VGI projects. The research in this area usually focuses on comparisons with authoritative data, measurements and quality statistics. In other papers, scholars have explored quality frameworks or studied the motivation and engagement of volunteers. This chapter examines OSM quality from a different point of view. The focus here is on examining how the qualitative elements of the micro-environment within OSM, such as data specifications and the OSM editors, have evolved over time. We discuss how their evolution can affect OSM data quality, taking into account a number of different factors and dimensions that directly affect the quality of the contributions.

How to cite this book chapter:

Antoniou, V and Skopeliti, A. 2017. The Impact of the Contribution Micro-environment on Data Quality: The Case of OSM. In: Foody, G, See, L, Fritz, S, Mooney, P, Olteanu-Raimond, A-M, Fonte, C C and Antoniou, V. (eds.) *Mapping and the Citizen Sensor*. Pp. 165–196. London: Ubiquity Press. DOI: https://doi.org/10.5334/bbf.h.

Keywords

Spatial data quality, OSM specifications, OSM editors, quality evaluation

1 Introduction

OpenStreetMap (OSM) is one of the first examples of Volunteered Geographic Information (VGI; Goodchild, 2007), and continues to be one of its prime examples. VGI has been defined as 'the widespread engagement of large numbers of private citizens, often with little in the way of formal qualifications, in the creation of geographic information' (Goodchild, 2007). A number of factors have helped this phenomenon to grow, including the removal of the selective availability of the Global Positioning System (GPS) in 2000 (Clinton, 2000), which has resulted in the proliferation of GPS-enabled devices, novel Web 2.0 practices and programming techniques as well as the development of spatial applications and products based on global-wide maps of satellite imagery by technology giants such as Google, Microsoft and Yahoo!. Since 2007, VGI has become intertwined with crowdsourcing, active local communities and social media, and thus can be found in many flavours and extracted from various sources (for more details, see Chapter 2 by See et al., 2017), such as web applications about toponyms, GPS tracks, sharing of geotagged photographs, synchronous micro-blogging, social networking sites, etc. A very interesting, and equally promising, interconnection of VGI is the one with the domain of citizen science (Haklay, 2013). As the latter gains momentum, the need for geotagged measurements and information is growing, and along with it the quest for solid answers about the caveats and challenges that VGI projects face, especially with respect to data quality. Thus, understanding how the most successful VGI project (i.e. OSM) has evolved in terms of quality will give insights valuable to other existing VGI projects or projects that will follow in the future, including those in the citizen science domain. Spatial data quality is the cornerstone of every spatial database, map, product or service. Measuring, understanding and documenting the quality of spatial data is of paramount importance for any kind of geodata, including VGI.

This chapter will examine OSM quality evolution from a new point of view. In Section 2, quality evaluation procedures, as described in the ISO quality framework, will be discussed. Then, in Section 3, the methodology for understanding the evolution of OSM quality will be introduced. The central focus will not be on the data themselves (as is usually the case in most OSM-based quality studies), but rather on the micro-environment inside which OSM is evolving. To this end, Section 4 will cover the evolution of OSM specifications, taking into account a number of different factors and dimensions that directly affect the quality of contributions; in Section 5, the evolution of OSM editors will be examined, as they are literally the entry point for all OSM contributions. Both

Sections will provide a critical view of the developments on these two fronts and of their impact on the overall quality of OSM. The chapter will conclude with a discussion of and conclusions on how all of these aspects can provide a useful context for OSM quality evaluation.

The purpose of this chapter is not to provide measurements or quantitative reports regarding the quality of OSM. Instead, the aim is to highlight new, important facets of OSM quality that have not been considered to date in what is otherwise a rich and growing literature on VGI quality. This chapter supports the idea that the evolution of OSM data quality is closely related to qualitative elements of the OSM micro-environment. These include the wiki-based and thus bottom-up build and constantly changing specifications, the digitisation software (i.e. the OSM editors), the mapping parties, the forums, the voting system, the local and global OSM communities, the few, yet most productive, contributors, and other seemingly small and unimportant factors that in reality determine to a great extent the evolution of the OSM initiative and consequently the quality of the data created. All of these factors are outside the traditional quality elements for spatial data (ISO, 2005) or even the new quality indicators suggested specifically for VGI (see Antoniou and Skopeliti, 2015 for an overview of these). This chapter focuses on two of these outside factors: OSM specifications and OSM editors.

2 Spatial Data Quality Evaluation Procedures

This book provides considerable material on the subject of spatial data quality. For example, in Chapter 7, Fonte et al. (2017) discuss VGI quality and review measures and indicators for this new breed of data. In Chapter 9, Skopeliti et al. (2017) discuss best practices and methods for visualising VGI quality, while Chapter 10, by Minghini et al. (2017), discusses best practices for data collection, including quality considerations. Finally, in Chapter 13, Olteanu-Raimond et al. (2017) examine the experience of European National Mapping Agencies (NMAs) with VGI data and discuss methods for obtaining contributions of high quality from volunteers.

Both in this book and in the literature available on the subject of VGI quality, most VGI cases or examples come from the OSM project. OSM is a prime example of VGI as it has managed to provide free, constantly updated, crowdsourced data for the globe. However, when research focuses on VGI data quality, scholars tend to examine some of the spatial quality elements for a given study area, e.g. cities, urban areas or nationwide (Antoniou, 2011; Girres and Touya, 2010; Haklay et al., 2010; Jokar Arsanjani et al., 2015). The studies usually follow a benchmark evaluation process, which involves creating a copy of what is a continuously changing dataset, and then evaluating this copy as if it were a static dataset. This method gives insight into the data quality at the time when the copy was created; thus, these efforts provide a good understanding of selected quality elements at a given point in time compared with corresponding authoritative datasets. However, spatial datasets, and especially VGI ones, are not static products and hence time is a critical factor that is not often considered. The starting point for a spatial product is the specifications that will be used to create the dataset. Yet these specifications can change over time for both authoritative and VGI datasets. In fact, the latter kind of Geographic Information (GI) is more susceptible to changes in specifications since bottom-up processes provide the flexibility for new rules to be established or existing ones deprecated more easily by the community of volunteers. While the path of evolution and change in the specifications of a product is inescapable, there is a fundamental difference in how each source of GI (i.e. authoritative or VGI) handles their dataset life-cycle. For example, authoritative data, collected by NMAs or Commercial Mapping Companies (CMCs), usually follow a versioning system. Users of such data are notified that a set of updates is available or, more relevant to our case, that a new dataset has been created based on new specifications. The product specifications can also be available to the interested parties. A case in point can be found in the practices of the UK's Ordnance Survey (OS). For the OS MasterMap product (OS 2001), for example, OS provides a detailed document that explains how each physical entity is conceived, modelled and stored and thus what accuracy and attributes should be expected. The important point here is that while a new dataset is developed, or during the migration from one form of specification to another, the datasets are not accessible to the users. This process takes place in-house, and only when the whole process has been concluded are the data available for use. This is in contrast with what takes place with VGI. In a sense, VGI datasets are following one of the main characteristics of Web 2.0 (O'Reilly, 2007), i.e. *perpetual beta*. This small phrase is usually applied to software development cycles, and means that there are no versioning cycles but rather a continuous effort of software development so as to match evolving user needs; here this notion spills over to datasets, and OSM is an excellent example for monitoring this. The perpetual editing of and changes to OSM specifications has made OSM evolve from a dataset with a handful of layers and physical features to an extremely detailed dataset, in many cases far more detailed than any NMA or CMC dataset. The difference between VGI and authoritative data is that in VGI while the evolution of datasets takes place the actual data are available without any guarantees or indications regarding the state or compliance of each feature in relation to a specification's version. It is not difficult to imagine that this process, while it has many advantages, can create a series of inconsistencies and, in fact, deteriorate the overall quality of the data.

Thus, while specification improvements might eventually be a necessary step for a better, more inclusive, detailed and meaningful dataset, during the transition time, the dataset is bound to suffer from inconsistencies, mixed feature versions and mixed typologies that exist in former and latter specifications. This is even more likely if there is a perpetual change in specifications without any rigorous provision on how to manage the data transition and compliance.

Returning the discussion to quality evaluation processes, benchmark comparisons are usually chosen not because they are necessarily the best way to evaluate the data quality of a VGI dataset but because they are the most practical to perform and report. ISO (2005) explains that benchmark procedures should be based on the establishment of a suitable reporting frequency. Sporadic and non-systematic evaluations, although perfectly acceptable in an academic environment, do not provide a clear view of OSM quality, or of the quality of any other VGI source. To this end, a different approach suggested by the ISO quality framework is to evaluate constantly changing datasets, as is the case of OSM data, using a continuous process. Here, the starting point could again be a benchmark test, but then there should be a continuous evaluation of the updates and of the impact that these updates might have on the overall dataset. However, there is no provision made for specification migration, perhaps because this sense of perpetual editing is not applicable to authoritative data.

3 Methodology

To evaluate OSM evolution from a quality point of view, we need to consider what process to use. A way forward is to follow one of the two ISO suggestions. This means that we need to develop a benchmarking method that will be able to examine an instance of the OSM data against an authoritative dataset on a regular basis (e.g. weekly, monthly, etc.). For a number of reasons, this is not straightforward. First, there is no global-scale authoritative dataset that could play the role of the reference data. Even if such datasets were available for academic research, it is not clear which one would be more detailed and at which places. For example, Vandecasteele and Devillers (2015) report that in many places OSM is far more detailed than any authoritative dataset available. Moreover, such an approach would require the implementation of considerable amounts of brute force computing on a regular basis. This approach would be possible in the context of confined academic experiments that would test either a few quality elements at a national level or all the quality elements for small areas, but it would be difficult to achieve and maintain both globally and regularly. The same applies to a continuous evaluation process, although the evaluation of the quality of OSM updates is a more straightforward task, given the fact that OSM provides regular updates in separate files and for various time intervals. However, the frequency of updates is inversely related to the number of changes, so, for practical reasons, evaluating the data quality continuously is beyond the means of most NMAs or CMCs.

Hence, an alternative approach is taken here, which is based on the evaluation of factors that directly affect OSM quality but are currently not studied by researchers, i.e. a study of the OSM specifications. The value of specifications in VGI has been discussed by Brando and Bucher (2010) and by Brando et al. (2011). The form of, and the rules included in, a product's specification, at any given point in time, is fundamental. This, along with metadata, is the starting point that allows potential users to understand the usability of the data. Monitoring and documenting the changes that have taken place in the specification of OSM over time could add another tool to the toolbox used for OSM quality evaluation, and could provide the necessary context for some of the academic efforts in this field.

Moreover, this approach will be coupled with an evaluation of the evolution of OSM editors. OSM contributions are uploaded through a number of OSM editors that have been developed and updated by the OSM community itself. The editing tools and the overall functionality of the editor, and, more importantly, the editor's conformance to the wiki specifications, play a significant role in the kind of edits submitted and consequently in the quality of the data contributed.

4 Evolution of OSM Specifications

4.1 General Changes to the Main OSM wiki Page

OSM specifications are described in a wiki-based process. The starting point is a MediaWiki[1] web page titled '*Map Features*' (OpenStreetMap, 2016). This page lists all of the physical features that should be included in the OSM database, along with some of the basic attributes that should describe each feature. The OSM community decides what is added or removed from this list through a voting system. In the OSM world, the features are called *keys* and the attributes *values*. In the '*Map Features*' web page, the physical features are grouped into categories and sub-categories depending on their semantics and nature. For each feature, additional information is available, such as the type of geometry that should be used (i.e. node, way or area), comments on what each feature represents, assisting documentation from Wikipedia, a photograph that shows how the feature appears on the OSM map and a photograph that functions as a photo-interpretation key. The latter photograph helps the contributors to better understand how to assign features on the ground to the OSM nomenclature. Moreover, each key/tag combination is further explained in other wiki pages, which themselves include more details about the way the feature should be digitised, additional attributes that could further describe the feature, and the possible combinations of the attributes.

For web pages created with MediaWiki, it is possible to access the pages' history and trace back what changes have been made, at which time and by whom. Moreover, a short summary of the changes is available, along with a classification of whether a change was a minor edit or not (computed based on whether the person who performs the edit has marked the edit as minor or not[2]). Thus, in order to understand how this (quasi) specification of OSM has evolved, we examined how the '*Map Features*' page has changed over time. At the time of writing (May 2016), there were 847 versions of this wiki page alone, with the first one dating back to 20 December 2005. This means that a major or minor edit has taken place approximately every 4.4 days since on average.

The first point of analysis was to examine when each version was released. Figure 1 shows the number of changes per year and the corresponding percentage. This provides a good understanding of whether OSM specifications are constantly changing or if there are any emerging patterns. Figure 1 shows that most of the changes (88%) have taken place in the first three years of OSM's life, while, from 2011 onwards, each year's overall changes do not exceed 2% of the total of changes. This is an interesting observation as it paints a picture of a crowdsourced product that has matured extremely fast compared to the breadth and length of its aims (i.e. to 'create and distribute free geographic data for the world'[3]).

The next step is to analyse the importance of these changes. Taking into account the automatic assignment of an edit into minor or not, we explored when and how many edits take place each year for each kind of change. It is understandable that the number of characters changed cannot be an entirely safe measure of a change's importance. However, it is considered as a good indicator that can give a basic understanding of the amount of work put forward in every change. Figure 2 presents the percentage of major and minor changes per year. Despite being a fast maturing product as noted above, major changes in the specifications take place constantly. This observation should be considered in combination with that of the flexibility provided to contributors, which is in line with the openness and spirit of inclusiveness that characterises the OSM project. For example, in the wiki-forums it is explicitly stated that the OSM community might introduce best practices, guidelines or even deprecated features and attributes and that nothing is banned. Contributors are free to add whatever they believe will better describe the physical world.

Thus, inconsistencies and mismatches in the keys and values used can come from both a 'formal' change in the specifications and the free key/tag combination choice available to users. Interestingly, in the case when changes in the specification are introduced, automatic correction of the existing features is highly discouraged; the rules state: 'Under no circumstances should you automatically (or semi-automatically) change "deprecated" tags to something else in the database on a large scale without conforming to the Automated Edits code of conduct. Any such edits will be reverted'[4].

4.2 Development of Feature Specifications

The analysis so far has provided an initial overview of OSM specification's development over time. Now the focus turns to the actual changes that took place. For practical reasons, a selection of some of the 847 '*Map Features*' page versions had to be made in order to use them for comparison. The

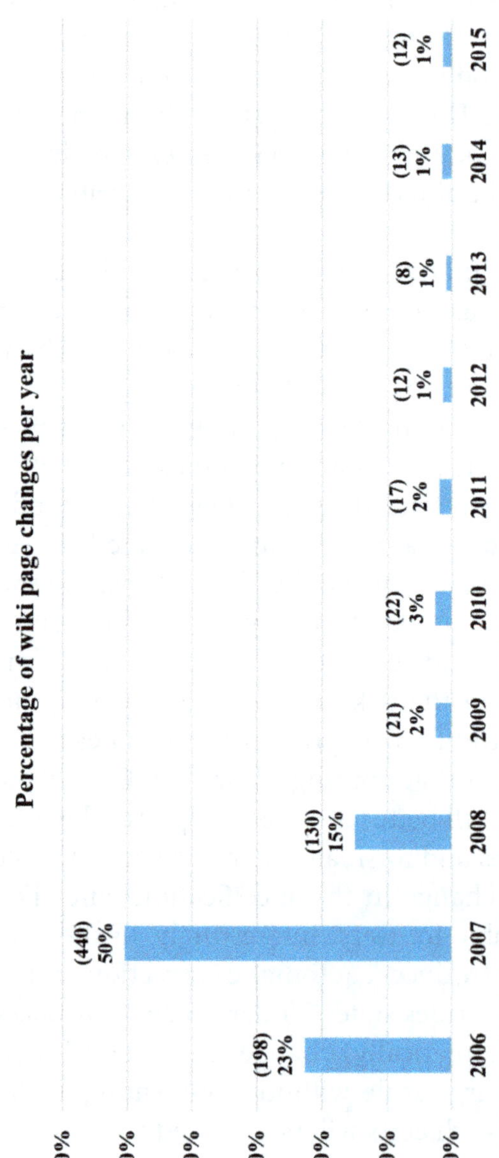

Fig. 1: Number of changes per year for the OSM Map Features wiki page.

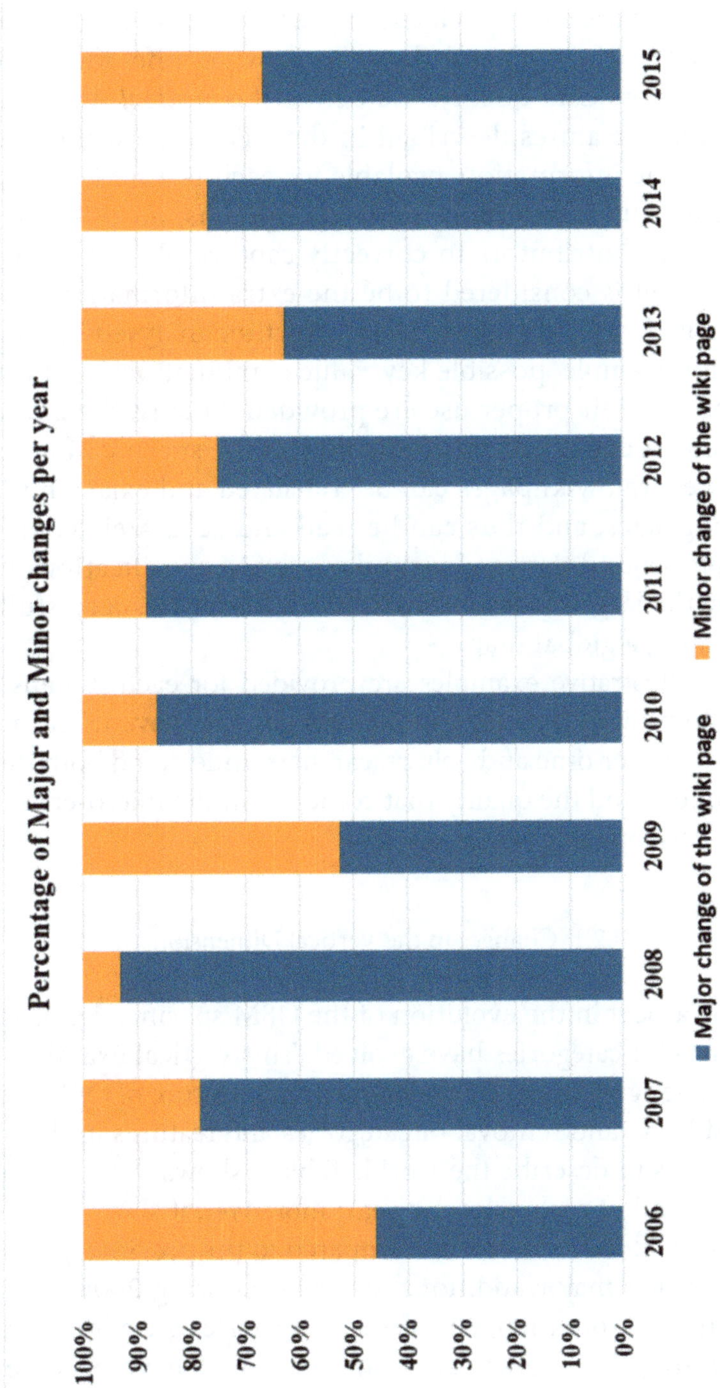

Fig. 2: Major and minor changes to the Map Features wiki page.

versions selected were those closest to the end of each calendar year from 2006 up until 2015. Then, in order to better monitor the development of the specification, we examined the alterations that took place in four dimensions: the *vertical, horizontal, in-depth* and *internationalisation* dimensions. All four dimensions are closely related to the OSM data (in fact are different aspects of the OSM content) and thus can provide a helpful point of view in the effort to assess data quality. We define the *vertical* dimension as the number of physical features described in the wiki page, while the *horizontal* dimension is the information available for each feature (i.e. keys, values, comments, rendering instructions and photographs; all of these are helpful in guiding the contributors to correctly capture physical features). The *in-depth* dimension is considered to be the extra information available for each feature: both keys and tags are usually further analysed in separate wiki pages where, for example, possible key/value combinations or more detailed instructions about their proper use are provided. Finally, the *internationalisation* dimension is defined as the availability of the specification in different languages. In general, wiki pages can be translated and exist simultaneously in different languages, and thus can be read and accurately comprehended by many people around the world; similarly, OSM specifications need to be understood by the largest possible audience in order to successfully achieve the aim of creating a global map.

A number of illustrative examples are provided for each dimension. These examples aim to provide a picture of the changes that have taken place in the OSM specification over time and help researchers understand both the volatility in the contributions and the quality that comes from the micro-environment in which OSM is developing.

4.2.1 Changes in the Vertical Dimension

One interesting aspect in the evolution of the OSM specification is to examine how the major OSM categories have evolved. This vertical examination of the '*Map Features*' page gives a sense of how the nomenclature of OSM has changed through the addition and removal of categories and features in the list of entities that OSM uses to describe the world. Table 1 shows the number of active categories at the end of each calendar year; moreover, it shows how many categories have been added or removed compared to the previous year.

It can be seen that major additions took place during 2008, where 48 categories were added. From then, new feature categories are added almost every year, but interestingly there are also categories that have been removed as independent typologies in the nomenclature of OSM and have been merged with others. Examples of the categories added include *power* and *shop* in 2007, *facilities, education* and *transportation* in 2008, *geological* in 2009, *emergency*,

Table 1: Additions and removals of OSM categories from the Map Features wiki page.

	2006	2007	2008	2009	2010	2011	2012	2013	2014	2015
Categories Present	28	32	78	83	90	97	91	93	96	93
Categories Removed*		0	2	0	0	0	4	0	0	3
Categories With a Name Change*		0	0	0	0	0	2	0	0	0
Categories Added*		4	48	5	7	7	0	0	3	0
**compared to the previous year*										

medical rescue and *firefighters* in 2010, *commercial* and *civil amenity* in 2011 and *traffic calming* in 2014. Examples of removals include the categories of *cycleway*, *tracktype*, *abutters* and *naming* in 2012.

Apart from the changes in the major OSM categories, there have also been changes recorded to the features in each category. Tables 2, 3 and 4 present illustrative examples of how selected features have evolved over time. More specifically, Table 2 shows the sub-categories of *Highways* and *Places* as well as the number of distinct features included in each of these sub-categories. It can be seen that, for these two major categories, which in fact include all road network and all gazetteer data, there have not been any changes since 2008. This does not mean that there have not been changes in the wiki pages that further explain the attributes of each distinct feature, but that at least at this high level the nomenclature has been stable since 2008. The flip side is that while the geometry (i.e. positional accuracy) of the road network or places might still be correct, since they have not been updated since 2007 it is likely that they might suffer from attribution inconsistencies that affect their thematic accuracy and logical consistency.

Table 3 shows how the *Buildings* category has evolved. Here again, at the sub-categories level and in terms of the number of features per sub-category, *Buildings* have been stable since 2011. The interesting point here is that this major category, which includes the footprints of buildings, was introduced in OSM in 2011. Thus, areas that have not been updated since 2011, either because there was a bulk upload in the past or because the area was mapped by a very productive user that did not return to update it (for more, see Antoniou and Schlieder, 2014), would probably not have this type of feature, since capturing buildings was out of the scope of OSM before 2011.

Table 2: The number of sub-categories and distinct features (keys) included in the Highways and Places main OSM categories from 2006 to 2015.

Primary Feature Category	Feature Sub Category	2015	2014-2009	2008	2007	2006
		Distinct Features (Keys)				
Highways	Roads	8	No change	8	47*	42*
	Link roads	5		5		
	Special road types	6		6		
	Paths	4		4		
	When sidewalk (or pavement) is tagged on the main roadway	1		1		
	When cycleway is drawn as its own way	1		1		
	Cycleway tagged on the main roadway or lane	8		8		
	Lifecycle	2		2		
	Attributes	27		27		
	Other highway features	18		18		
Places	Administratively declared places	7	No change	7	15*	15*
	Populated settlements, urban	7		7		
	Populated settlements, urban and rural	6		6		
	Other places	6		6		
	Additional attributes	6		6		

* *Different groupings and typologies used for OSM Keys*

Finally, Table 4 shows the changes in the *Additional Properties* category. This category was introduced in 2012 as a successor to the *Naming* category, and includes important features and information such as *Addresses*, *Annotation* and *Name*. However, it can be seen that there are frequent and important changes in OSM typology that make it difficult for contributors to follow all the specification's provisions. For example, *Addresses* did not exist until 2008; it was later added to the *Naming* category, and then, in 2012, it was re-assigned to *Additional Properties*. Similarly, *Place* was removed from the *Additional Properties* category and formed a new one.

Table 3: The number of sub-categories and distinct features (keys) included in the Buildings main OSM category from 2006 to 2015.

Primary Feature Category	Feature Sub Category	2015	2014	2013	2012	2011	2010	2009	2008	2007	2006
		Distinct Features (Keys)									
Buildings	Accommodation	11	11	11	11	11					
	Commercial	4	4	4	4	4					
	Civic/Amenity	15	15	15	15	15	***Did not exist***				
	Other Buildings	23	23	23	23	23					
	Additional Attributes	4	4	4	4	4					

Table 4: The number of sub-categories and distinct features (keys) included in the Additional Properties main OSM category from 2006 to 2015.

Primary Feature Category	Feature Sub Category	2015	2014	2013	2012*	2011	2010	2009	2008	2007	2006
		Distinct Features (Keys)									
Additional properties	Addresses										
	Tags for individual houses	10	10	10	10	10**	10**	10**	***Did not exist***		
	For countries using hamlet, subdistrict, district, province, state	6	6	6	6	6**	6**	6**			
	Tags for interpolation ways	3	3	3	3	3**	3**	3**			
	Annotation	23	23	23	23	23	23	23	23	*11****	*9****
	Name	13	13	13	13	13	13	13	13	7	6
	Properties	36	36	36	36	*36***	*36***	*36***	*36***	*16****	*10****
	References	12	12	12	12	12	12	12	12	12	8
	Restrictions	40	40	40	40	*40***	*40***	*40***	*40***	*26****	*28****
	Places	***New Primary Feature Category***				32	32	32	32	11	15
	Editor keys	***Did not exist***								2	***Did not exist***

** Change in the name of the Primary Feature Category*
*** Included in another Primary Feature Category*
**** As stand-alone Primary Feature Category*

Apart from the distinct feature keys that have been added or removed over time, major changes in how the OSM community models the world took place in 2008 and 2012. In 2006, the world, according to OSM, was divided into a number of major categories: *Physical, Non Physical, Abutters, Accessories, Properties, Restrictions, Naming* and *Annotation*. During the next year, these major categories were further enriched with sub-categories, and then, in the following year, there was another typology. Indeed, in 2008 there were only three major categories: *Physical, Non-Physical* and *Naming*. The first category went from including 17 sub-categories to including 59, while the second included as sub-categories all the major categories of 2007 apart from those specifically related to the naming process (e.g. *Name, References, Places, Annotation*, etc.), which were assigned to the last main category.

In 2012, the features were re-assigned into two new major categories: *Primary Features* and *Additional Properties*. The *Physical* sub-categories were added to the former category, but it also included sub-categories from the *Non-Physical*, such as *Route, Boundary* and *Sport*. The latter category remained with six main sub-categories: *Addresses, Annotation, Name, Properties, References* and *Restrictions*. Also, in 2012, some major changes took place regarding the grouping of the physical entities in various sub-categories and classes. For example, the entity *Places*, which used to be a class under the *Naming* sub-category in 2011, became an independent sub-category in 2012 below the *Primary Features*, while the *Naming* sub-category was assigned to the *Additional Properties* category. Furthermore, during the study period (i.e. 2006–2015), considerable volatility was recorded in some sub-categories. A case in point is the *Naming* sub-category, which listed 3 features in 2007, 9 features in 2008 and 13 features in 2009 (before it was split again in 2012).

While these are only some illustrative, and perhaps confusing, examples of the changes recorded in the OSM specification, two things are evident with respect to the commitment of contributors. First, for OSM contributors that have been consistently contributing during the entire period, it should have been difficult to meticulously follow all of the changes; thus, it should not come as a surprise that even experienced users might have introduced errors and inconsistencies in the data. On the other hand, there are either occasional contributors or contributors that have just a short active period and never contribute again; for both of these types of contributors, the best case scenario would be that contributors have consulted the active specification at a specific point in time and collected the data based on this version. In the worst case, the contributions were based on previous knowledge and understanding of the specification. In any case, and taking into account the fact that automatic corrections are discouraged, it is highly likely that a considerable number of contributions are out of date in terms of specification compliance. This also puts quality frameworks that are based on contributor evaluation under fresh scrutiny (see e.g. D'Antonio et al., 2014; van Exel et al., 2010).

4.2.2 Changes in the Horizontal Dimension

The '*Map Features*' page, apart from the addition and removal of new categories, sub-categories and features, has also changed in terms of the available information for each of these categories and features. While modest changes have been recorded compared to the vertical dimension, this horizontal dimension still plays a significant role in the rules and information that volunteers are equipped with when collecting data and contributing to the project.

Two illustrative examples are presented to show the evolution in the horizontal dimension. The first example (Figures 3 and 4) shows one of the major physical entities: *Highways*. Even from the early days of the OSM project, it was made clear that volunteers needed as much information as possible in order to be able to unequivocally distinguish between and capture various physical entities. However, the actual information available was not enough for safely guiding volunteers. For example, at the end of 2006 (Figure 3), the main feature-attribute combination, which is a description of what each feature name represents and how features are portrayed on the OSM map, became available. Thus, in practice, a volunteer could use only the short description as a guide for interpreting the entity before digitising and assigning it to the correct category. For more information, the volunteer would have had to follow a link attached to the *Highway* key. At the end of 2006, a small number of photographs and basic information was available so as to guide the contributors. It is obvious that the incomplete description of each feature, although it does not stop contributors collecting the data, makes the collection error prone in terms of thematic and logical consistency, and especially so at a time when satellite imagery was not so common and was of low resolution when it was available.

In contrast, Figure 4 shows the current specification section of *Highways*. The available information for each physical feature has expanded to include a photo-interpretation key that can more easily guide contributors. Furthermore, apart from the link attached to the *highway* key, which links to a page more detailed than the 2006 one, each *value* also has its own wiki page (see also Section 4.2.3). In these pages, more details are provided regarding what is preferable for the volunteers to follow and what to avoid. Moreover, a wide list of possible key-value combinations is provided, with explanations and examples.

A similar example is provided by contrasting the 2006 and 2015 wiki pages on *aerialways* (Figures 5 and 6). As this feature is not one of the fundamental entities of a base map, there was only a basic description of it in 2006 (Figure 5; note also that the structure of the table is different from that of the table for the *highways* of 2006). In contrast, in 2015 (Figure 6), the available information is as complete as that of the *highways*. Moreover, the comments are supported by Wikipedia articles and some basic instructions are given about the key-value information.

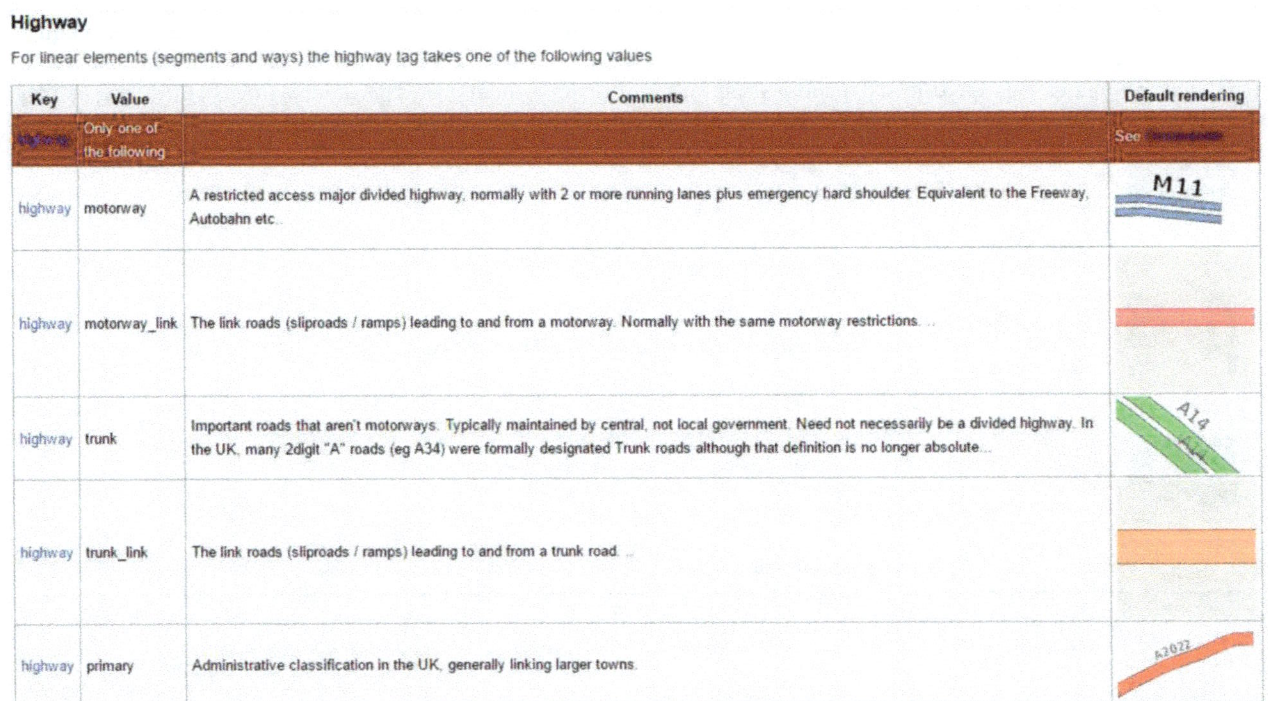

Highway

For linear elements (segments and ways) the highway tag takes one of the following values

Key	Value	Comments	Default rendering
highway	Only one of the following		See
highway	motorway	A restricted access major divided highway, normally with 2 or more running lanes plus emergency hard shoulder. Equivalent to the Freeway, Autobahn etc..	M11
highway	motorway_link	The link roads (sliproads / ramps) leading to and from a motorway. Normally with the same motorway restrictions. ..	
highway	trunk	Important roads that aren't motorways. Typically maintained by central, not local government. Need not necessarily be a divided highway. In the UK, many 2digit "A" roads (eg A34) were formally designated Trunk roads although that definition is no longer absolute...	A14
highway	trunk_link	The link roads (sliproads / ramps) leading to and from a trunk road. ..	
highway	primary	Administrative classification in the UK, generally linking larger towns.	A2022

Fig. 3: Part of the Map Features wiki page (end of 2006) that specifies various types of roads that should be captured (includes key, value, comments and default rendering of the entity on the OSM map). @OpenStreetMap contributors.

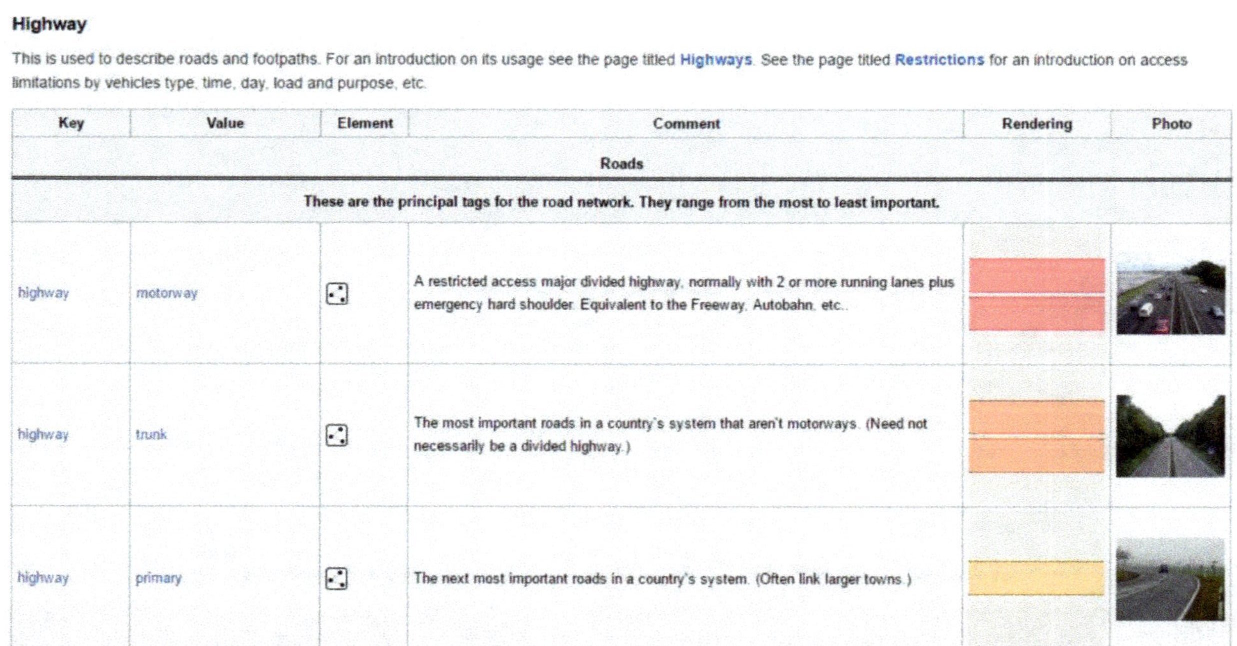

Highway

This is used to describe roads and footpaths. For an introduction on its usage see the page titled **Highways**. See the page titled **Restrictions** for an introduction on access limitations by vehicles type, time, day, load and purpose, etc.

Key	Value	Element	Comment	Rendering	Photo
Roads					
These are the principal tags for the road network. They range from the most to least important.					
highway	motorway		A restricted access major divided highway, normally with 2 or more running lanes plus emergency hard shoulder. Equivalent to the Freeway, Autobahn, etc..		
highway	trunk		The most important roads in a country's system that aren't motorways. (Need not necessarily be a divided highway.)		
highway	primary		The next most important roads in a country's system. (Often link larger towns.)		

Fig. 4: Part of the Map Features wiki page (end of 2015) that specifies various types of roads that should be captured (includes key, value, geometric element to be used, comments, default rendering of the entity on the OSM map and a photo-interpretation key). @OpenStreetMap contributors.

Aerialway

Feature	Feature type	Key	Value	Comments
Linear	Physical	aerialway	Only one of the following	
Linear	Physical	aerialway	cable_car	
Linear	Physical	aerialway	chair_lift	
Linear	Physical	aerialway	drag_lift	
Linear	Physical	aerialway	overhead_lines	
Linear	Physical	aerialway	User Defined	
Linear	Physical	User Defined	Any of the following	
Linear	Physical	User Defined	User Defined	

Man Made

Fig. 5: Part of the Map Features wiki page (end of 2006) that specifies how aerialways should be captured (includes geometry to be used, category, key, value and an empty comments column). @OpenStreetMap contributors.

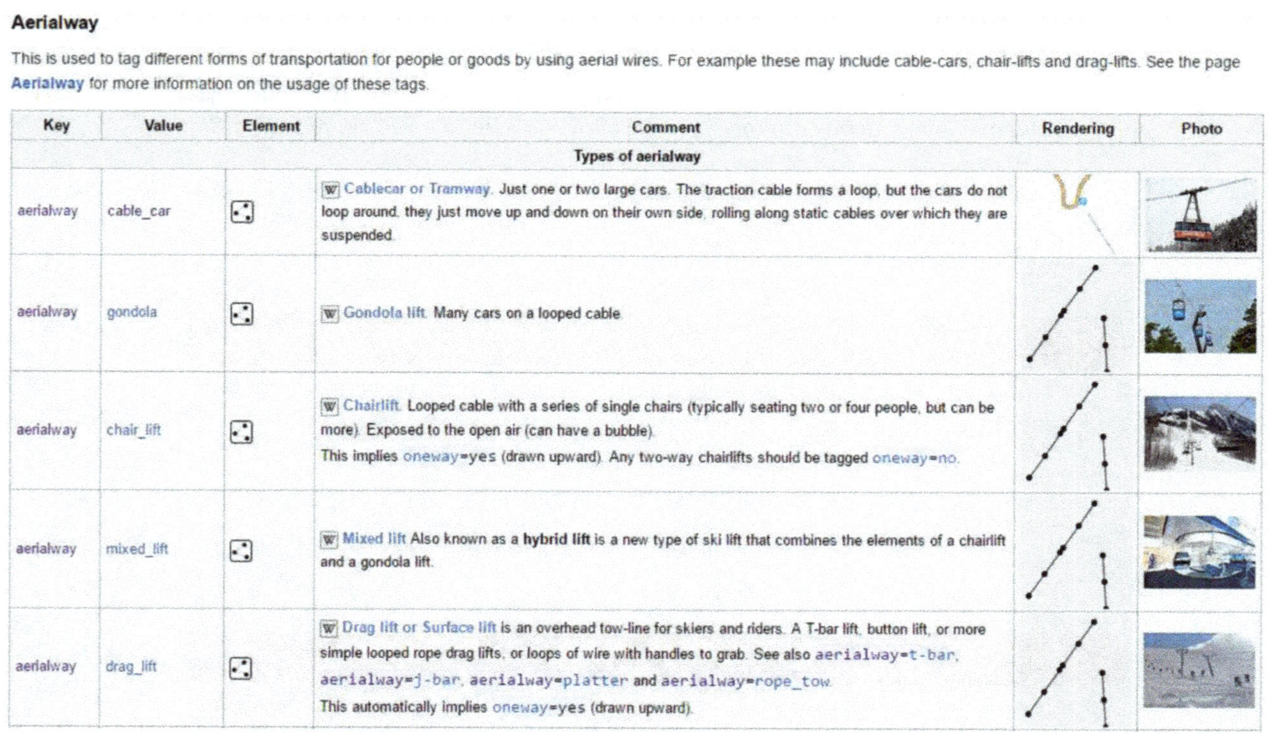

Aerialway

This is used to tag different forms of transportation for people or goods by using aerial wires. For example these may include cable-cars, chair-lifts and drag-lifts. See the page **Aerialway** for more information on the usage of these tags.

Key	Value	Element	Comment	Rendering	Photo
			Types of aerialway		
aerialway	cable_car		Cablecar or Tramway. Just one or two large cars. The traction cable forms a loop, but the cars do not loop around, they just move up and down on their own side, rolling along static cables over which they are suspended.		
aerialway	gondola		Gondola lift. Many cars on a looped cable.		
aerialway	chair_lift		Chairlift. Looped cable with a series of single chairs (typically seating two or four people, but can be more). Exposed to the open air (can have a bubble). This implies oneway=yes (drawn upward). Any two-way chairlifts should be tagged oneway=no.		
aerialway	mixed_lift		Mixed lift Also known as a **hybrid lift** is a new type of ski lift that combines the elements of a chairlift and a gondola lift.		
aerialway	drag_lift		Drag lift or Surface lift is an overhead tow-line for skiers and riders. A T-bar lift, button lift, or more simple looped rope drag lifts, or loops of wire with handles to grab. See also aerialway=t-bar, aerialway=j-bar, aerialway=platter and aerialway=rope_tow. This automatically implies oneway=yes (drawn upward).		

Fig. 6: Part of the Map Features wiki page (end of 2015) that specifies how aerialways should be captured (includes key, value, geometric element to be used, comments with references from Wikipedia, default rendering of the entity on the OSM map and a photo-interpretation key). @OpenStreetMap contributors.

We have used these two examples to highlight the evolution of the OSM specification. From 2006 to 2015, each feature followed its own pace regarding the available information provided to the OSM community. Thus, the quality of the contributions for each feature could have varied accordingly. The mobilisation of thousands of enthusiastic, yet mostly inexperienced, contributors has inevitably led to 'learning-by-doing' in the face of incomplete and changing specifications.

4.2.3 Changes in the In-depth Dimension

The in-depth dimension of the '*Map Features*' has been briefly discussed in the previous section. It refers to the available information for each key/value combination and the attribution process that contributors should follow. As explained, each physical entity has developed independently and the level of detail might vary considerably at different time periods. Here we provide one example to illustrate changes: *unclassified* roads. Figure 7 shows the *unclassified* roads wiki page at the end of 2008, which included the basic information regarding the mapping of the *highway=unclassified* combination.

In contrast, the same page at the end of 2015 (Figure 8) includes more detailed information about the preferable attributes that can be assigned to this entity as well as instructions about how to map the entity, when it is applicable, situations where other tags should be used, examples of determining applicability and even disambiguation instructions when the public/private status is unclear.

4.2.4 Changes in Internationalisation

Right from the beginning of the project, OSM aspired to create a global and free map. It is obvious that this could not be achieved without global participation. When examining the internationalisation of OSM, we can see that the '*Map Features*' page is currently (i.e. in May 2016) available in 49 languages (Table 5). Although there has been no calculation regarding the percentage of the global population covered, it is clear that the basic rules of OSM can be understood by a broad audience. However, this was not always the case. Until the end of 2009, the '*Map Features*' page was only available in English. From the end of 2010, however, until 2015, the number of available languages was 45.

Apart from the '*Map Features*' page, which is the starting point of the specification, there are documentation pages for each OSM key and value in order to better explain the use cases and the most appropriate combinations. These pages should also be available in as many languages as possible. However, their availability varies and, in general, there are considerably fewer available languages than for the '*Map Features*' page. For example, the key *aerialway* is available in 10 languages (čeština, deutsch, english, italiano, magyar, polski, português do Brasil, русский, 한국어 and 日本語) while the combination *amenity=cafe*

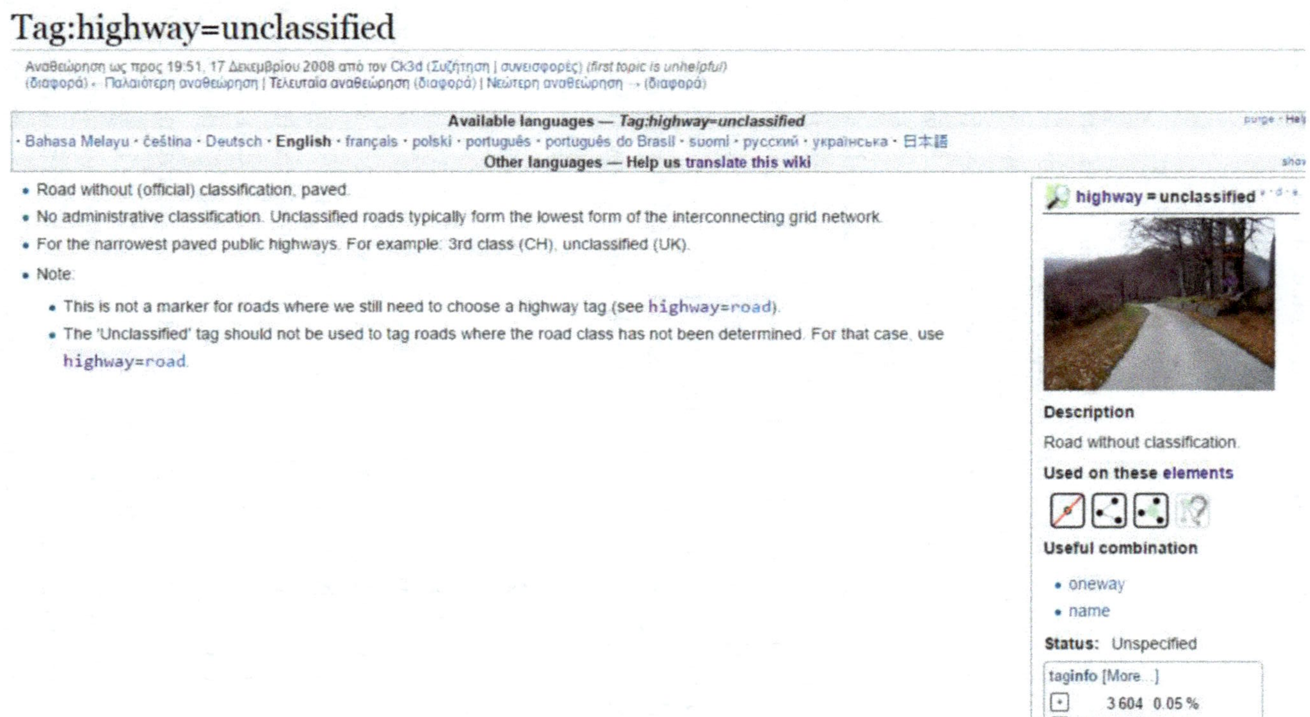
Tag:highway=unclassified

Αναθεώρηση ως προς 19:51, 17 Δεκεμβρίου 2008 από τον Ck3d (Συζήτηση | συνεισφορές) (first topic is unhelpful)
(διαφορά) ← Παλαιότερη αναθεώρηση | Τελευταία αναθεώρηση (διαφορά) | Νεώτερη αναθεώρηση → (διαφορά)

Available languages — *Tag:highway=unclassified*
· Bahasa Melayu · čeština · Deutsch · **English** · français · polski · português · português do Brasil · suomi · русский · українська · 日本語
Other languages — Help us translate this wiki

- Road without (official) classification, paved.
- No administrative classification. Unclassified roads typically form the lowest form of the interconnecting grid network.
- For the narrowest paved public highways. For example: 3rd class (CH), unclassified (UK).
- Note:
  - This is not a marker for roads where we still need to choose a highway tag (see `highway=road`).
  - The 'Unclassified' tag should not be used to tag roads where the road class has not been determined. For that case, use `highway=road`.

highway = unclassified

Description
Road without classification.

Used on these elements

Useful combination
- oneway
- name

Status: Unspecified

taginfo [More...]
3 604 0.05 %

Fig. 7: The wiki page that specifies the use of the highway=unclassified combination (end of 2008). @OpenStreetMap contributors.

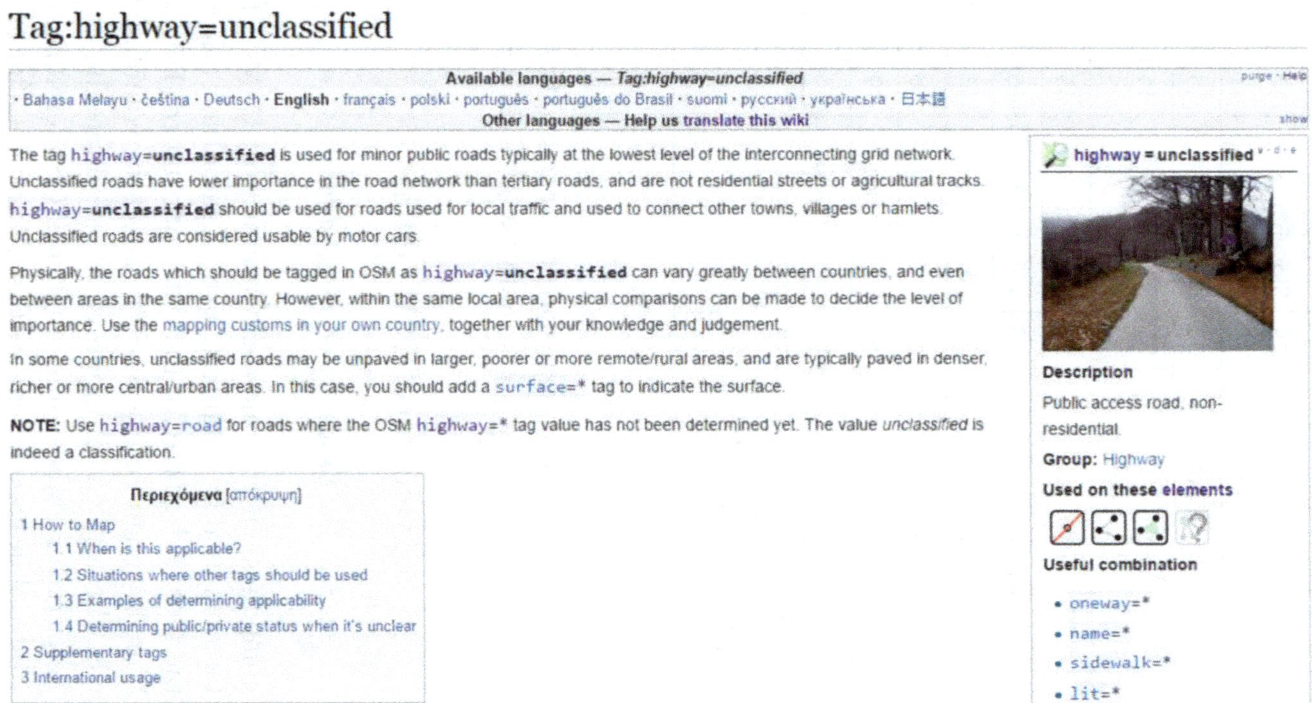

Tag:highway=unclassified

Available languages — *Tag:highway=unclassified*
· Bahasa Melayu · čeština · Deutsch · **English** · français · polski · português · português do Brasil · suomi · русский · українська · 日本語
Other languages — Help us translate this wiki

The tag highway=**unclassified** is used for minor public roads typically at the lowest level of the interconnecting grid network. Unclassified roads have lower importance in the road network than tertiary roads, and are not residential streets or agricultural tracks. highway=**unclassified** should be used for roads used for local traffic and used to connect other towns, villages or hamlets. Unclassified roads are considered usable by motor cars.

Physically, the roads which should be tagged in OSM as highway=**unclassified** can vary greatly between countries, and even between areas in the same country. However, within the same local area, physical comparisons can be made to decide the level of importance. Use the mapping customs in your own country, together with your knowledge and judgement.

In some countries, unclassified roads may be unpaved in larger, poorer or more remote/rural areas, and are typically paved in denser, richer or more central/urban areas. In this case, you should add a surface=* tag to indicate the surface.

NOTE: Use highway=road for roads where the OSM highway=* tag value has not been determined yet. The value *unclassified* is indeed a classification.

Περιεχόμενα [απόκρυψη]

1 How to Map
1.1 When is this applicable?
1.2 Situations where other tags should be used
1.3 Examples of determining applicability
1.4 Determining public/private status when it's unclear
2 Supplementary tags
3 International usage

highway = unclassified

Description
Public access road, non-residential.
Group: Highway
Used on these elements
Useful combination
- oneway=*
- name=*
- sidewalk=*
- lit=*

Fig. 8: The wiki page that specifies the use of highway=unclassified combination (end of 2015). @OpenStreetMap contributors.

Table 5: Available languages for the Map Features wiki page (as of May 2016).

#	Language	#	Language	#	Language	#	Language
1	asturianu	14	Hrvatski	**27**	Română	**40**	Ελληνικά
2	azərbaycanca	15	Íslenska	**28**	Shqip	**41**	ქართული
3	Bahasa indonesia	16	Italiano	**29**	Slovenčina	**42**	தமிழ்
4	bosanski	17	kreyòl ayisyen	**30**	Slovenščina	**43**	한국어
5	català	18	kréyòl gwadloupéyen	**31**	Suomi	**44**	日本語
6	čeština	19	Latviešu	**32**	Svenska	**45**	中文（简体）
7	dansk	20	Lietuvių	**33**	Tiếng Việt	**46**	中文（繁體）
8	Deutsch	21	Magyar	**34**	Türkçe	**47**	עברית
9	eesti	22	nederlands	**35**	српски/srpski	**48**	العربية
10	english	23	norsk bokmål	**36**	Български	**49**	فارسی
11	español	24	Polski	**37**	македонски		
12	esperanto	25	Português	**38**	Русский		
13	français	**26**	português do Brasil	**39**	Українська		

is available in 12 languages (čeština, deutsch, eesti, english, français, italiano, nederlands, português do Brasil, русский, ελληνικά, 日本語，中文（简体).

5 Evolution of OSM Editors

5.1 The Usage of the OSM Editors

An important component of the micro-environment of OSM is the editing tools. The OSM editors used by volunteers play an important role as they primarily dictate the type and quality of the data contributed. For example, an embedded functionality in an OSM editor can direct the volunteer to or avert them from specific choices that can improve or deteriorate the quality of the contribution. There are currently a large number of OSM editors available for various media, from online browser editors (e.g. iD and Potlatch 2), to desktop and offline editors such as JOSM and Merkaartor, to GIS software add-ons, e.g. for QGIS and ArcGIS, through to editors for mobile devices, like the Vespucci and OsmAndFrom. By reviewing the history of the OSM wiki pages dedicated to editors[5], it becomes clear that the number of available editors has increased as the project has developed (Figure 9).

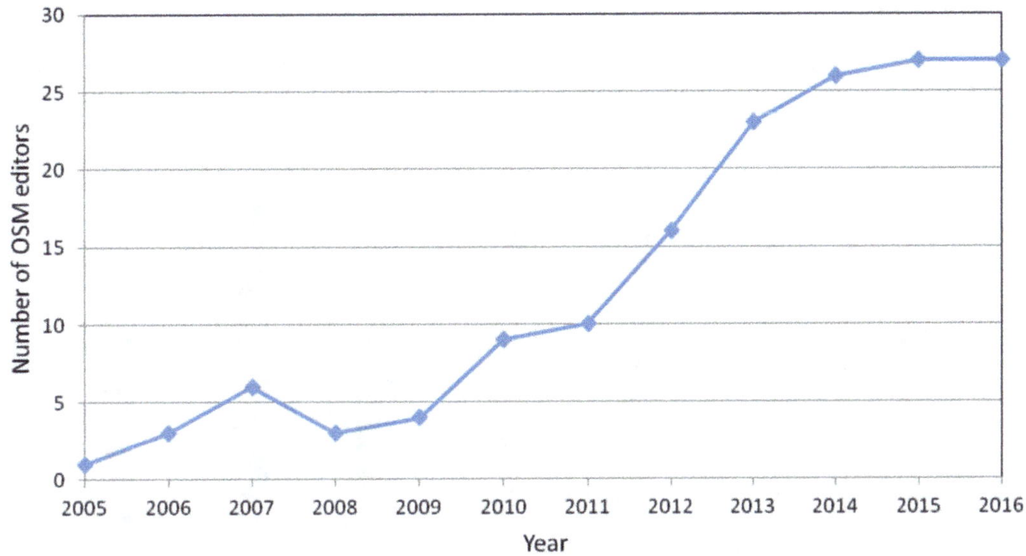

Fig. 9: Number of OSM editors.

The variety and the large number of OSM editors currently in use indicates the degree of interest in the OSM project. However, this wide range of OSM editors diversifies the data sources and can possibly affect the coherence and homogeneity of the contributions. Indeed, at the time of writing (i.e. May 2016), there were 27 editors available for the OSM community to choose from. This freedom, while in line with the ideology of a crowdsourced project, might undermine the overall effort for a usable dataset of high quality. However, the flip side of this observation might reside in the penetration that selected editors have in the OSM community. Indeed, by examining the statistics from the OSM wiki pages[6] regarding the most popular editors, a more encouraging picture is painted. By using the number of *changesets* as a criterion for the years 2009 to 2015 (Figure 10), it can be seen that the most popular editors in 2015 are iD, JOSM and Potlatch 2. An OSM *changeset* is a group of changes made by a single user over a short period of time. One *changeset* might include a number of *edits* (see below) such as the addition of new elements and tags or a change in values.

While the OSM community seems to have settled on using primarily 3 out of the 27 editors available, the findings in Figure 10 raise concerns regarding the quality and homogeneity of the contributions submitted with other editors in the past. For example, Potlach 1, which used to be one of the most popular editors in 2009, is now abandoned, and Potlach 2 has been completely rewritten. Similarly, Merkaartor, which provided 4–5% of changesets each year from 2009 until 2011, has now almost entirely disappeared. Interestingly, purpose-built editors for mobile devices have not managed to diffuse into the OSM community. For example, Vespucci has a small percentage, i.e. around 1%. The most popular editor between 2009 and 2012 was JOSM, followed by the online

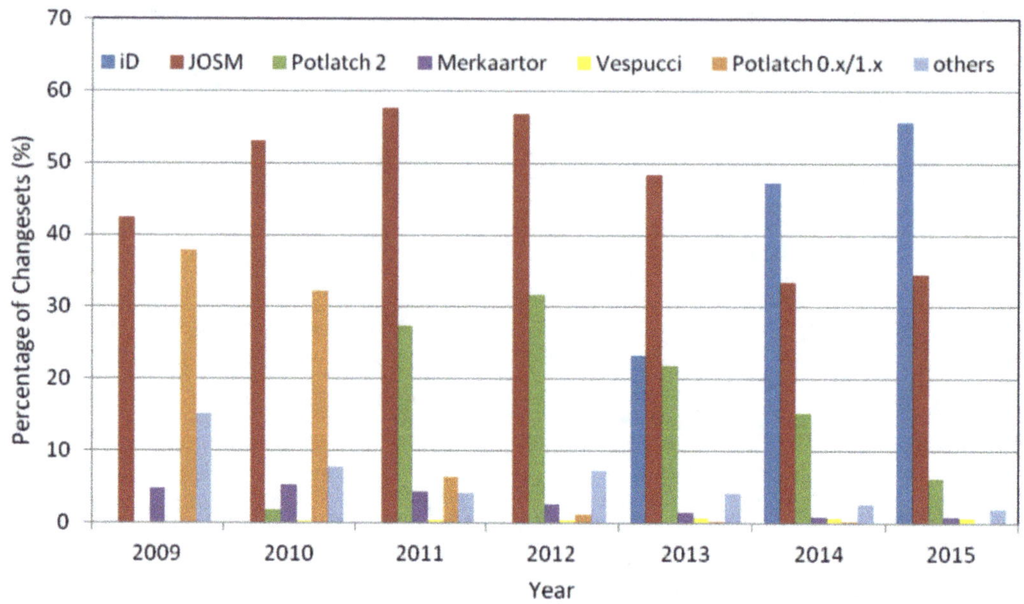

Fig. 10: Percentage of changesets per OSM editor.

editors on the OSM website: initially Potlatch 1, and then Potlatch 2 and iD. However, from 2014, iD has become the most frequently used editor when counting *changesets*. Yet when measuring the number of *edits*, JOSM has been the most popular editor since 2010 (Figure 11). Nevertheless, in 2015, JOSM use decreased by 5.6% while iD use has increased by 4.1%.

From what has been presented so far, it is evident that there is a strong volatility in the choices of the OSM community. The majority of the *changesets* and *edits* take place through a small number of editors that succeed each other over time. While the aim of this chapter is not to compare and evaluate the functionality of each editor, it is to be noted that the potential differences in their functionality or abidance to the OSM specifications might cause inconsistencies and deteriorate the overall quality of the data submitted. However, on the positive side, the strength and devotion of the OSM community in creating new editors that adapt to new challenges and requirements can be seen.

5.2 The Functionality of the Editors

Apart from the number of OSM editors available, what has also changed is their functionality. The existence of a set of rules that function as a product specification also needs to be supported by the available tools for the task. Thus, the level and efficiency of the editors at any given point in time plays a crucial role in the quality of the contributions. Here we present the evolution of the functionality across the active editors from 2006 to the present:

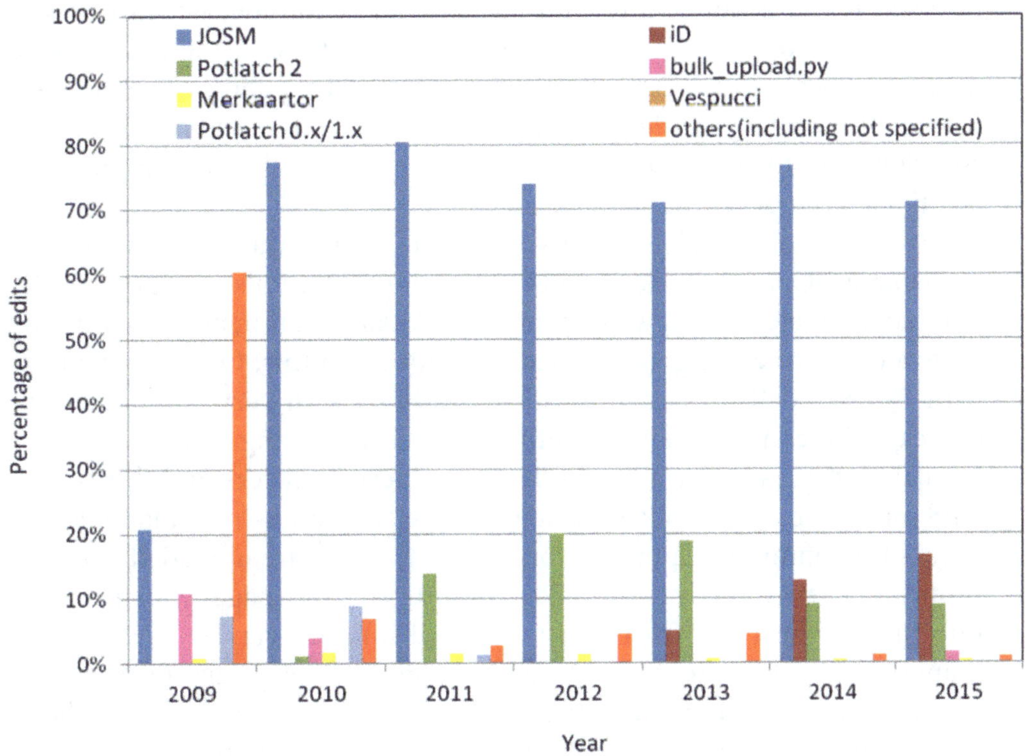

Fig. 11: Percentage of edits per OSM editor.

In 2006, the OSM editors serve only to upload GPS tracks. Only the online editing applet provides a Landsat photo, and thus GPS tracks cannot be verified in comparison with a satellite image.

- In 2007, Landsat overlay becomes available in JOSM 1.0 and some editing facilities are offered. Merkaartor, a small editor for OSM with some unique features like anti-aliased displaying and transparent display of map features, also appears.
- In 2007, the online editor applet displays Yahoo! Aerial Imagery under the GPS trackpoints while editing. This is very useful, and in fact more accurate than GPS data in the areas where coverage is most detailed (cities). In other areas it may sometimes assist in correcting GPS tracks.
- In 2008, photomapping is added in JOSM, which allows users to retrieve photographs and work with them on screen, positioned alongside the map data in the editor. In addition, if GPS location information is included in the photograph files or a GPS track is available, JOSM's photograph mapping features can be used to see them in context, and perhaps position new elements based on the recorded photograph positions.
- In 2008, Merkaartor can use satellite imagery from Yahoo! or any other Web Map Service (WMS).

- In 2009, JOSM acquires fast fluid panning and zooming, which provides for precise mapping. It is now possible to work offline using downloaded data files, local photo and GPX files. Offline editing can help volunteers work more carefully in a less rushed manner, and thus could provide better contributions. In addition, advanced editing functionality, which improves positional accuracy, becomes available.
- In 2010, Yahoo! Aerial Imagery, Bing and other aerial imagery become available in JOSM as backgrounds for tracing. JOSM also supports audio mapping. Potlatch 2, a new version of the Potlatch editor, appears, offering quite a different editing experience. In addition to this, OSM cooperation with QGIS and Esri's ArcGIS leads to add-ons with very comprehensive GIS capabilities and advanced editing, further improving quality.
- In 2015, JOSM provides a large selection of aerial imagery and third-party GPS traces as backgrounds for tracing, as well as a built-in validator, which checks for common mapping errors before the data are uploaded. Tags are shown to users directly with links to the OSM wiki page, which returns information for a tag. In iD, custom aerial imagery can be used, photographs are directly available in the editor from Mapillary[7], and OSM editors have access to billions of GPS tracks recorded by Strava[8] users, which allows for very precise mapping of twisted roads and trails. Potlatch 2 develops advanced features, including vector backgrounds, a merging/conflation functionality for specialists and several aerial imagery backgrounds, which are preconfigured, as well as the introduction of an option for custom Tile Map Service (TMS) imagery.
- At the time of writing (May 2016), JOSM seems to be the most promising editor in terms of quality assurance based on the tools offered, such as advanced geometry and topology editing; the resolving of conflicts; the tagging of presets; a validator that checks for common mapping errors before data upload; selection of background images and custom TMS, WMS and Web Map Tile Service (WMTS); selection of third-party GPS traces immediately available as backgrounds for tracing; etc.

6 Discussion and Conclusions

It is not common for a discussion section to begin with what the study has not done. Yet, in this case, it is necessary. We only scratched the surface of what could be done. We sampled only a few of the 847 versions of just one wiki page, albeit an important one, and we used these to examine selected cases of the changes recorded. The entire OSM specification consists of hundreds more wiki pages with information about each feature and the possible key/value combinations. Each of these extra pages have their history, which might, in turn, consist of hundreds of versions. The workload required to monitor each and every change would be immense. The other thing that we did not do is examine the OSM editor's evolution from a data quality viewpoint. This would require comparing the evolving functionality of all available editors against the active OSM specification at each point in time across a timeline; again, this is a task that would be next to impossible.

The value of this chapter is in its context and orientation. Regarding the former, the methodology chosen did not try to provide quantitative descriptions of different quality elements or indicators but rather to provide context and to expand the discussion on OSM quality by delving into the micro-environment of OSM. Indeed, we treated the '*Map Features*' wiki page, the main OSM specification page and the OSM editors as living organisms and chose to examine how they have grown and evolved over time. By not studying and thus not fully understanding the environment within which OSM data are created, studies on the subject of data quality do not have a solid context, i.e. they deal with the symptoms and ignore the cause. This, in turn, leads us to orientation. VGI quality has become a popular subject of study among researchers. Much of the literature has focused on the nature of the phenomenon (Antoniou, 2011), on the contributors (Ciepłuch et al., 2011; Nedović-Budić and Budhathoki, 2010) and on the social engineering behind it (Haklay, 2010; Haklay et al., 2010; Zielstra and Zipf, 2010). Other, more technical papers have delved into statistics and measures of various quality elements and indicators (Barron et al., 2014; Keßler and de Groot, 2013), usually by comparing OSM data with authoritative products. In this chapter, the idea was to re-orient the discussion towards the fundamentals of spatial products. The specifications of a product and the tools available to produce it largely define the outcome, regardless of the effort, the workload or the enthusiasm put into producing it. OSM is clearly much more than a spatial product, and the value of VGI, in general, is orders of magnitude greater than the achieved quality (Antoniou, 2016). However, if the goal is to improve the quality of VGI, then we need to have a better understanding of the micro-environment within which each VGI project grows.

Notes

1 https://www.mediawiki.org
2 https://meta.wikimedia.org/wiki/Help:Minor_edit
3 http://wiki.openstreetmap.org/wiki/Main_Page
4 http://wiki.openstreetmap.org/wiki/Deprecated_features
5 http://wiki.openstreetmap.org/wiki/Editors
6 http://wiki.openstreetmap.org/wiki/Editor_usage_stats
7 https://www.mapillary.com
8 https://www.strava.com

Reference list

Antoniou, V., 2016. Volunteered Geographic Information: Measuring Quality, Understanding the Value. *Geomedia* 20.

Antoniou, V., 2011. User Generated Spatial Content: An Analysis of the Phenomenon and its Challenges for Mapping Agencies. Unpublished PhD Thesis. University College London (UCL), London, UK.

Antoniou, V., Schlieder, C., 2014. Participation patterns, VGI and gamification, in: Proceedings of AGILE 2014. Presented at the AGILE 2014, Castellón, Spain, pp. 3–6.

Antoniou, V., Skopeliti, A., 2015. Measures and indicators of VGI quality: An overview, in: ISPRS Annals of the Photogrammetry, Remote Sensing and Spatial Information Sciences. Presented at the ISPRS Geospatial Week 2015, ISPRS Annals, La Grande Motte, France, pp. 345–351.

Barron, C., Neis, P., Zipf, A., 2014. A comprehensive framework for intrinsic OpenStreetMap quality analysis. *Transactions in GIS* 18, 877–895. DOI: https://doi.org/10.1111/tgis.12073

Brando, C., Bucher, B., 2010. Quality in user generated spatial content: A matter of specifications, in: Proceedings of the 13th AGILE International Conference on Geographic Information Science. Springer Verlag, Guimar aes, Portugal, pp. 11–14.

Brando, C., Bucher, B., Abadie, N., 2011. Specifications for user generated spatial content, in: Geertman, S., Reinhardt, W., Toppen, F. (Eds.), *Advancing Geoinformation Science for a Changing World, Lecture Notes in Geoinformation and Cartography*. Springer Berlin Heidelberg, pp. 479–495.

Ciepłuch, B., Mooney, P., Winstanley, A., 2011. Building generic quality indicators for OpenStreetMap, in: Proceedings of the 19th Annual GIS Research UK (GISRUK) Conference. Portsmouth, England, 27–29 April 2011. Available at: http://eprintsprod.nuim.ie/2483/1/Gisruk2011-edit2-Blazej.pdf [Last accessed 16 May 2017].

Clinton, W., 2000. *Improving the civilian global positioning system (GPS)*. The White House, Office of Science & Technology Policy: Washington, D.C., USA. Available at: https://clinton4.nara.gov/WH/EOP/OSTP/html/0053_4.html [Last accessed 16 May 2017].

D'Antonio, F., Fogliaroni, P., Kauppinen, T., 2014. VGI edit history reveals data trustworthiness and user reputation, in: Huerta, J., Schade, S., Granell, G. (eds.), *Proceedings of the 17th AGILE International Conference on Geographic Information Science: Connecting a Digital Europe through Location and Place*, Castellón, Spain, 3–6 June 2014, Available at: https://agile-online.org/conference_paper/cds/agile_2014/agile2014_140.pdf Last accessed 16 May 2017].

Fonte, C C, Antoniou, V, Bastin, L, Estima, J, Arsanjani, J J, Bayas, J-C L, See, L and Vatseva, R. 2017. Assessing VGI Data Quality. In: Foody, G, See, L, Fritz, S, Mooney, P, Olteanu-Raimond, A-M, Fonte, C C and Antoniou, V. (eds.)

Mapping and the Citizen Sensor. Pp. 137–163. London: Ubiquity Press. DOI: https://doi.org/10.5334/bbf.g.

Girres, J.-F., Touya, G., 2010. Quality assessment of the French OpenStreetMap dataset. *Transactions in GIS* 14, 435–459.

Goodchild, M.F., 2007. Citizens as sensors: the world of volunteered geography. *GeoJournal* 69, 211–221. DOI: https://doi.org/10.1007/s10708-007-9111-y

Haklay, M., 2013. Citizen science and volunteered geographic information: Overview and typology of participation, in: Sui, D., Elwood, S., Goodchild, M. (Eds.), *Crowdsourcing Geographic Knowledge*. Springer Netherlands, Dordrecht, Netherlands, pp. 105–122.

Haklay, M., 2010. How good is volunteered geographical information? A comparative study of OpenStreetMap and Ordnance Survey datasets. *Environment and Planning B: Planning and Design* 37, 682–703. DOI: https://doi.org/10.1068/b35097

Haklay, M., Basiouka, S., Antoniou, V., Ather, A., 2010. How many volunteers does it take to map an area well? The validity of Linus' Law to volunteered geographic information. *The Cartographic Journal* 47, 315–322.

ISO, 2005. *ISO 9000: Quality Management Systems - Fundamentals and Vocabulary*. International Organisation for Standardisation (ISO), Geneva, Switzerland.

Jokar Arsanjani, J., Mooney, P., Zipf, A., Schauss, A., 2015. Quality assessment of the contributed land use information from OpenStreetMap versus authoritative datasets, in: Jokar Arsanjani, J., Zipf, A., Mooney, P., Helbich, M. (Eds.), *OpenStreetMap in GIScience, Lecture Notes in Geoinformation and Cartography*. Springer International Publishing, Cham, pp. 37–58.

Keßler, C., de Groot, R.T.A., 2013. Trust as a proxy measure for the quality of Volunteered Geographic Information in the case of OpenStreetMap, in: Vandenbroucke, D., Bucher, B., Crompvoets, J. (Eds.), *Geographic Information Science at the Heart of Europe*. Springer International Publishing, Cham, pp. 21–37.

Minghini, M, Antoniou, V, Fonte, C C, Estima, J, Olteanu-Raimond, A-M, See, L, Laakso, M, Skopeliti, A, Mooney, P, Arsanjani, J J, Lupia, F. 2017. The Relevance of Protocols for VGI Collection. In: Foody, G, See, L, Fritz, S, Mooney, P, Olteanu-Raimond, A-M, Fonte, C C and Antoniou, V. (eds.) *Mapping and the Citizen Sensor*. Pp. 223–247. London: Ubiquity Press. DOI: https://doi.org/10.5334/bbf.j.

Nedović-Budić, Z., & Budhathoki, U. N. R., 2010. Motives for VGI Participants, in: '*VGI for SDI' Workshop*, Wageningen University, Netherlands, 16 April 2010. Available at: https://www.wur.nl/upload_mm/9/b/1/4deca512-7005-4974-8974-2d9b234c6eb0_Zorica_Nama_VGI_for_SDI.pdf [Last accessed 16 May 2017].

Olteanu-Raimond, A-M, Laakso, M, Antoniou, V, Fonte, C C, Fonseca, A, Grus, M, Harding, J, Kellenberger, T, Minghini, M, Skopeliti, A. 2017. VGI in National Mapping Agencies: Experiences and Recommendations.

In: Foody, G, See, L, Fritz, S, Mooney, P, Olteanu-Raimond, A-M, Fonte, C C and Antoniou, V. (eds.) *Mapping and the Citizen Sensor*. Pp. 299–326. London: Ubiquity Press. DOI: https://doi.org/10.5334/bbf.m.

OpenStreetMap, 2016. Map Features, 18 November 2016. Available at http://wiki.openstreetmap.org/wiki/Map_Features [Last accessed 4 April 2017]

Ordnance Survey (OS), 2001. OS MasterMap™ real-world object catalogue, 2001. Available at https://www.ordnancesurvey.co.uk/docs/legends/os-mastermap-real-world-object-catalogue.pdf [Last accessed 4 April 2017]

O'Reilly, T., 2007. What is Web 2.0: Design patterns and business models for the next generation of software. *Communications & Strategies* 65, 17–37.

See, L, Estima, J, Pőd.r, A, Arsanjani, J J, Bayas, J-C L and Vatseva, R. 2017. Sources of VGI for Mapping. In: Foody, G, See, L, Fritz, S, Mooney, P, Olteanu-Raimond, A-M, Fonte, C C and Antoniou, V. (eds.) *Mapping and the Citizen Sensor*. Pp. 13–35. London: Ubiquity Press. DOI: https://doi.org/10.5334/bbf.b.

Skopeliti, A, Antoniou, V and Bandrova, T. 2017. Visualisation and Communication of VGI Quality. In: Foody, G, See, L, Fritz, S, Mooney, P, Olteanu-Raimond, A-M, Fonte, C C and Antoniou, V. (eds.) *Mapping and the Citizen Sensor*. Pp. 197–222. London: Ubiquity Press. DOI: https://doi.org/10.5334/bbf.i.

van Exel, M., Dias, E., Fruijtier, S., 2010. The impact of crowdsourcing on spatial data quality indicators, in: Wallgrün, J.O., Lautenschütz, A.K. (eds.), *Proceedings of the GIScience 2010 Doctoral Colloquium*, Zurich, Switzerland, 14–17 September 2010, Akademische Verlagsgesellschaft Aka GmbH (IOS Press), Heidelberg, Germany. Available at: http://www.giscience2010.org/pdfs/paper_213.pdf [Last accessed 16 May 2017].

Vandecasteele, A., Devillers, R., 2015. Improving volunteered geographic information quality using a tag recommender system: The case of OpenStreetMap, in: Jokar Arsanjani, J., Zipf, A., Mooney, P., Helbich, M. (Eds.), *OpenStreetMap in GIScience, Lecture Notes in Geoinformation and Cartography*. Springer International Publishing, Cham, Switzerland, pp. 59–80.

Zielstra, D., Zipf, A., 2010. A comparative study of proprietary geodata and volunteered geographic information for Germany, in: Painho, M., Santos, M., Y., Pundt, H. (eds.), *Proceedings of the 13th AGILE International Conference on Geographic Information Science: Geospatial Thinking*, Guimarães, Portugal, 11–14 May 2010, pp. 1–15. Available at: https://agile-online.org/Conference_Paper/CDs/agile_2010/ShortPapers_PDF/142_DOC.pdf [Last accessed 16 May 2017].

CHAPTER 9

Visualisation and Communication of VGI Quality

Andriani Skopeliti*, Vyron Antoniou† and Temenoujka Bandrova‡

*School of Rural and Surveying Engineering, National Technical University of Athens, 9 H. Polytechniou, Zografou, 15780, Greece, askop@survey.ntua.gr

†Hellenic Army General Staff, Geographic Directorate, PAPAGOU Camp, Mesogeion 227-231, Cholargos, 15561, Greece

‡University of Architecture, Civil Engineering and Geodesy, 1, Chr. Smirnenski blvd., Sofia, Bulgaria

Abstract

The flourishing of VGI projects has transformed the average web user into an eager geographic data user and contributor. As it is difficult for the crowd to perceive VGI quality, visualisation can play a critical role in communicating data quality. At the same time, although VGI quality has been a prominent research topic for scientists, quality visualisation has not been exploited to its full potential. Since the crowd encompasses a diverse pool of users, VGI quality visualisation caters for different needs and exhibits variable functionality, operating as an awareness tool for the novice user as well as an exploration tool for the expert user / scientist. The scope of this chapter is to present a framework for VGI quality visualisation that takes into account factors such as methods for quality visualisation of spatial data, the nature of VGI data quality, user profiles and the visualisation environment. In addition, a review of the available methods for data quality visualisation, which have emerged from cartography,

How to cite this book chapter:

Skopeliti, A, Antoniou, V and Bandrova, T. 2017. Visualisation and Communication of VGI Quality. In: Foody, G, See, L, Fritz, S, Mooney, P, Olteanu-Raimond, A-M, Fonte, C C and Antoniou, V. (eds.) *Mapping and the Citizen Sensor*. Pp. 197–222. London: Ubiquity Press. DOI: https://doi.org/10.5334/bbf.i.

is presented, and a number of guidelines for VGI quality visualisation are proposed, taking into account user characteristics.

Keywords

quality, VGI, visualisation, VGI quality awareness, VGI quality exploration, visualisation framework

1 Introduction

Quality visualisation of geospatial data is as important as the data themselves (Pang, 2001). The recent development in VGI projects, such as OpenStreetMap (OSM) and Geonames, makes this topic even more critical and challenging, as novice users now access, use and create geographic information. The novice user does not question the quality of VGI data, as he/she is either unaware of the quality issue or erroneously believes that quality problems do not exist in the dataset. The source of geographic data (i.e. VGI vs. proprietary/authoritative) is not perceived as an important factor when determining the credibility of a map (Parker, 2014). A nicely designed map in terms of cartography and an operational map environment, e.g. OSM, is considered as a reliable source. Judgement is based on peripheral signals such as visual design and symbology (e.g. 'if it looks good and attractive, then it is good'; Idris et al., 2011). Quality reporting in text and tables may be easily understood by experts but not by the diverse pool of VGI users. Since visualisation can communicate data quality to all users (Buttenfield, 1983; Drecki, 2002; MacEachren et al., 2005), it is proposed to use visualisation to reveal VGI data quality.

VGI quality has been given particular attention by scientists. Much of the work concentrates on assessing and reporting VGI quality in diverse outlets, but only a few studies include visualisations. According to the OSM wiki, there are a number of online web pages characterised as 'Visualisation tools'[1] related to 'Quality assurance'. However, these mainly refer to error and bug reporting tools with maps and do not constitute an actual quality visualisation environment. Visualisation has not been exploited to its full potential and scientists have not taken full advantage of its capabilities. As a result, researchers miss aspects of VGI quality that visualisation could reveal. One may assume that in the early days of VGI, VGI quality measures and indicators were not mature enough to be visually represented: past research has suggested that without a good understanding of quality, effective approaches to visualisation remain elusive (MacEachren et al., 2005). However, a review of the literature indicates the existence of a plethora of measures and indicators that now manage to successfully express VGI quality (see e.g. Antoniou and Skopeliti, 2015; Senaratne et al., 2016).

1.1 The Role of VGI Quality Visualisation

Visualisation can be used to communicate VGI quality to the crowd (Figure 1). Visualisation transforms VGI quality from an issue that is rather ignored and difficult to perceive into a perceptible and vivid data characteristic. As the crowd consists of a diverse pool of users in terms of knowledge and experience with spatial data, VGI quality visualisation needs to satisfy different requirements. Visualisation is applicable to two distinct but related activities: visual thinking, which is exploratory and engages scientists; and visual communication, which is explanatory and refers to the distribution of existing knowledge (DiBiase et al., 1992). Thus VGI quality visualisation can have multiple functionalities: it can be considered as an awareness tool for the novice user as well as an exploration tool for the expert user / scientist. Users with intermediate knowledge and experience can take advantage of the different functionalities depending on their abilities. In more detail, VGI data quality visualisation can be considered:

- *An awareness tool for the novice user* that can be used to draw the attention of the crowd to VGI quality; force the crowd to question VGI quality; communicate quality in a way that can be understood by the layperson; stimulate contribution improvements; etc. Many research projects (MacEachren et al., 1995; Leitner and Buttenfield, 2000; Cliburn et al., 2002; Deitrick, 2007) have demonstrated that quality visualisation supports the process of

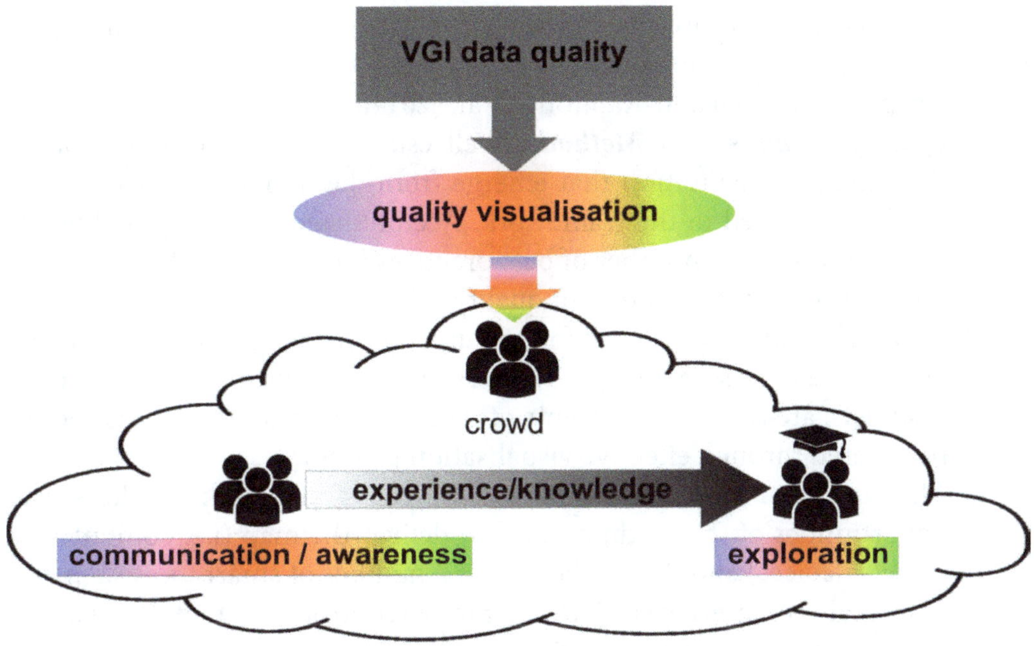

Fig. 1: VGI data quality, visualisation, users and functionality.

decision-making and leads to significantly better decisions. Consequently, it is important to inform users about data quality in order to select VGI data that are appropriate for a specific purpose. Although experts do not find uncertainty visualisation overwhelming, confusing or useless (Kunz, 2011), with so many non-expert VGI users, there is a need to make sure that visualisation is understandable by all users, not only expert ones (Jones, 2011). This can be achieved by exploring the full potential of data quality visualisation and selecting the appropriate methods.

- *And an exploration tool for the expert user / scientist* that can aid researchers to study the appropriateness and the ability of measures and indicators to express quality; to discover dependencies to extrinsic socio-economic or demographic factors; to explore the spatial distribution and heterogeneity of VGI quality; etc.

1.2 A Framework for VGI Quality Visualisation

In the previous paragraph, the role of VGI quality visualisation as an awareness and as an exploration tool has been discussed. However, although VGI quality visualisation is acknowledged as necessary, it is also considered as a big challenge (Sester et al., 2014). As a result, a framework for VGI quality visualisation that can facilitate and guide the successful design of VGI quality visualisation is much welcomed; this framework acknowledges four interactive parameters that influence VGI quality visualisation (Figure 2):

i) ***VGI Data Quality***: The framework takes into account the nature of VGI datasets, the applicable data quality elements and the measures and indicators used to measure quality – see Chapter 7 by Fonte et al. (2017) and Chapter 13 by Olteanu-Raimond et al. (2017a).
ii) ***Quality Visualisation Methods***: Well established methods for spatial data quality visualisation that emerge from the domain of cartography can be integrated in the framework. Accumulated cartographic knowledge can provide a number of best practices for a successful visual communication and exploration of quality (see Section 4).
iii) ***Users***: The framework caters for end users of all backgrounds. The members of the diverse pool of VGI users, who range from novice users to scientists, are the final recipients of data quality, and their needs should be covered through effective visualisation processes.
iv) ***Medium/Visualisation Environment***: The framework exploits the opportunities of the medium used to deliver the map (i.e. computer or mobile devices) and the availability of a number of smart tools such as a graphical user interface (GUI), interactive controls, etc. that create a rich and effective visualisation environment.

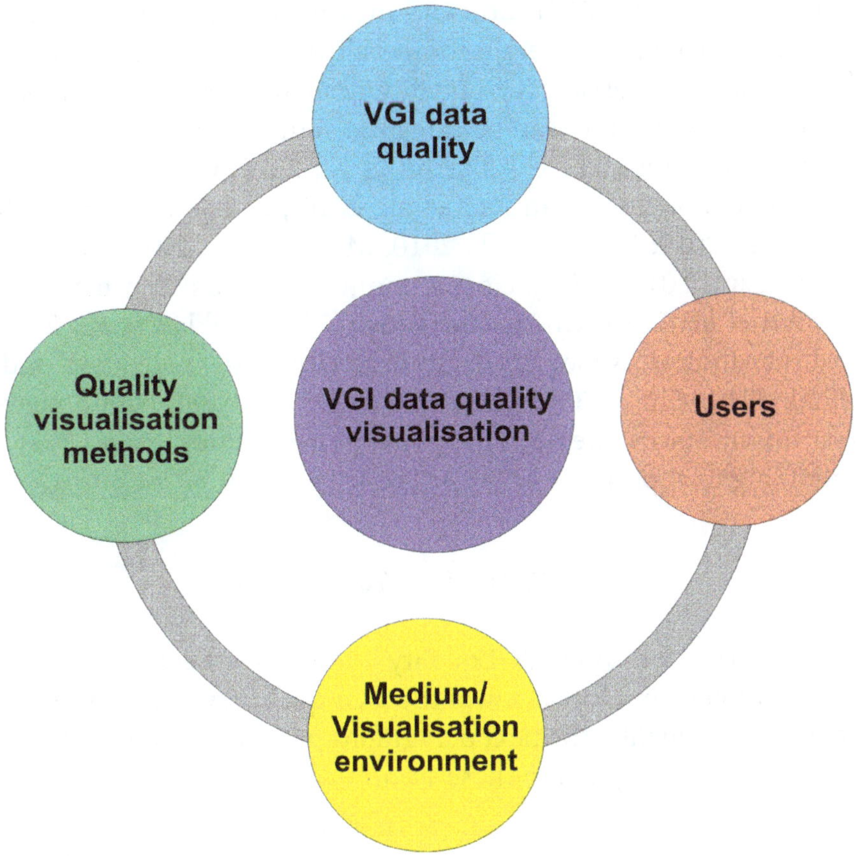

Fig. 2: A framework for VGI quality visualisation.

The above factors of the VGI quality visualisation framework are discussed in detail in Section 3.

In this context, the paper is structured as follows: Section 2 provides an overview of the present status of VGI quality visualisation, Section 3 describes in detail the elements of the framework for VGI quality visualisation and Section 4 presents the state of the art in data quality visualisation methods, providing specific guidelines for VGI data quality visualisation. The chapter ends with conclusions and proposals for future work.

2 Present Status of VGI Quality Visualisation

2.1 Measures and Indicators for VGI Quality

Scientists assess VGI quality with measures and indicators (see Chapter 7 by Fonte et al., 2017). A number of studies have tried to estimate VGI quality by comparing VGI with proprietary data (e.g. Girres and Touya, 2010; Haklay,

2010; Zielstra and Zipf, 2010), utilising measures that emerge from quality assessment, data matching, generalisation evaluation, etc. Because measures are not sufficient for characterising VGI quality, academic research focuses on data quality indicators. Indicators can be categorised into (Antoniou and Skopeliti, 2015): i) data indicators (see e.g. Barron et al., 2014; Ciepłuch et al., 2010a; Keßler and de Groot, 2013; van Exel et al., 2010); ii) demographic indicators (see e.g. Haklay, 2010; Haklay et al., 2010; Mullen et al., 2015; Tulloch, 2008; Zielstra and Zipf, 2010); iii) socio-economic indicators (see e.g. Antoniou, 2011; Elwood et al., 2013; Girres and Touya, 2010; Haklay et al., 2010); and iv) contributor indicators (see e.g. D'Antonio et al., 2014; Nedović-Budić and Budhathoki, 2010). Since VGI quality is currently assessed with a plethora of measures and indicators, the need for visual representation makes VGI quality visualisation highly topical.

2.2 VGI Quality Visualisation

Once meta-information about VGI quality is available, there are different ways to portray it graphically. Only a few of the VGI quality studies have provided a visualisation of the quality; the next paragraphs present a detailed review of the visualisation methods applied in these studies.

2.2.1 Measures

A number of studies access VGI quality with measures based on the comparison of VGI and proprietary data and provide quality visualisation (e.g. Antoniou, 2011; Fan et al., 2014; Forghani and Delavar, 2014; Haklay, 2010). Values of quality measures (e.g. distance between features, length difference of the road network, the area and density difference of buildings, etc.) are calculated for a grid that covers the study area, and are portrayed utilising colour schemes based on hue and value.

2.2.2 Contributor Indicators

Other studies assess the 'perceived quality' instead of the 'measured quality', i.e. user perception about the data quality, which is based on personal opinion and commentary and feedback from other users, is portrayed. Inspired by the popular web rating system that is utilised in sites such as Amazon, eBay, iTunes, etc. and that assesses quality on a 1 to 5 rating system, the quality visualisation proposed by Jones (2011) results in a Virtual Globe with glyphs (e.g. star 2D, star 3D), where visual variables such as size and colour portray the magnitude of quality. Schiewe (2013) records the opinion of the user for the current region of interest in OSM with a 'like' or 'dislike' button and visualises it with pictograms such as smiling faces, targets, etc.

2.2.3 Data Indicators

In recent studies, a number of data indicators have been proposed and visualised. Two different approaches are observed: indicators can be computed and visualised at the feature level or using grid cells that cover the study area. In the first approach, nodes, points and lines are used. For example, Trame and Keßler (2011) visualised the number of versions for OSM POIs (Points of Interest) by using a colour spectral scheme (heat map[2]) and overlaid the representation onto OSM. In another study (van Exel, 2011a), contour lines were used to visualise the average number of version (updates) of any node in the OSM database. Contours of different values were visualised with different hues. Van Exel (2011b) also proposed a combined visualisation of two metrics for the linear OSM features: (i) the time passed since a feature has last been updated by the community is visualised using a hue colour scheme and (ii) the number of versions, indicating how many updates a feature has received since its creation, is visualised using the width of the linear symbol. In another study (Keßler and de Groot, 2013), the trustworthiness of selected features was assessed by the numbers of versions, users, confirmations, corrections and rollbacks and was then visualised with different hue colour schemes. Two cases of interactive visualisation have also been recorded. Antoniou (2011) used an interactive map, which could alternate between data and quality visualisation, to visualise conceptual compliance to the OSM wiki-based specifications for each feature, using a hue colour scheme. In iOSMAnalyzer (Barron et al., 2013), 25 intrinsic measures referring to 'General Area Information', 'Routing & Navigation', 'Address-Search', 'Points of Interest-Search', 'Map-Applications' and 'User-Information & -Behavior' were calculated and portrayed in maps using hue colour schemes.

Other studies in the literature take the second approach, which is the grid-based approach. The densities of points and other indicators (Ciepłuch et al., 2010b) for OSM data have been computed for a grid and visualised utilising a colour spectrum scheme. In Roick et al. (2012), OSM data for Europe were divided into hexagonal cells and a number of spatio-temporal quality metrics (user activity, topicality and number of features) were calculated and visualised with hue and value colour schemes in a web application. The conceptual compliance (Ballatore and Zipf, 2015) of tags was calculated on a 10 km^2 grid and portrayed using a value colour scheme. In another study (Camboim et al., 2015), completeness (number of buildings/km^2, road density, road length, percentage of unclassified roads) and temporal quality (number of editors and days since last edition) were computed for administrative regions and visualised utilising a number of hue and value colour schemes.

2.3 Evaluation of Existing VGI Quality Visualisations

From the above analysis, it becomes evident that VGI quality assessment has been conducted per feature or per area (grid cell or administrative area) and that this pattern is followed for VGI quality visualisation as well. The visualisation of VGI quality, as it appears in the studies mentioned above, can be characterised as cartographically poor. Although a number of methods for quality visualisation exist in the cartographic literature (see Section 4), only a few of them have been applied. Most cases use only colour schemes based on hue and value. Additionally, quality visualisation is notably presented separately, independently from the data, offline and asynchronously. Thus, it does not permit quality judgement while looking at the data, and it obscures data visualisation, as attribute information is lost. With poor symbolisation or design choices, quality visualisation leads to more, rather than less, uncertainty about the data depicted (MacEachren et al., 2005). Practices for VGI quality visualisation need to be revised and updated based on a framework for VGI quality visualisation.

3 A Framework for VGI Quality Visualisation

The scope of this section is to discuss in detail the components of the framework for VGI quality visualisation presented in Section 1. Each component is analysed in order to present its contribution to quality visualisation. Finally, a number of guidelines are proposed that can help the design of a VGI quality visualisation environment.

3.1 VGI Data Quality

The nature of VGI datasets – see Chapter 2 by See et al. (2017) and Chapter 3 by Mooney and Minghini (2017) – and their quality aspects play an important role in the choices regarding visualisation. Past research (Buttenfield and Beard, 1994; Buttenfield and Weibel, 1988; MacEachren, 1992; MacEachren, 1995) has proved that the selection of a visualisation method should be related to the quality element represented and the measure/indicator used. The main information that users need about VGI quality focuses on fitness-for-use. Since fitness-for-use depends on a number of quality elements (such as positional accuracy, completeness, currency, etc.) and on criteria related to the planned use of the data, users may need to be presented with visualisations for a number of data quality measures and indicators in order to reach a decision on the suitability of a dataset. As a result, in order for users to fully benefit from the provision of various measures and indicators, a wide variety of visualisation methods should be provided, enhanced with interactivity to maximise functionality.

The nature of the quality indicator or measure affects the functionality of the visualisation as an awareness tool or as an exploratory tool. For instance, quality measures that are computed through comparison with authoritative data, although descriptive, cannot be used to support the *quality awareness* role: they are computed offline, post-processing is needed and they depend on the existence of reference data, which is not always the case. On the contrary, they are considered valuable for VGI *quality exploration* by scientists. Visualisation, as a VGI quality awareness tool, requires quality indicators that can only be calculated in real time from the VGI data or other available data, for simultaneous provision to the user.

Therefore, in order to provide for good understanding of quality and fitness-for-use judgement, one should provide a number of data quality measures and indicators along with visualisation support. Specific visualisation functionality, e.g. quality awareness or quality exploration, is made possible by selecting the appropriate quality descriptors, as explained above.

3.2 Quality Visualisation Methods

Quality visualisation can be handled as the cartographic portrayal of any other spatial phenomenon. Thus, the analysis of the measure/indicator and the values that describe it, of the classification according to geometry (point, line, area), and of the measurement scale (continuous or discrete; ordinal or categorical) will lead to the selection of the appropriate visualisation method. VGI data visualisation and quality visualisation should work together as a whole (holistic/symbiotic approach) and balance simplicity, detail, richness of visualisation and ease of understanding. Technical feasibility should also be considered. Methods should not be too complex, so that they can be applied easily within the framework of a VGI project.

One of the most attractive developments in cartography, which are based on modern technologies, is 3D mapping. 3D maps pose new challenges to cartographers, as these representations must be very well adapted to the context of the user and must provide understandable and easy-to-perceive information and messages. Some VGI data can be mapped in 3D. The 'third dimension is a growing topic in OSM (OpenStreetMap Wiki, 2017), for example, a number of web pages provide maps with 3D rendering of buildings. Data quality visualisation methods are considered to be adaptable to the 3D context, yet the subject hides big challenges (Bandrova et al., 2012; Jones, 2011; Pang et al., 1997).

A detailed review of available quality visualisation techniques emerging from cartography, as well as guidelines to select the appropriate methods taking into account usability and user experience, is presented in Section 4.

3.3 Users

An important factor for successful map design is to know who the audience is. Regarding VGI, there will always be a group of unknown users despite the effort of producers to register volunteers and involve them in user groups (Vullings et al., 2015). Since cartographic representations can only be optimised if end users and data types are known (Kunz et al., 2011), it is impossible to provide successful VGI quality visualisations for all users. Users with no knowledge of visualisation quality will work with a map differently than a professional who has been dealing with the issue for some time (Brus and Pechanec, 2015). Fortunately, the dual role of visualisation as a communication and as an exploration tool (DiBiase et al., 1992) can serve all VGI user needs. The idea of levels of uncertainty visualisation in relation to the experience and needs of the user is discussed in Beard and Mackaness (1993). Three levels are distinguished: the first level is simply a notification of poor data quality, with 'poor' defined on the basis of a predetermined threshold; the second level adds detail, such as the location and type of quality conflict, etc.; and the third level focuses on giving users methods for investigating the reasons for uncertainty. A VGI quality visualisation environment should provide for all users and take into account different user needs and characteristics. Based on this context, VGI quality visualisation design should address the profiles of at least two user groups, which are opposites in terms of experience and knowledge: the novice user profile and the expert user / scientist profile.

3.4 Medium/Visualisation Environment

Among the quality visualisation methods addressed in the literature, a frequently repeated idea is that users need control over depictions of quality (MacEachren et al., 2005). Cliburn et al. (2002) proposed to help users cope with the complexity of the display by providing interactivity. Interactive functionality can facilitate the interpretation of visualisation and cater for the different needs of heterogeneous user groups. A number of choices can be available in interactive functionality: selection among different cartographic methods for the visualisation (see Section 4); or customisation of the selected visualisation method according to user needs, e.g. configuration of visual variables such as colour schemes based on hue and value, symbol sizes, and data quality value classification, among others. Once the visualisation meets the requirements of the user (Kunz et al., 2011), the cartographic representation can be analysed visually, or, in addition, explored with the help of further functionality (e.g. a tooltip window displaying detailed information). Of course only expert users can make good use of strong interactivity, whereas novice users may be restricted to graphic modification of visualisations.

Graphical user interfaces (GUIs) are a powerful tool in visualisation support as they enhance functionality, through e.g. the graphic modification of visualisations, screen division and simultaneous display of data and quality visualisation in neighbouring windows, interactive tools such as a 'quality slider' that controls the appearance of the data in relation to quality, buttons that control whether different components – data or quality – should be visually dominant, etc. Functionality classification, based on Cron et al. (2007), includes: general functions, functions for navigation, didactic functions, cartographic and visualisation functions and GIS functions. Cartographic and visualisation functionality (Cron et al., 2007) refers to map manipulation, redlining (addition of drawings, labelling, and comments) and exploratory data analysis.

Apart from the need for the ability of a visualisation method to be understandable by any user, another important factor is the technical feasibility of the visualisation method's implementation (Jones, 2011). Technological advances can now provide geospatial applications with interactivity, flexibility and user friendliness so as to create the perfect environment for VGI quality visualisation. The integration of these qualities in the GUIs of a VGI project (irrespective of the device used) will further enhance the effort to communicate quality.

As a result, the design of the visualisation environment should strike a balance between interactivity, cartographic and visualisation functionality, and technical feasibility, taking into account the expected functionality, e.g. quality awareness or quality exploration, and the user profile, e.g. novice user or expert user/scientist.

3.5 Guidelines for VGI Quality Visualisation Implementation

From the above analysis of the framework, a number of guidelines may arise that can help the design of VGI quality visualisation:

- Various data quality measures and indicators should be provided to the user in order to achieve successful communication of quality and permit a successful fitness-for-use judgement;
- The nature of the VGI pool of users should be addressed and user needs and characteristics taken into account; in particular, user profiles on the opposite ends of the experience and knowledge spectrum (the novice user and the expert user / scientist) should be taken into account;
- Visualisation functionality e.g. quality awareness or quality exploration should be provided by selecting the appropriate quality descriptors or measures;
- Visualisation techniques and guidelines emerging from cartography that take usability and user profile into account should be applied; and
- A visualisation environment that balances interactivity with cartographic and visualisation functionality and technical feasibility should be designed.

4 A Review of Methods for Quality Visualisation

Research in the field of quality visualisation for geospatial data has been ongoing for the last 30 years (Aerts et al., 2003; Buttenfield and Beard, 1994; Buttenfield and Weibel, 1988; Drecki, 2002; Goodchild et al., 1994; Leitner and Buttenfield, 2000; MacEachren, 1992; MacEachren et al., 2005; McGranaghan, 1993; Van der Wel et al., 1998; Wittenbrink et al., 1996; Zuk and Carpendale, 2006). In this section, papers about geographic data uncertainty and quality visualisation are reviewed and summarised, in order to acquire a catalogue of methods/techniques that can be applied to VGI quality visualisation. This review may act as an informative guide for designing a VGI quality visualisation.

The main challenge of any visualisation effort is to select the most appropriate method. Symbolisation is based on visual variables introduced by Bertin (1983). These include location; size; shape; orientation; colour hue; colour value (or brightness (Wilkinson, 2005), or lightness (Slocum et al., 2003)); texture (grain); colour saturation; arrangement (Morrison, 1974); clarity (fuzziness); resolution (of boundaries and images); and transparency (MacEachren, 1992). MacEachren (1995) describes the syntax for the above visual variables, giving a three-step rating of good, marginal and poor, for use with numerical, ordinal and categorical data (Roth, 2015).

In this paper, visualisation methods are presented in tables according to the classification that appears in the bibliography (Gershon, 1998; Kinkeldey et al., 2014a; MacEachren et al., 2005). First, intrinsic visualisation methods are presented in Table 1. Intrinsic visualisation methods (Howard and MacEachren, 1996) alter the symbology used to portray data values to additionally represent quality, through manipulation of a visual variable that has not been used to portray data values, e.g. the colour value. Table 1 presents the visual variables that can be used to portray quality. In order to make the functionality of visual variables understandable to non-experts, the notion of a visualisation metaphor was introduced by MacEachren (1992), was adopted by other researchers (e.g. Kardos et al., 2006) and is also integrated in Table 1. A number of the visual variables presented in Table 1 can be used in combination with hue (Hengl, 2003; Howard and MacEachren, 1996), resulting in combinations such as hue, saturation and value or value and hue, in order to form colour schemes, e.g. sequential colour schemes, diverging colour schemes, and qualitative colour schemes (Brewer, 1994; Harrower and Brewer, 2003). Such schemes can be applied in bivariate representations, which depict data and quality together, treating quality as a second variable (Kunz et al., 2011; MacEachren et al., 2005). All intrinsic approaches have in common the fact that slight changes in uncertainty can be difficult to identify, especially for datasets with great variability (Kunz et al., 2011). However, this can be mitigated with the help of interactive functionality.

Table 1: Intrinsic visualisation methods.

Visual Variable	Good Quality	Bad Quality	Metaphor: Quality is/has:	Examples in
Colour saturation	high saturation (pure hue)	low saturation (graying)	pure colour	Hengl (2003); Leitner and Buttenfield (2000); MacEachren et al. (1998); Schweizer and Goodchild (1992)
Colour value	high value or low value	reverse value of the good quality value	dark or light (both metaphors are applicable depending on value chosen for good quality)	Aerts et al. (2003); Brewer (1994); Edwards and Nelson (2001); Jiang et al. (1995); Leitner and Buttenfield (2000)
Transparency (fog)	transparent (clear)	opaque (foggy)	transparent (clear)	Drecki (2002); MacEachren (1992)
Clarity / fuzziness (fill clarity or edge crispness)	in focus areas or crisp border	out of focus areas or fuzzy border	in focus and crisp	MacEachren (1992)
Resolution of raster images or vector line work	high resolution	coarse resolution	high resolution	Finger and Bisantz (2002); Leitner and Buttenfield (2000); MacEachren et al. (1998)
Texture (integrated into existing symbols and thus becoming intrinsic)	fine (small) texture	coarse (large) texture	clear (no texture overlay – texture works as a barrier to see the data where quality is bad)	Kunz et al. (2011); Leitner and Buttenfield (2000); MacEachren et al. (1998); Monmonier (1990); Retchless (2012)

Table 2: Extrinsic visualisation methods.

Method	Description	Visual variable to portray quality	Examples in
Glyphs	graphical objects with 2D or 3D geometry, such as circle, sphere, vertical bar, pyramid, square etc.	size, colour value, saturation etc.	McKenzie et al. (2016); Pang (2001); Slocum et al. (2003)
Contours	lines that represent same values (isolines) of quality	size (thickness), colour value (brightness), connectedness, colour hue, texture etc.	DiBiase et al. (1992); Howard and MacEachren (1996); Pang (2008)
Grids / Tessellations	a grid or other tessellation e.g. hexagons overlaid to the data	size (grid size), texture (grid pattern), grid outline (boundaries) etc.	Cedilnik and Rheingans (2000); Kardos et al. (2008); Kinkeldey et al. (2014b); Mullins (2014); Pang (2008)

Extrinsic techniques (Howard and MacEachren, 1996), which introduce new objects to depict quality, e.g. glyphs, grids, etc., that work independently of the existing symbols for data values, are presented in Table 2. These new objects portray quality using appropriate visual variables such as size, colour value, texture, etc.

In terms of visual organisation, extrinsic visualisation methods (Gershon, 1998; Howard and MacEachren, 1996) can be *coincident*, if data and quality are represented in one map, or *adjacent*, if they are represented in adjacent maps. (Intrinsic visualisations are, by definition, coincident.)

Finally, quality visualisation methods can be static, like the ones already presented, or dynamic. Dynamic representations are presented in Table 3. Animation is related to three basic design elements, or 'dynamic variables': scene duration, rate of change between scenes and scene order (DiBiase et al., 1992). The range of possible dynamic approaches is wide because elements from animation and interaction can be combined in numerous ways. Intrinsic and extrinsic visualisation methods are static, but they can also be transformed into dynamic methods through animation.

4.1 Quality Visualisation Methods and VGI Data

A number of studies that present methods for quality visualisation have also studied their usability (Aerts et al., 2003; Cliburn et al., 2002; Fisher, 1993;

Table 3: Dynamic visualisation methods.

Dynamic Variable	Quality is represented by	Metaphor	Examples in
Sound	sonic variables	a low pitch sound depicts good quality and a high pitch sound, bad quality. It can be cursor-driven.	Fisher (1994) 1994; Krygier (1994); Lodha et al. (1996)
Animation	scene duration	long duration of an object on the screen depicts good quality	Fisher (1993); MacEachren et al. (1998)
	rate of change between scenes	questionable quality is portrayed with rapid blinking	Evans (1997); Fisher (1993); Monmonier and Gluck (1994); Kardos et al. (2006)
	spatially variable blurring	questionable quality is portrayed with very blurred regions	Gershon (1992); MacEachren et al. (2005)
	scene order	multiple representations: a number of possible data values are represented, and the existence of many different values creates questions on quality	Bastin et al. (2002); Ehlschlaeger et al. (1997)

Gershon, 1992; Kardos et al., 2006; Kinkeldey et al., 2014a; Lodha et al., 1996; MacEachren et al., 1998; Pang, 2001; Schweizer and Goodchild, 1992). In the following paragraphs, a number of guidelines for VGI quality visualisation in relation to user experience are discussed, once again taking the two main user profiles into account: the novice user and the expert user/scientist.

Which method to use (intrinsic vs. extrinsic): Slocum et al. (2003) found that intrinsic techniques give a better overview of uncertainty, but that in-depth analysis is easier with extrinsic techniques. This is in agreement with Kunz et al. (2011), who noted that none of the intrinsic approaches can successfully portray the variability in quality. As a result, it is proposed to use intrinsic methods as awareness tools for novice users and extrinsic methods as exploratory tools for the experts.

Which visual variable to use in intrinsic visualisations: Regarding the intuitiveness needed for novice users (MacEachren et al., 2012), colour value, fog (transparency) and clarity (fuzziness) visual metaphors are preferable. On the other hand, expert users prefer transparency or saturation (Kunz, 2011). In terms of user performance, Kinkeldey et al. (2014a) conclude that colour saturation is not recommended, while colour hue and value as well as transparency provide better alternatives. Also, texture on colour fill and resolution lead to good results and thus can be used with intrinsic visualisations.

Which variable to use in extrinsic methods: Studies on extrinsic displays (Kinkeldey et al., 2014a) highlight the potential of glyph and grid-based techniques for quality representation. According to a different usability study (Senaratne et al., 2012), contours are considered the best method.

Which technique (coincident vs. adjacent) to use: Research suggests that both coincident and adjacent approaches have their applications. According to Kinkeldey et al. (2014a), coincident maps can be seen as the preferable option because the integration of uncertainty into the display makes it easier to retrieve data and quality simultaneously. This is why they are advised for the novice users in order to ensure that quality information will not escape their attention. The problem of advanced complexity, which may be an obstacle for the novice user, can be minimised with good cartographic design and interactivity (e.g. use of on/off buttons). Expert users can work with both techniques and should be able to decide which one to use.

Static or dynamic: There is evidence (Kinkeldey et al., 2014a) that animated views have a potential to successfully represent quality when static solutions are not feasible, but there is little evidence that they perform equally or better than more traditional static depictions when these are available. Regarding dynamic techniques, animations are the most promising ones as they can be used to attract the attention of the user (Gershon, 1992; Blenkinsop et al., 2000). Thus, dynamic visualisations can be used with novice users in order to highlight VGI quality issues and increase awareness. Expert users can again work with all of the methods, and they should be able to decide which one to use.

Scale: Finally, one should consider the dynamic scale of the VGI display environment, e.g. the OSM web page. The scale plays an important role in the selection of an appropriate visualisation method, as intrinsic methods are best for larger scales and extrinsic methods such as grid and contours are preferable for a global quality visualisation at smaller scales.

5 Conclusions and Future Plans

From the above analysis, it is clear that there is an emerging need for VGI data quality visualisation. A number of measures and indicators for VGI quality (Antoniou and Skopeliti, 2015) have been proposed, there is knowledge on quality visualisation (MacEachren et al., 2005; Kinkeldey et al., 2014a) and the technology is now available. Since the crowd encompasses a diverse pool of users, VGI quality visualisation should cater for different needs and exhibit variable functionality, operating as an awareness tool for the novice user as well as an exploration tool for expert users / scientists. A framework for successful VGI quality visualisation was presented, incorporating factors such as the nature of VGI data quality, user profiles, methods for quality visualisation of spatial data, and the visualisation environment.

Effective VGI quality visualisation will have a positive impact on a VGI project's overall quality: quality visualisation will help users decide on fitness-for-use, the quality of contributions will improve, the reputation of VGI will rise as quality is better communicated through visualisation, quality awareness will increase, sceptical users will change their opinion (since most of the time VGI quality is better than expected) and quality metadata hidden in data will be revealed, e.g. by utilising information from history files or elapsing tags in the case of OSM. Thus there are only merits to VGI quality visualisation for both VGI data and VGI projects.

VGI quality visualisation is also of interest to National Mapping and Cadastral Agencies (NMCAs) that embrace VGI. Today many NMCAs encourage and welcome VGI contributions in their geoportals (see Chapter 13 by Olteanu-Raimond et al., 2017a). Volunteers are playing an increasingly important role in ensuring that authoritative sources of geographic information are accurate and kept up-to-date. VGI data and authoritative data can be visualised in the geoportal of NMCAs and one of the aforementioned methods can be employed to portray quality. Data will be enhanced, but at the same time the user will be informed about data quality. Whereas authoritative data can be better in terms of quality elements such as homogeneity (Olteanu-Raimond et al., 2017b), VGI may prove to be better in terms of completeness (Vandecasteele and Devillers, 2015), currency (Goodchild and Glennon, 2010) and positional accuracy (Haklay, 2010). These differences in quality may only become apparent, especially to non-experts, through visualisation.

For the future development of this research topic, it is proposed to create a prototype for VGI quality visualisation, combining existing measures and indicators (Antoniou and Skopeliti, 2015) of VGI quality with a variety of visualisation methods (MacEachren et al., 2005; Kinkeldey et al., 2014a). For the choice of suitable visualisation methods for the crowd, it is important to confirm the usability and effectiveness of methods with the pool of VGI users. The prototype can be used to conduct a user survey that records and evaluates the crowd response on VGI quality visualisation and verifies methods in practice. Knowledge about VGI quality visualisation as it relates specifically to the crowd acquired through a user survey can then be implemented in the development of an interactive visualisation environment in the framework of any VGI project.

Notes

1 http://wiki.openstreetmap.org/wiki/Quality_assurance#Visualisation_tools

2 A heat map utilizes a colour scheme that is part of the colour spectrum; it is called heat map because this colour scheme is traditionally used in cartography for the visualisation of temperature.

Reference list

Aerts, J.C.J.H., Clarke, K.C., Keuper, A.D., 2003. Testing popular visualization techniques for representing model uncertainty. *Cartography and Geographic Information Science* 30, 249–261. DOI: https://doi.org/10.1559/152304003100011180

Antoniou, V., 2011. User Generated Spatial Content: An Analysis of the Phenomenon and its Challenges for Mapping Agencies. Unpublished PhD Thesis. University College London (UCL), London, UK.

Antoniou, V., Skopeliti, A., 2015. Measures and indicators of VGI quality: An overview, in: ISPRS Annals of the Photogrammetry, Remote Sensing and Spatial Information Sciences. Presented at the ISPRS Geospatial Week 2015, ISPRS Annals, La Grande Motte, France, pp. 345–351.

Ballatore, A., Zipf, A., 2015. A conceptual quality framework for Volunteered Geographic Information, in: Fabrikant, S.I., Raubal, M., Bertolotto, M., Davies, C., Freundschuh, S., Bell, S. (Eds.), *Spatial Information Theory*. Springer International Publishing, Cham, pp. 89–107.

Bandrova, T., Zlatanova, S., Konecny, M., 2012. Three-dimensional maps for disaster management. *ISPRS Annals of Photogrammetry, Remote Sensing and Spatial Information Sciences I-4, 245–250.* DOI: https://doi.org/10.5194/isprsannals-I-4-245-2012

Barron, C., Neis, P., Zipf, A., 2014. A comprehensive framework for intrinsic OpenStreetMap quality analysis. *Transactions in GIS* 18, 877–895. DOI: https://doi.org/10.1111/tgis.12073

Barron, C., Neis, P., Zipf, A., 2013. iOSMAnalyzer – ein umfassendes Werkzeug für intrinsische OSM Qualitätsuntersu*chungen, in: Proceedings of the AGIT 2013 - Symposium und Fachmesse Angewandte Geoinformatik*, Wichmann, Salzburg, 3–5 July, Austria, pp. 142–151.

Bastin, L., Fisher, P.F., Wood, J., 2002. Visualizing uncertainty in multi-spectral remotely sensed imagery. *Computers & Geosciences* 28, 337–350. DOI: https://doi.org/10.1016/S0098-3004(01)00051-6

Beard, K., Mackaness, W., 1993. Visual access to data quality in Geographic Information Systems. *Cartographica: The International Journal for Geographic Information and Geovisualization* 30, 37–45. DOI: https://doi.org/10.3138/C205-5885-23M7-0664

Bertin, J., 1983. *Semiology of graphics: Diagrams, networks, maps* (Translation from a French 1967 edition by W.Berg, editor). University of Wisconsin Press, Madison, WI, USA.

Blenkinsop, S., Fisher, P., Bastin, L., Wood, J., 2000. Evaluating the perception of uncertainty in alternative visualization strategies. *Cartographica: The International Journal for Geographic Information and Geovisualization* 37, 1–14. DOI: https://doi.org/10.3138/3645-4V22-0M23-3T52

Brewer, C.A., 1994. Color use guidelines for mapping and visualization, in: MacEachren, A.M., Taylor, D.R.F. (Eds.), *Visualization in Modern Cartography*. Elsevier Science Limited, Tarrytown, NY, USA, pp. 123–148.

Brus, J., Pechanec, V., 2015. The user centered framework for visualization of spatial data quality, in: Brus, J., Vondrakova, A., Vozenilek, V. (Eds.), *Modern Trends in Cartography*. Springer International Publishing, Cham, pp. 325–338.

Buttenfield, B.P., 1983. Representing data quality. *Cartographica* 30, 1–7.

Buttenfield, B.P., Beard, M.K., 1994. Graphical and geographical components of data quality, in: Unwin, D.J., Hearnshaw, H. (Eds.), *Visualization in Geographic Information Systems*. John Wiley & Sons, London, UK, pp. 150–157.

Buttenfield, B.P., Weibel, R., 1988. Visualizing the quality of cartographic data, in: *Proceedings of the Third International Geographic Information Systems Symposium (GIS/LIS 88)*, San Antonio, Texas, USA, 30 November – 2 December.

Camboim, S.P., Bravo, J.V.M., Sluter, C.R., 2015. An investigation into the completeness of, and the updates to, OpenStreetMap data in a heterogeneous area in Brazil. *ISPRS International Journal of Geo-Information* 4, 1366–1388. DOI: https://doi.org/10.3390/ijgi4031366

Cedilnik, A., Rheingans, P., 2000. Procedural annotation of uncertain information, in: Visualization 2000. Proceedings. Presented at the Visualization 2000. Proceedings, pp. 77–84. DOI: https://doi.org/10.1109/VISUAL.2000.885679

Ciepłuch, B., Jacob, R., Mooney, P., Winstanley, A., 2010a. Comparison of the accuracy of OpenStreetMap for Ireland with Google Maps and Bing Maps, in: Proceedings of the Ninth International Symposium on Spatial Accuracy Assessment in Natural Resources and Environmental Sciences, University of Leicester, UK, 20–23 July, pp. 37–40. Available at: http://eprints.maynoothuniversity.ie/2476/1/Accuracy2010-Ciepluch-Submitted.pdf

Ciepłuch, B., Jacob, R., Winstanley, A., Mooney, P., 2010. Comparison of the accuracy of OpenStreetMap for Ireland with Google Maps and Bing Maps, in: Proceedings of the Accuracy 2010 Symposium, Leicester, UK, 20–23 July. Available at: http://www.spatial-accuracy.org/system/files/img-X07133419_0.pdf

Cliburn, D.C., Feddema, J.J., Miller, J.R., Slocum, T.A., 2002. Design and evaluation of a decision support system in a water balance application. *Computers & Graphics* 26, 931–949. DOI: https://doi.org/10.1016/S0097-8493(02)00181-4

Cron, J., Sieber, R., Hurni, L., 2007. Guidelines to optimized graphical user interfaces of interactive atlases, in: Proceedings of the 23rd International Cartographic Conference, Moscow, Russia, 4–10 August. Available at: http://icaci.org/files/documents/ICC_proceedings/ICC2007/documents/doc/THEME%2015/Oral%205/GUIDELINES%20TO%20OPTIMIZED%20GRAPHICAL%20USER%20INTERFACES%20OF%20INTERAC.doc [Last accessed 16 May 2017].

D'Antonio, F., Fogliaroni, P., Kauppinen, T., 2014. VGI edit history reveals data trustworthiness and user reputation, in: *Proceedings of 17th AGILE International Conference on Geographic Information Science*, Castellón, Spain,

3–16 June. Available at: https://agile-online.org/conference_paper/cds/agile_2014/agile2014_140.pdf [Last accessed 16 May 2017].

Deitrick, S.A., 2007. Uncertainty visualization and decision making: Does visualizing uncertain information change decisions?, in: Proceedings of the XXIII International Cartographic Conference. Presented at the XXIII International Cartographic Conference, pp. 4–10.

DiBiase, D., MacEachren, A.M., Krygier, J.B., Reeves, C., 1992. Animation and the role of map design in scientific visualization. *Cartography and Geographic Information Science* 19, 201–214. DOI: https://doi.org/10.1559/152304092783721295

Drecki, I., 2002. Visualisation of uncertainty in geographical data, in: Shi, W., Fisher, P., Goodchild, M. (Eds.), *Spatial Data Quality*. Taylor & Francis, London, UK, pp. 140–159, NaN-143.

Edwards, L.D., Nelson, E.S., 2001. Visualizing data certainty: A case study using graduated circle maps. *Cartographic Perspectives* 38, 19–36.

Ehlschlaeger, C.R., Shortridge, A.M., Goodchild, M.F., 1997. Exploratory Cartograpic VisualisationVisualizing spatial data uncertainty using animation. *Computers & Geosciences* 23, 387–395. DOI: https://doi.org/10.1016/S0098-3004(97)00005-8

Elwood, S., Goodchild, M.F., Sui, D., 2013. Prospects for VGI research and the emerging fourth paradigm, in: Sui, D., Elwood, S., Goodchild, M. (Eds.), *Crowdsourcing Geographic Knowledge*. Springer Netherlands, Dordrecht, pp. 361–375.

Evans, B.J., 1997. Dynamic display of spatial data-reliability: Does it benefit the map user? *Computers & Geosciences* 23, 409–422. DOI: https://doi.org/10.1016/S0098-3004(97)00011-3

Fan, H., Zipf, A., Fu, Q., Neis, P., 2014. Quality assessment for building footprints data on OpenStreetMap. *International Journal of Geographical Information Science* 28, 700–719. DOI: https://doi.org/10.1080/13658816.2013.867495

Finger, R., Bisantz, A.M., 2002. Utilizing graphical formats to convey uncertainty in a decision-making task. *Theoretical Issues in Ergonomics Science* 3, 1–25. DOI: https://doi.org/10.1080/14639220110110324

Fisher, P.F., 1994. Randomization and sound for the visualization of uncertain spatial information, in: Unwin, D.J., Hearnshaw, H. (Eds.), *Visualization in Geographic Information Systems*. John Wiley & Sons, London, UK, pp. 181–185.

Fisher, P.F., 1993. Visualizing uncertainty in soil maps by animation. *Cartographica: The International Journal for Geographic Information and Geovisualization* 30, 20–27. DOI: https://doi.org/10.3138/B204-32P4-263L-76W0

Fonte, C C, Antoniou, V, Bastin, L, Estima, J, Arsanjani, J J, Bayas, J-C L, See, L and Vatseva, R. 2017. Assessing VGI Data Quality. In: Foody, G, See, L, Fritz, S, Mooney, P, Olteanu-Raimond, A-M, Fonte, C C and Antoniou, V.

(eds.) *Mapping and the Citizen Sensor*. Pp. 137–163. London: Ubiquity Press. DOI: https://doi.org/10.5334/bbf.g.

Forghani, M., Delavar, M., 2014. A quality study of the OpenStreetMap dataset for Tehran. *ISPRS International Journal of Geo-Information* 3, 750–763. DOI: https://doi.org/10.3390/ijgi3020750

Gershon, N.D., 1998. Visualization of an imperfect world. *IEEE Computer Graphics and Applications* 18, 43–45. DOI: https://doi.org/10.1109/38.689662

Gershon, N.D., 1992. Visualization of fuzzy data using generalized animation. *Visualization, 1992. Visualization '92, Proceedings, IEEE Conference on*, Boston, MA, pp. 268–273. DOI: https://doi.org/10.1109/VISUAL.1992.235199

Girres, J.-F., Touya, G., 2010. Quality assessment of the French OpenStreetMap dataset. *Transactions in GIS* 14, 435–459.

Goodchild, M.F., Chih-Chang, L., Leung, Y., 1994. Visualizing fuzzy maps, in: Unwin, D.J., Hearnshaw, H. (Eds.), *Visualization in Geographic Information Systems*. John Wiley & Sons, London, UK, pp. 158–167.

Goodchild, M.F., Glennon, J.A., 2010. Crowdsourcing geographic information for disaster response: a research frontier. *International Journal of Digital Earth* 3, 231–241. DOI: https://doi.org/10.1080/17538941003759255

Haklay, M., 2010. How good is volunteered geographical information? A comparative study of OpenStreetMap and Ordnance Survey datasets. *Environment and Planning B: Planning and Design* 37, 682–703. DOI: https://doi.org/10.1068/b35097

Haklay, M., Basiouka, S., Antoniou, V., Ather, A., 2010. How many volunteers does it take to map an area well? The validity of Linus' Law to volunteered geographic information. *The Cartographic Journal* 47, 315–322.

Harrower, M., Brewer, C.A., 2003. ColorBrewer.org: An online tool for selecting colour schemes for maps. *The Cartographic Journal* 40, 27–37. DOI: https://doi.org/10.1179/000870403235002042

Hengl, T., 2003. Visualisation of uncertainty using the HIS colour model: Computations with colours, in: Proceedings of the 7th International Conference on GeoComputation. Southampton, UK, p. 8–17.

Howard, D., MacEachren, A.M., 1996. Interface design for geographic visualization: Tools for representing reliability. *Cartography and Geographic Information Science* 23, 59–77. DOI: https://doi.org/10.1559/152304096782562109

Idris, N.H., Jackson, M.J., Abrahart, R.J., 2011. Colour coded traffic light labelling: An approach to assist users in judging data credibility in map mashup applications, in: Proceedings of the 7th International Symposium on Spatial Data Quality, Coimbra, Portugal, 12 - 14 October, p. 201–206.

Jiang, B., Ormeling, F., Kainz, W., 1995. Visualization support for fuzzy spatial analysis, in: *Auto Carto* 12. Charlotte, NC, USA, pp. 291–300.

Jones, K., 2011. Communicating perceived geospatial data quality of 3D objects in virtual globes. MSc Thesis. School of Graduate Studies, Department of Geography, Memorial University of Newfoundland.

Kardos, J., Moore, A., Benwell, G., 2008. Exploring tessellation metaphors in the display of geographical uncertainty, in: Moore, A., Drecki, I. (Eds.), *Geospatial Vision, Lecture Notes in Geoinformation and Cartography*. Springer Berlin Heidelberg, pp. 113–140.

Kardos, J., Moore, A., Benwell, G., 2006. Expressing attribute uncertainty in spatial data using blinking regions, in: Caetano, M., Painho, M. (Eds.), Proceedings of the 7th International Symposium on Spatial Accuracy Assessment in Natural Resources and Environmental Sciences, Lisbon, Portugal, 5–7 July, pp. 814–824. Available at: http://www.spatial-accuracy.org/system/files/Kardos2006accuracy.pdf [Last accessed 16 May 2017].

Keßler, C., de Groot, R.T.A., 2013. Trust as a proxy measure for the quality of Volunteered Geographic Information in the case of OpenStreetMap, in: Vandenbroucke, D., Bucher, B., Crompvoets, J. (Eds.), *Geographic Information Science at the Heart of Europe*. Springer International Publishing, Cham, pp. 21–37.

Kinkeldey, C., MacEachren, A.M., Schiewe, J., 2014a. How to assess visual communication of uncertainty? A systematic review of geospatial uncertainty visualisation user studies. *The Cartographic Journal* 51, 372–386. DOI: https://doi.org/10.1179/1743277414Y.0000000099

Kinkeldey, C., Mason, J., Klippel, A., Schiewe, J., 2014b. Evaluation of noise annotation lines: using noise to represent thematic uncertainty in maps. *Cartography and Geographic Information Science* 41, 430–439. DOI: https://doi.org/10.1080/15230406.2014.949868

Krygier, J., 1994. Sound and geographic visualization, in: Unwin, D.J., Hearnshaw, H. (Eds.), *Visualization in Geographic Information Systems*. John Wiley & Sons, London, UK, pp. 149–166.

Kunz, M.H., 2011. *Interactive visualizations of natural hazards data and associated uncertainties*. Doctoral dissertation, ETH, Zurich, Switzerland. Available at: http://e-collection.library.ethz.ch/eserv/eth:2927/eth-2927-01.pdf [Last accessed 16 May 2017].

Kunz, M.H., Grêt-Regamey, A., Hurni, L., 2011. Visualization of uncertainty in natural hazards assessments using an interactive cartographic information system. *Natural Hazards* 59, 1735–1751. DOI: https://doi.org/10.1007/s11069-011-9864-y

Leitner, M., Buttenfield, B.P., 2000. Guidelines for the display of attribute certainty. *Cartography and Geographic Information Science* 27, 3–14. DOI: https://doi.org/10.1559/152304000783548037

Lodha, S.K., Pang, A., Sheehan, R.E., Wittenbrink, C.M., 1996. UFLOW: visualizing uncertainty in fluid flow. ACM, pp. 249–254. DOI: https://doi.org/10.1109/VISUAL.1996.568116

MacEachren, A.M., 1995. *How Maps Work: Representation, Visualization and Design*. The Guildford Press, New York, NY, USA.

MacEachren, A.M., 1992. Visualizing uncertain information. *Cartographic Perspectives* 13, 10–19.

MacEachren, A.M., Brewer, C.A., Pickle, L.W., 1995. Mapping health statistics: Representing data reliability, in: Proceedings of the 17th International Cartographic Conference. Barcelona, Spain, pp. 311–319.

MacEachren, A.M., Brewer, C.A., Pickle, L.W., 1998. Visualizing georeferenced data: representing reliability of health statistics. *Environment and Planning A* 30, 1547–1561. DOI: https://doi.org/10.1068/a301547

MacEachren, A.M., Robinson, A., Hopper, S., Gardner, S., Murray, R., Gahegan, M., Hetzler, E., 2005. Visualizing geospatial information uncertainty: What we know and what we need to know. *Cartography and Geographic Information Science* 32, 139–160. DOI: https://doi.org/10.1559/1523040054738936

MacEachren, A.M., Roth, R.E., O'Brien, J., Li, B., Swingley, D., Gahegan, M., 2012. Visual semiotics & uncertainty visualization: An empirical study. *IEEE Transactions on Visualization and Computer Graphics* 18, 2496–2505. DOI: https://doi.org/10.1109/TVCG.2012.279

McGranaghan, M., 1993. A cartographic view of spatial data quality. *Cartographica: The International Journal for Geographic Information and Geovisualization* 30, 8–19. DOI: https://doi.org/10.3138/310V-0067-7570-6566

McKenzie, G., Hegarty, M., Barrett, T., Goodchild, M., 2016. Assessing the effectiveness of different visualizations for judgments of positional uncertainty. International *Journal of Geographical Information Science* 30, 221–239. DOI: https://doi.org/10.1080/13658816.2015.1082566

Monmonier, M., 1990. Strategies for the Interactive Exploration of Geographic Correlation. Proceedings of the 4th International Symposium on Spatial Data Handling, IGU:512–521., in: Brassel, K., Kishimoto, H. (Eds.), Proceedings of the 4th International Symposium on Spatial Data Handling. Department of Geography, University of Zurich, Switzerland, pp. 381–389.

Monmonier, M., Gluck, M., 1994. Focus groups for design improvement in dynamic cartography. *Cartography and Geographic Information Systems* 21, 37–47. DOI: https://doi.org/10.1559/152304094782563948

Mooney, P and Minghini, M. 2017. A Review of OpenStreetMap Data. In: Foody, G, See, L, Fritz, S, Mooney, P, Olteanu-Raimond, A-M, Fonte, C C and Antoniou, V. (eds.) *Mapping and the Citizen Sensor*. Pp. 37–59. London: Ubiquity Press. DOI: https://doi.org/10.5334/bbf.c.

Morrison, J., 1974. A theoretical framework for cartographic generalization with the emphasis on the process of symbolization. *International Yearbook of Cartography* 14, 115–127.

Mullen, W.F., Jackson, S.P., Croitoru, A., Crooks, A., Stefanidis, A., Agouris, P., 2015. Assessing the impact of demographic characteristics on spatial error in volunteered geographic information features. *GeoJournal* 80, 587–605. DOI: https://doi.org/10.1007/s10708-014-9564-8

Mullins, R.S., 2014. Interpretive uncertainty and the evaluation of symbols and a taxonomy of symbol evaluation methods and mobile evaluation tool. MSc Thesis. The Pennsylvania State University, The Graduate School, College of Earth and Mineral Sciences.

Nedovic-Budic, Z., Budhathoki, N. R., 2010. Motives for VGI participants, in: Workshop "VGI for SDI?", Wageningen University and the Subcommission Geo-Information Infra-structure of the Netherlands Geodetic Commission, , Wageningen, Netherlands, 16 April. Available at: http://www.ncgeo.nl/phocadownload/Zorica_Nama_VGI_for_SDI.pdf [Last accessed 16 May 2017].

Olteanu-Raimond, A-M, Laakso, M, Antoniou, V, Fonte, C C, Fonseca, A, Grus, M, Harding, J, Kellenberger, T, Minghini, M, Skopeliti, A. 2017a. VGI in National Mapping Agencies: Experiences and Recommendations. In: Foody, G, See, L, Fritz, S, Mooney, P, Olteanu-Raimond, A-M, Fonte, C C and Antoniou, V. (eds.) *Mapping and the Citizen Sensor*. Pp. 299–326. London: Ubiquity Press. DOI: https://doi.org/10.5334/bbf.m.

Olteanu-Raimond, A.-M., Hart, G., Foody, G., Touya, G., Kellenberger, T., Demetriou, D., 2017b. The scale of VGI in map production: A perspective of European National Mapping Agencies. *Transactions in GIS* 21, 74–90. DOI: https://doi.org/10.1111/tgis.12189

OpenStreetMap Wiki, 2017. 3D, 17 February 2017. Available at http://wiki.openstreetmap.org/wiki/3D [Last accessed: 3 April 2017]

Pang, A.T., 2008. Visualizing uncertainty in natural hazards, in: Bostrom, A., French, P.S., Gottlieb, S. (Eds.), *Risk Assessment, Modeling and Decision Support, Risk, Governance and Society*. Springer Berlin Heidelberg, pp. 261–294.

Pang, A.T., 2001. Visualizing uncertainty in geo-spatial data. *Computer Science and Telecommunications Board*, Arlington, VA.

Pang, A.T., Wittenbrink, C.M., Lodha, S.K., 1997. Approaches to uncertainty visualization. *The Visual Computer* 13, 370–390. DOI: https://doi.org/10.1007/s003710050111

Parker, C.J., 2014. A framework of neogeography, in: *The Fundamentals of Human Factors Design for Volunteered Geographic Information, SpringerBriefs in Geography*. Springer International Publishing, Cham, pp. 11–22.

Retchless, D., 2012. Mapping Climate Change Uncertainty: Effects on Risk Perceptions and Decision Making. Poster. Presented at the AGU Fall Meeting, San Francisco, CA.

Roick, O., Loos, L., Zipf, A., 2012. Technical Framework for Visualizing Spatio-Temporal Quality Metrics of Volunteered Geographic Information, in: Proceedings of GEOINFORMATIK 2012-Mobility and Environment, Braunschweig, Germany, 28 – 30 March. Available at: http://koenigstuhl.geog.uni-heidelberg.de/publications/2012/Roick/Roick_OSMatrix_Geoinformatik2012.pdf [Last accessed 16 May 2017].

Roth, R.E., 2015. Visual variables, in: *The International Encyclopedia of Geography: People, The Earth, Environment, and Technology*. Wiley-Blackwell.

Schiewe, J., 2013. Improving collective intelligence and exploration in a VGI like context through communication of uncertainty information. Presented at the GeoVIz, Hamburg, Germany.

Schweizer, D.M., Goodchild, M.F., 1992. Data quality and choropleth maps: An experiment with the use of color, in: Proceedings of GIS/LIS'92. Presented at the GIS/LIS'92, ACSM and ASPRS, San Jose, CA, USA, pp. 686–699.

See, L, Estima, J, Pőd.r, A, Arsanjani, J J, Bayas, J-C L and Vatseva, R. 2017. Sources of VGI for Mapping. In: Foody, G, See, L, Fritz, S, Mooney, P, Olteanu-Raimond, A-M, Fonte, C C and Antoniou, V. (eds.) *Mapping and the Citizen Sensor*. Pp. 13–35. London: Ubiquity Press. DOI: https://doi.org/10.5334/bbf.b.

Senaratne, H., Gerharz, L., Pebesma, E., Schwering, A., 2012. Usability of spatio-temporal uncertainty visualisation methods, in: Gensel, J., Josselin, D., Vandenbroucke, D. (Eds.), *Bridging the Geographic Information Sciences*. Springer Berlin Heidelberg, Berlin, Heidelberg, pp. 3–23.

Senaratne, H., Mobasheri, A., Ali, A.L., Capineri, C., Haklay, M. (Muki), 2016. A review of volunteered geographic information quality assessment methods. I*nternational Journal of Geographical Information Science* 1–29. DOI: https://doi.org/10.1080/13658816.2016.1189556

Sester, M., Jokar Arsanjani, J., Klammer, R., Burghardt, D., Haunert, J.-H., 2014. Integrating and generalising Volunteered Geographic Information, in: Burghardt, D., Duchêne, C., Mackaness, W. (Eds.), *Abstracting Geographic Information in a Data Rich World*. Springer International Publishing, Cham, pp. 119–155.

Slocum, T.A., McMaster, R.B., Kessler, F.C., Howard, H.H., 2003. *Thematic Cartography and Geovisualization*. Pearson Prentice Hall, Upper Saddle River NJ.

Trame, J., Keßler, C., 2011. Exploring the lineage of Volunteered Geographic Information with heat maps in: Proceedings of GeoVIz 2011, Hamburg, Germany, 10–11 March. Available at: http://www.geomatik-hamburg.de/geoviz11/abstracts/28_TrameKessler_Abstract_GeoViz2011.pdf [Last accessed 16 May 2017].

Tulloch, D.L., 2008. Is VGI participation? From vernal pools to video games. *GeoJournal* 72, 161–171. DOI: https://doi.org/10.1007/s10708-008-9185-1

Van der Wel, F.J., Van der Gaag, L.C., Gorte, B.G., 1998. Visual exploration of uncertainty in remote-sensing classification. *Computers & Geosciences* 24, 335–343. DOI: https://doi.org/10.1016/S0098-3004(97)00120-9

van Exel, M., 2011a. A new OpenStreetmap visualization: Version contour lines. *Oegeo Martijn van Exel and his OpenStreetMap adventures* [blog] 20 June 2011. Available at: https://oegeo.wordpress.com/2011/06/20/a-new-openstreetmap-visualization-version-contour-lines/ [Last accessed 16 May 2017]

van Exel, M., 2011b. A new OpenStreetmap visualization: Taking the temperature of local OpenStreetMap communities. *Oegeo Martijn van Exel and his OpenStreetMap adventures* [blog] 19 September 2011. Available at: https://oegeo.wordpress.com/2011/09/19/taking-the-temperature-of-local-openstreetmap-communities/ [Last accessed 16 May 2017] van Exel, M., Dias, E., Fruijtier, S., 2010. The impact of crowdsourcing on spatial data quality indicators, in: Wallgrün, J.O., Lautenschütz, A.K. (eds.), Proceedings of the GIScience 2010 Doctoral Colloquium, Zurich, Switzerland, 14–17 September 2010. Available at: http://www.giscience2010.org/pdfs/paper_213.pdf [Last accessed 16 May 2017]

Vandecasteele, A., Devillers, R., 2015. Improving volunteered geographic information quality using a tag recommender system: The case of OpenStreetMap, in: Jokar Arsanjani, J., Zipf, A., Mooney, P., Helbich, M. (Eds.), *OpenStreetMap in GIScience, Lecture Notes in Geoinformation and Cartography*. Springer International Publishing, Cham, Switzerland, pp. 59–80.

Vullings, L.A.E., Bulens, J.D., Rip, F.I., Boss, M., Meijer, M., Hazeu, G., Storm, M., 2015. Spatial data quality: What do you mean? ?, in: Proceedings of 18th AGILE International Conference on Geographic Information Science, Lisbon, Portugal, 9–12 June 2015, pp. 9–12. Available at: https://agile-online.org/conference_paper/cds/agile_2015/shortpapers/87/87_Paper_in_PDF.pdf [Last accessed 16 May 2017]

Wilkinson, L., 2005. *The Grammar of Graphics (Statistics and Computing)*. Springer-Verlag, New York.

Wittenbrink, C.M., Pang, A.T., Lodha, S.K., 1996. Glyphs for visualizing uncertainty in vector fields. *IEEE Transactions on Visualization and Computer Graphics* 2, 266–279. DOI: https://doi.org/10.1109/2945.537309

Zielstra, D., Zipf, A., 2010. A comparative study of proprietary geodata and volunteered geographic information for Germany. Presented at the 13th AGILE International Conference on Geographic Information Science 2010, Guimarães, Portugal, pp. 1–15.

Zuk, T., Carpendale, S., 2006. Theoretical analysis of uncertainty visualizations, in: Proceedings of the International Society for Optics and Photonics (SPIE) - Visualization and Data Analysis, 15–19 January 2006, San Jose, California, USA, p. 606007. DOI: https://doi.org/10.1117/12.643631

CHAPTER 10

The Relevance of Protocols for VGI Collection

Marco Minghini*, Vyron Antoniou†, Cidália Costa Fonte‡, Jacinto Estima§, Ana-Maria Olteanu-Raimond¶, Linda See‖, Mari Laakso**, Andriani Skopeliti††, Peter Mooney‡‡, Jamal Jokar Arsanjani§§, Flavio Lupia¶¶

*Department of Civil and Environmental Engineering, Politecnico di Milano, Piazza Leonardo da Vinci 32, 20133 Milano, Italy, marco.minghini@polimi.it

†Hellenic Army General Staff, Geographic Directorate, PAPAGOU Camp, Mesogeion 227-231, Cholargos, 15561, Greece

‡Department of Mathematics, University of Coimbra, 3001-501 Coimbra, Portugal / INESC Coimbra, Rua Sílvio Lima, Pólo II, 3030-290 Coimbra, Portugal

§NOVA IMS, Universidade Nova de Lisboa, 1070-312, Lisbon, Portugal

¶Paris-Est, LASTIG COGIT, IGN, ENSG, F-94160 Saint-Mande, France.

‖International Institute for Applied Systems Analysis (IIASA), Schlossplatz 1, 2361 Laxenburg, Austria

**Finnish Geospatial Research Institute, Kirkkonummi 02430, Finland

††School of Rural and Surveying Engineering, National Technical University of Athens, 9 H. Polytechniou, Zografou, 15780, Greece

‡‡Department of Computer Science, Maynooth University, Maynooth, Co. Kildare, Ireland

§§Department of Planning and Development, Aalborg University Copenhagen, A.C. Meyers Vænge 15, DK-2450 Copenhagen, Denmark

¶¶Council for Agricultural Research and Economics (CREA), Via Po, 14 00198 Roma, Italy

How to cite this book chapter:

Minghini, M, Antoniou, V, Fonte, C C, Estima, J, Olteanu-Raimond, A-M, See, L, Laakso, M, Skopeliti, A, Mooney, P, Arsanjani, J J, Lupia, F. 2017. The Relevance of Protocols for VGI Collection. In: Foody, G, See, L, Fritz, S, Mooney, P, Olteanu-Raimond, A-M, Fonte, C C and Antoniou, V. (eds.) *Mapping and the Citizen Sensor.* Pp. 223–247. London: Ubiquity Press. DOI: https://doi.org/10.5334/bbf.j.

Abstract

Volunteered Geographic Information (VGI) has become a rich and well established source of geospatial data. From the popular OpenStreetMap (OSM) to many citizen science projects and social network platforms, the amount of geographically referenced information that is constantly being generated by citizens is burgeoning. The main issue that continues to hamper the full exploitation of VGI lies in its quality, which is by its nature typically undocumented and can range from very high quality to very poor. A crucial step towards improving VGI quality, which impacts on VGI usability, is the development and adoption of protocols, guidelines and best practices to assist users when collecting VGI. This chapter proposes a generic and flexible protocol for VGI data collection, which can be applied to new as well as to existing projects regardless of the specific type of geospatial information collected. The protocol is meant to balance the contrasting needs of providing VGI contributors with precise and detailed instructions while maintaining and growing the enthusiasm and motivation of contributors. Two real-world applications of the protocol are presented, which guide the collection of VGI in respectively the generation and updating of thematic information in a topographic building database; and the uploading of geotagged photographs for the improvement of land use and land cover maps. Technology is highlighted as a key factor in determining the success of the protocol implementation.

Keywords

Volunteered Geographic Information, protocol, best practices, data collection, data quality.

1 Introduction and Background

Volunteered Geographic Information (VGI) represents an important new source of citizen-contributed data (Goodchild, 2007), as outlined in detail in Chapter 2 (See et al., 2017). VGI can be a complementary source of information to authoritative data such as detailed road networks and building footprints, and may be the only source of map data usable after a natural disaster or crisis event has occurred, for example in the case of mapping efforts by the Humanitarian OpenStreetMap Team (HOT)[1]. Yet the main barrier to the widespread use of VGI remains the assessment and documentation of data quality (Johnson and Sieber, 2013; Olteanu-Raimond et al., 2017a). This is particularly true when quality compliance is an essential requirement for VGI exploitation, such as for its exploitation by governments, National Mapping Agencies (NMAs), public bodies (fire fighters, civil protection etc.) and private companies, which make use of geospatial data to take decisions. From this perspective, an analysis of VGI exploitation by NMAs is made in Chapter 13 (Olteanu-Raimond et al., 2017b), while some guidance on VGI data quality assessment is provided in Chapter 7 (Fonte et al., 2017). The latter chapter describes measures and indicators that are generally applied to VGI after the data have been collected. Instead, more attention should be placed on how to ensure high-quality data collection during the data capture phase. One approach for doing this is to develop and adopt generic and flexible guidelines, best practices and protocols for VGI collection. While guidelines and best practices refer to a set of rules, instructions, suggestions, recommendations or situations that indicate how VGI should be collected, perhaps by reference to examples or ideal cases, protocols can be defined as strict sequences of instructions regulating VGI collection. Specific attention should be paid to the structure and complexity of such guidelines, best practices and protocols; in particular, they should not discourage citizens from contributing, while simultaneously ensuring that the collected data are of an acceptable quality for the purpose of the specific VGI project. Not secondarily, they should ease or facilitate the reuse of VGI for projects and applications other than the one(s) it was originally collected for.

The relevance of establishing protocols in VGI projects and the potential problems for communities and society that arise when these protocols are absent have been highlighted by many authors, including Sui (2007), Johnson and Sieber (2013) and See et al. (2016). In Europe, only a few NMAs have experience with using or integrating VGI in their authoritative datasets (Olteanu-Raimond et al., 2017a), while protocols for VGI within NMAs, governments or Commercial Mapping Companies (CMCs) are lacking (Johnson and Sieber, 2013). Conversely, as mentioned above, many authors have developed methodologies to study the quality of VGI (after it has been collected) and have undertaken VGI comparison, integration or conflation with data from NMAs and CMCs to build more up-to-date, accurate and complete datasets (Girres and Touya, 2010; Haklay, 2010; Ludwig et al., 2011; Al-Bakri and Fairbairn, 2012; Du et al., 2012; Pourabdollah et al., 2013; Touya et al., 2013; Gao et al., 2014; Jokar Arsanjani et al., 2015b; Brovelli et al., 2016a; Fan et al., 2016).

To instruct users in the production of data that are fit-for-purpose, some VGI projects provide detailed guidelines instead of defining a real protocol. OpenStreetMap (OSM)[2] is the most popular VGI project and one of the most studied in the literature (Jokar Arsanjani et al., 2015c); it is extensively described in Chapter 3 (Mooney and Minghini, 2017). Over its more than ten years of life, there has been a progressive development of guidelines about the types of geographic features that users can create and the attributes (or tags) that can be attached to them. The updated version of these guidelines is maintained in a page[3] on the OpenStreetMap Wiki, while their development and enrichment over time is discussed in Chapter 8 (Antoniou and Skopeliti, 2017). It is worth mentioning that, although a real, strict protocol for creating OSM data does not exist and indeed there is considerable freedom left to the contributors, several studies have documented the high quality of OSM crowdsourced datasets (see e.g. Neis et al., 2011; Fan et al., 2014; Dorn et al., 2015; Jokar Arsanjani et al., 2015a). Another example of VGI project that provides guidelines is the National Map Corps[4], a mapping crowdsourcing programme similar to OSM that supports the Geospatial Information Office of the U.S. Geological Survey (USGS) in gathering rapidly-changing landscape feature data for The National Map (Bearden, 2007).

In other cases, protocols have been designed to assist volunteers in contributing high-quality data that could fit the VGI project's needs and purposes. A well known example is that of Geo-Wiki (Fritz et al., 2012), which is an online crowdsourcing platform where volunteers – provided with a strict and detailed protocol – are asked to use very fine spatial resolution imagery to gather information on land cover and land use to improve global land cover maps. Similarly, an extensive and detailed protocol for digitising old French maps was created and enriched through user collaboration on a dedicated platform[5], which allowed for consistent data records to be maintained (Perret et al., 2015). In the same way, the GéoPeuple project used protocols to create topographic vector datasets from old French maps for analysing population growth (Ruas et al., 2014). The Degree Confluence Project[6] is an example of a project applying a protocol to collect photographs of the landscape from all the intersection points (or confluences) of one degree latitude-longitude around the globe. Volunteers are asked to take either photographs in the four cardinal compass directions (north, south, east, west) or one or more panoramic views from the intersection, one general photograph taken within 100 metres of the confluence, and one photograph of the GPS used. Users then upload all the photographs, along with a text describing the landscape as well as their journey to the confluence point (Fritz et al., 2009). In principle, these photographs may then be reused in another VGI project to yield reference data for map validation (Foody and Boyd, 2012).

The addition of such protocols in VGI projects usually comes with trade-offs; in other words, as the complexity or length of the protocol increases, the participation or retention rate may become lower (see Chapter 5 (Fritz et al., 2017) on motivation and participation for examples). A contrary example to the Degree Confluence Project in the same domain of VGI photograph-based initiatives is represented by Flickr and Panoramio. These are VGI photograph sharing sites that do not provide any protocols regarding how the photographs should be taken or what information should be added. Users can add a title, a comment/description, one or more tags and the location, but these are optional. The lack of protocols is reflected in the very high participation rates (Michel, 2015; Panorank, 2016), but also in the variable quality of the contributions when considering them for applications such as land cover and land use mapping (see e.g. Leung and Newsam, 2012; Estima and Painho, 2014; Antoniou et al., 2016).

To show an example of the variability of the photographs in terms of tags, a random sample of around 130,000 geotagged photographs that were uploaded to Flickr and Panoramio for the London region in May 2015 was analysed. The frequency of the number of tags associated with the photographs was computed and plotted in Figure 1 as a function of increasing numbers of tags. Clearly the vast majority of photographs (almost 1/3 of the total) have no tags associated to them. In addition, the number of photographs with one to seven tags are within the limits of random variation (although some trends can be spotted; for instance if a user decides to include tags, they usually prefer to append from two to six tags instead of just one). Conversely, the frequency of photographs with eight or more tags shows an almost progressive decrease. This can be seen as a proxy for the following relationship: the more freedom users have in terms of contributions, the more heterogeneous the contributions will be, accompanied with a likely decrease in average quality in terms of their use in further applications. Hence the role of guidelines and protocols could substantially increase the exploitation of VGI for applications not even considered by the person collecting the data.

The definition of protocols is more common in other established citizen science activities where many examples can be found. Accurate data collection by citizens depends on the provision of three elements: clear data collection protocols, simple and logical data forms, and support for participants on protocol use and information submission (Bonney et al., 2009). Pocock et al. (2014) argue that volunteers are more likely to provide information following a given standard if the value of their contribution is recognised. However, if the project requires a complex standard for gathering data, strategies for supporting participants must be deployed and protocols need to be thoroughly tested (Tweddle et al., 2012). Acknowledgement of participants, even simply demonstrating the usefulness of the data, plays a central role in encouraging participation (Pilz et al., 2006).

As discussed in more detail in Chapter 2 (See et al., 2017), VGI can be collected either actively or passively. While in active projects users collect data in a conscious way, passive data collection happens when contributions are gathered without any active engagement (Haklay, 2013). Similarly, Harvey (2013) has made a distinction between truly volunteered versus contributed geographic information (CGI). While the former refers to data that are collected with permission (such as an edit in the OSM database), the latter refers to data collected as part of an automated, open-ended or uncontrollable process (such as the tracking of mobile phones). Information contributed to a passive VGI project typically demands much more processing to result in meaningful information. It is possible to impose a set of protocols in active VGI, but this is usually not possible when using passive VGI or CGI, where the data volumes are often larger than in active sources and hence the data need to be filtered if they are to be used. For example, Bordogna et al. (2015) demonstrated how input data can be filtered based on minimum quality criteria specified by the user, for example to remove geotagged photographs downloaded from repositories such as Flickr and Panoramio.

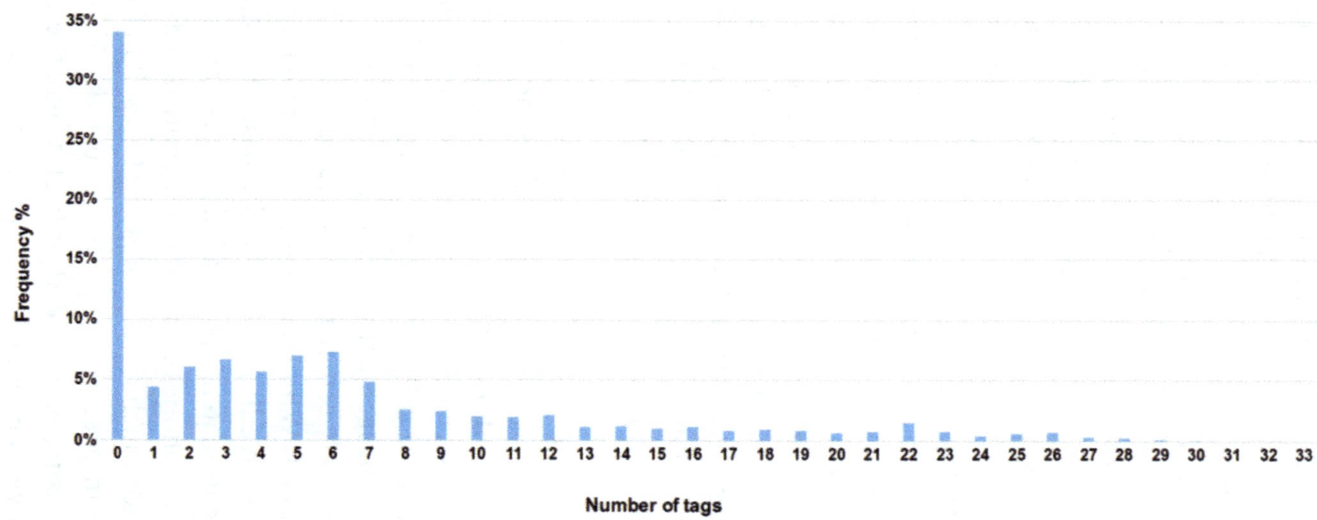

Fig. 1: Frequency of the number of tags for a sample of geotagged photographs uploaded to Flickr and Panoramio (total number of photographs: 130,408; total number of tags: 700,789).

Hence this chapter limits its focus to active VGI projects, where the role played by protocols can be crucial for the quality of the data collected. The chapter seeks to emphasise the need for data collection protocols in VGI projects, and explores how technology can be seamlessly exploited to facilitate collection of suitable data. The chapter takes its origin in a previous work by Mooney et al. (2016), who defined a general and flexible protocol for collecting VGI vector data.

In Section 2 this protocol is briefly presented with the idea of generalising it to all types of VGI projects and VGI data collected. In Section 3 attention is placed on which protocols are required to meet minimum data quality requirements and how technology can play a role in helping to enforce protocols in a user-friendly way. Section 4 presents examples of how the protocol can be applied to two real-world applications, one related to the collection of VGI vector data and the other to geotagged photographs, and reflects upon the relationship between protocols and volunteer motivation. Section 5 concludes the chapter and explores open questions as well as the needs and directions for future research.

2 A Reference Protocol for VGI Collection

A generic protocol has been proposed and developed by Mooney et al. (2016), which can be applied by new VGI projects focused on vector data collection. It can also be used retrospectively on existing data in current VGI projects. This protocol aims to be inclusive of all participants to VGI projects, from new to experienced VGI contributors. By guiding contributors in the process of VGI data collection, the protocol seeks to improve the quality of data in order to both fit the purpose of the specific VGI project for which they are collected and to facilitate their reuse within other, future and potentially unintended, applications. The protocol assumes only a basic working knowledge of geographic information science with basic file and data handling skills from information technology. The protocol has been developed in a bidirectional fashion, i.e. the authors have carefully considered mapping practices in bottom-up approaches (VGI, for example) and top-down approaches (like those used by some NMAs). In this way the protocol is positioned at the intersection between these two opposing approaches for the generation and collection of geographic vector information.

The protocol should be reasonably general and potentially usable by any VGI project based on the collection of vector data through digitisation, field survey or bulk import. The authors have been careful not to relate to any specific VGI initiative, like, for example, OSM, so as to ensure the protocol has potential for further/future customisation or improvement for other specific VGI projects. On the other hand, it gives concrete technical recommendations to easily guide users into a replicable step-by-step data collection process using the tools and processes that they currently possess and use. The protocol is formalised into five main stages as follows:

- Initialisation
- Data Collection
- Self-Assessment/Quality Control
- Data Submission
- Feedback to the Community.

Initialisation – This involves the users of the protocol becoming familiar with the VGI project and its specific goals and objectives. Familiarisation with the proper devices or technologies for the tasks to be accomplished is required. Users are encouraged to conduct tests of the data collection process to familiarise themselves with the process in general.

Data Collection – Users must carefully plan the data collection process. Data collection in this protocol can be considered as one of the following: digitisation, field survey, or bulk import of existing vector data. Obstacles, problems and technical issues with the specific type of data collection method must be carefully considered before proceeding. At all times data collection must be performed according to the VGI project specifications.

Self-Assessment/Quality Control – This step involves users making their own checks and assessments of their data collection process and the data that have been collected. The users should clearly state if problems were encountered (for instance if there was a GPS signal loss during field collection, licence issues in bulk import, or poor resolution imagery used in digitisation).

Data Submission – In this step users submit, potentially using specific application software, all the data to the project website or application. Submission must be successful and a post-submission check should outline any issues that were encountered during this process.

Feedback to the Community – The protocol encourages users to use all available channels to provide feedback on their experiences. According to Perret et al. (2015), controlling, tracking and reporting all aspects of the process is recommended in VGI. Feedback includes any problems that were encountered, issues that the user resolved, tips or guidance for other users in the project etc.

Despite these five main stages of data collection being intended to be sequential, it is sometimes not easy to establish a well defined limit between them. For example, during data collection the VGI contributors may need to get back to the initialisation stage to get more insight on the project specifications; similarly, contributors may realise that quality control is required again after data submission.

Currently, the protocol described is available to participants in VGI projects in the form of a printed or soft copy manual or document. The future goal of this work is to communicate the concepts of the proposed protocol in order to also influence and guide future software implementations for VGI vector data collection. As will be shown through examples in Section 4, in order for the protocol to be effectively adopted by VGI projects, the role of technology – and hence of VGI software developers – is fundamental. If this protocol can be directly implemented in software within VGI projects, the protocol can be communicated to more users and lead to overall improvements in VGI vector data collection.

3 The Role of Protocols for VGI Quality

While for authoritative data the evaluation of data quality is a well established subject, in VGI it remains rather elusive and vague. What is fundamentally different between authoritative data and VGI is the data collection process. For NMAs and CMCs, rigorous protocols and well defined procedures are in place that must be followed by surveyors. The management of surveyors, the updating of the protocols and the specifications, and the migration from a data scheme to another are fully controlled. A totally different landscape exists for VGI projects, in which the enthusiasm of an enormous but disparate set of volunteers is the driving force. In the case of NMAs and CMCs the logic is simple: production protocols and specifications need to be followed, since the final product will be examined for its quality using various measures (such as the ISO/TC211 quality framework). Similarly, in VGI volunteers should have to fully understand that following or ignoring guidelines, best practices and protocols will have a direct impact on the final spatial product and consequently on its usability. VGI projects can learn a lot from the advances in citizen science. In many cases, the quality of data in citizen science is attained through carefully designed and standardised protocols for participation (Kasperowski and Kullenberg, 2015). Standardisation ensures the validity and accuracy of contributions and classifications performed by citizens (Cohn, 2008: 194). In this context, the following subsections examine, in detail, each of the five data collection stages described above against protocol and best practice instructions.

3.1 Initialisation

One aspect that may influence the quality of the collected information is the type of instructions provided to the volunteers in the ***initialisation*** stage. While the initial impulse of most trained surveyors is to employ the standard data quality methods from their field, when designing citizen science projects a different approach for ensuring data quality may be necessary, taking into consideration the degree of participation and the expectations around contributors' skills (Wiggins et al., 2011). If the VGI collection is made for a particular purpose, then the instructions should be detailed enough so that volunteers understand exactly what they are expected to provide. However, instructions with too much detail should be avoided, or at least it should not be mandatory for the volunteer to go through all the detail, because this may be demotivating. The appropriate level of detail of the instructions is, in some circumstances, not easy to establish. Therefore, for some types of VGI projects, studies that identify how volunteers react to several types of instructions should be undertaken, as this reaction may have an important impact on the quality of the generated data (Kerle and Hoffman, 2013). Two practical examples of the importance of instructions for the quality of generated data are the following: if the volunteers need to collect georeferenced photographs, then it should be indicated what must be georeferenced: for example, is it the place where the photograph was taken from or the phenomena shown on the photograph?; and when providing a classification of land cover or disaster damage, how much detailed explanation is required, e.g. the thematic resolution of land cover classes or the choice of one among several damage classes, should be determined.

3.2 Data Collection

Familiarising contributors with the project's aims and goals may enhance their awareness, which, in turn, can help to improve the overall quality of the contributions. Nevertheless, crowdsourced participation inherently suffers from biases, inconsistencies and errors; thus the focus is on how to exclude these inherent characteristics from the ***data collection*** stage. Participation biases can result from various causes. The digital divide, socio-economic factors, demographic distribution and individual perceptions can all have an influence on volunteer contributions (Haklay, 2010; Brovelli et al., 2016b). Here protocols should act preemptively and hinder the appearance of biases. For example, it should be taken for granted that individuals have their own understanding and conceptualisation of the world that might not coincide with a VGI project's mission or specifications. Protocols should clearly state the point of view that volunteers should hold and which processes they should follow to collect the data. In an effort to relieve volunteers from extremely detailed protocols, projects might provide a minimalistic approach on the procedures to follow (Batini et al., 2009). However, this hides two dangers: first, setting the bar lower will probably result in data that are of lower quality. Secondly, more active and experienced volunteers might be discouraged by the approach taken. Thus, the challenge is to provide protocols and best practices that will balance data quality with participation.

3.3 *Self-Assessment/Quality Control*

Data collection might be influenced by factors that make the process error-prone, leading to errors and inconsistencies in the data. For example, weather, landscape, collaboration with other individuals or the instruments used are just a few factors that might affect in-situ measurements. Here the stage of ***self-assessment and quality control*** has much to offer. Thus, before uploading data, each volunteer should self-assess the quality of their data and perform all possible quality controls. Protocols should provide enough guidance and explain common pitfalls that can lead to inconsistencies and errors and how to avoid them.

3.4 *Data Submission*

The next stage for which protocols should provide detailed guidance is ***data submission***. Inevitably, individual contributions are generally small, sparse and fragmented, and yet valuable for the evolution of a crowdsourced project. Active and meticulous data collection followed by indifferent data submission (e.g. just pressing the 'upload' button) might not be sufficient. Protocols should stress that data submitted should, when possible, be validated against existing observations or measurements so that no vague or inconsistent cases appear. Even more important is that an individual's work does not harm or destroy other volunteer contributions. This does not mean that updates or alterations should be avoided, but rather that it is important to have a balance between contributor efforts, a way to evaluate the need for change, and a versioning system capable of roll-back to the previous state of the project if needed. Furthermore, submission should not be confined only to data: protocols should require the addition of metadata and supporting/documentation material when possible. For example, filling a form or submitting a geotagged image might be valuable for quality control by other volunteers or moderators. Similarly, any pitfall, problem or simple concern encountered during the data submission stage should be appropriately added to the contributed data.

3.5 *Feedback to the Community*

Finally, the ***feedback to the community*** may include the participation in discussion forums, which may help other volunteers to create higher quality data. Perret et al. (2015) highlighted the fact that VGI projects should continuously evolve through the feedback each contributor gets from and gives to others, for instance in terms of how a certain problem encountered while collecting data was solved or any other recommendations or guidance. Communication channels with the VGI project managers and administrators should be provided as well so that the project itself can evolve based on the user feedback. Thus, a continuous circle is formed that improves the protocol and enhances the overall VGI project quality. This way, common mistakes will hopefully start to disappear and overall data quality will be improved.

4 Applying the Protocol to Real-World Examples

In this section we present two hypothetical, extended examples of real-world applications of the VGI vector data protocol described above. In the first example, the protocol is applied to the updating and collection of new thematic information in a topographic building database. In the second example the protocol is applied to a different domain, that is the collection of photographs for land use / land cover (LULC) mapping.

4.1 Updating and Collecting New Thematic Information in a Topographic Building Database

In this example, an NMA is interested in exploiting crowdsourced vector data to improve their topographic building database. This improvement includes enriching and updating existing building objects (their geometry and thematic information) and capturing new building objects and associated thematic information. Buildings are typically very well mapped by NMAs, but the rapid pace of urban change can mean that keeping their database up-to-date is challenging in terms of resources. Additionally, the thematic information within these databases is often very poor. Typical information which is often missing includes: the function of the building, the number of floors in the building, cultural heritage information related to the building, the entrance(s), etc. As an additional challenge and motivation for VGI contributors, the NMA seeks to create a new layer from scratch to represent the entrances to buildings. This will be a multi-point layer, since a building might have more than one entrance. In this example, the NMA decides to develop a Web-based application to allow citizens to collect data. The implementation and presence of a protocol for this application will greatly assist in reducing the potential submission of low-quality data. Specifically, the Web-based application will use digitisation and field surveys as the means of collecting vector data. The application will present contributors with three layers: a base layer consisting of up-to-date orthoimagery of the region represented in the database; an overlay layer of the existing topographic building object database; and a layer for the entrances to buildings. Contributors will be encouraged to create and/or update the geometry and/or thematic information of building objects to reflect recent changes to building function, structure, etc. Additionally, contributors will be able to add vector point data to building objects to indicate the position of building entrances along with their door numbers. The implementation of the vector data protocol for this application will ensure that helpful advice and guidance is provided to all contributors in an attempt to maintain and ensure good quality. Guidance is provided for a number of categories:

- *Scale:* Select the appropriate cartographic scale for building level of detail, and preserve it over the collection and contribution process;
- *Shape:* Preserve building shape as much as possible (for instance keep the building corners squared whenever convenient) and digitise minimum details appropriate to the scale;
- *Logical Consistency:* Ensure that new buildings contributed or existing ones that are changed are always closed polygons and do not overlap;
- *Geometric Consistency:* Ensure that multiple entry points to buildings are represented as a multi-point object rather than creating a new point object for each individual entrance in the same building, and that door numbers for each entry point are different;
- *Thematic Quality Control:* Propose a list of thematic attributes and values to the user;
- *Metadata:* Allow free text comments on the visual quality (such as cloud cover, tree cover, shadows or resolution) of the imagery.

The five steps of the protocol workflow outlined in Section 2 are applied to this example as follows:

Initialisation – Citizens will need to register themselves on the Web-based application to use it and contribute vector data and information. Before collecting data, every contributor will need to complete all of the steps in a tutorial demonstration to understand which tasks are required and to familiarise themselves with the processes and tasks in general and with what the goals and objectives of the project are. Depending on the resources available, the NMA may develop a protected 'sandbox' version of the application, where contributors can test out the functionality of the application on a small subset of the topographic buildings database without actually making changes to the real database. This form of training will aid learning and help volunteers contribute effectively while still preserving their motivation.

Data Collection – Contributors will be encouraged to carefully plan their collection of new or updated data/information for the application. The application will specifically allow the digitisation of building objects on top of the orthoimagery, the addition of vector point data on building entrances, and the provision of new or updated thematic information associated with building objects. The software application will give prompts and tips to the contributors as they are working.

Self-Assessment/Quality Control – The application will provide functionality to allow contributors to make an initial assessment of the quality of the new data or changes to existing data that they are submitting. For example, if a contributor creates a new building footprint and does not supply any thematic information, the application would indicate this to the contributor. The contributor would then be presented with a generic list of thematic information from which they can choose the appropriate annotations. This would help emphasise the importance of thematic information in the application in the situation where many users may attach greater importance to geometrical data.
Data Submission – In this step, contributors submit their contributed vector data and/or thematic information to the application. The application will provide a space where contributors can provide metadata or descriptive information about their contribution. This could be used by the NMA to assess the overall quality of the contribution, as this information would describe the processes that the contributors used to make their contributions.
Feedback to the Community – The NMA will create a number of information channels to encourage contributors to provide feedback and discussions on their experiences of using the application and contributing vector data using the application. This feedback can include discussions on problems encountered with specific building types or structures, with certain thematic areas, etc. Through these channels, the NMA can provide assistance and feedback to the contributors in the community by offering suggestions on how problems may be fixed or resolved within the application. This creates a complete feedback loop within the vector protocol, which will allow for the protocol to be continuously improved.

4.2 Using Geotagged Photographs for LULC Mapping

In this example, an NMA is interested in exploiting geotagged photographs to improve their LULC maps, and in particular to provide much more data for training their classification algorithms and also to validate the map, if possible. The NMA has already experimented the use of photographs from existing photo-sharing sites such as Flickr and Panoramio, but it was observed that there was too much inconsistency in the tags and in the content of the photographs and thus that not all photographs were usable for the purpose of LULC mapping. Also, there was a strong spatial bias in the distribution of the photographs and not all required LULC types were captured.

Instead, the NMA decides to develop its own national-level photograph-sharing site specifically for the purpose of collecting photographs for LULC mapping, which will have a stricter protocol and ensure higher usable content and tags. At the same time, the data collection protocol should not hamper creativity or the spontaneous enthusiasm that drives contributors while aiming for the huge volumes of data that are a characteristic of popular social media sites. The NMA decides to develop a customised mobile-based photograph-sharing application, which can use technology to help ensure that specific parts of the data collection protocol are adhered to. The application should have the following features:

- Contributors will be taken through a step-by-step procedure for each location photographed;
- This procedure will require the contributor to take either a set of photographs in the four cardinal directions or a single 360-degree photograph. If the participant chooses the option of taking photographs in four different directions, then the compass in the mobile device will only allow the user to take a photograph when facing the correct cardinal direction;
- The application will prevent participants from using the zoom function, ensuring that the photographs show content closest to their geographic position;
- A 'guide line' will be added to the application so that the contributors can line up the horizon with the 'guide line', so that photographs containing one-third sky and two-thirds landscape are taken;
- The photograph should be dominated by landscape but without restricting the addition of other elements (such as people and animals); moderators or automated methods can be used to assign weights to these photographs for the purpose of LULC creation/validation;
- Once the photographs are taken, the participant will be presented with the possibility to assign tags from a pre-specified list (drawn from the LULC nomenclature used by the NMA) to the photographs, which will be mandatory, along with the possibility to add free form tags, which will be optional;
- The final step in the procedure will be to ask contributors to estimate the distance at which the LULC changes, to indicate how homogeneous or heterogeneous the landscape is;
- There will be at least two modes of operation in the protocol. In the first mode, participants can take photographs at any location, so the geotagged photographs will be useful for creating LULC training datasets; in the second mode, participants will be sent to specific locations, or 'quests' in the form of photograph-caching, which can be used to satisfy the sampling needs of the NMA for LULC map validation and reduce the spatial bias that is common in geotagged photographs from social media photograph-sharing sites.

As much as possible, elements of the protocol will be hidden or incorporated seamlessly into the workflow of the application through technology. In other cases, the protocol will be implemented via elements of gamification, which will be added to maintain, if not grow, the pool of participants and to create a certain level of competition among them, particularly for the photo-caching mode of the application.

Following the vector protocol outlined in Section 2, the five steps are applied as follows:

Initialisation – This first stage will be achieved by providing contributors with a guided tour of the project, including information on how each step contributes to the overall objectives of the project. In addition, step-by-step instructions will be provided to contributors when they first use the application. The guided tour will be mandatory yet short and easy to follow. Once the user has 'passed' through this stage and become familiar with the function of the application, they will be able to take further photographs.

Data Collection – This will be implemented via field survey, which will be facilitated by the mobile application. As outlined above, there will be two main modes of data collection where participants can: (i) photograph landscapes in any location or (ii) be directed to specific locations. Optionally, a third mode will be possible in which participants can turn off the protocol and photograph freely. The purpose of these three modes will be clearly explained to the participants. The mode employed will also allow the NMA to categorise the photographs for a specific use: the first mode may be more suitable for LULC map creation; the second for LULC map validation; while the third can be either omitted or used for training after careful checking.

Self-Assessment/Quality Control – In this step the mobile application will record the positional accuracy and other related parameters (such as dilution of precision (DOP) and type of GPS receiver) as an additional source of information to accompany the photographs. Through the application, the contributor will also estimate the heterogeneity of the LULC, which will provide the NMA with an indication of whether the photograph is in a homogeneous or mixed land cover class. There will be a mechanism implemented that will allow contributors to review the photographs in order to make sure that they comply with the protocol and are of sufficient quality. Contributors will be given the option to retake photographs that are of poorer quality. For instance, in this stage the app will display the position of the photographs taken on top of orthoimagery in order to easily spot positions recorded with low accuracy.

Data Submission – The application will not require data connection in the field but will automatically synchronise the photographs when connected to wifi, so that poor mobile signals will not be an issue. Once photographs are submitted, the online application will allow contributors to view, share and manage their photographs, for instance to correct the tagging of their photographs and thereby improve the labels needed for LULC classification.

Feedback to the Community – The final step will consist in sending out regular information/rich newsletters to contributors, giving them information about levels of improvement in LULC mapping, highlighting those areas that have been better mapped and featuring the contributions of active contributors. It will also highlight what areas are missing and guide participants to go out and photograph these areas. At this stage, the online application will also allow contributors to rate the contributions of other participants and start conversations and discussions in order to exchange and share suggestions that would lead to an overall improvement in the project's data quality.

Although some research on using geotagged photographs for LULC training and validation has been undertaken in the past (see e.g. Antoniou et al., 2016), this example is still largely hypothetical. However, a similar protocol for collecting geotagged photographs for LULC-related purposes is currently being tested by the FotoQuest Europe student campaign[7]. This initiative asks volunteers to survey specific locations with the purpose of validating the official EU LULC datasets derived from the Land Use and Coverage Area frame Survey (LUCAS) performed by EUROSTAT[8]. For more information on what geotagged photographs can offer, see Chapter 4 (Touya et al., 2017) on using geotagged photographs for examining OSM quality and for verifying the applicability and suitability of various cartographic processes.

5 Discussion and Conclusions

VGI has become a mainstream presence in the GIScience domain. By its own nature, the driving force behind VGI lies in the crowd. The progressive mitigation of the digital divide – not just the traditional one that considers Internet access, but also the second-level digital divide that looks at the real capacity of people to make use of available technology (Hargittai, 2002) – will likely result in an ever increasing amount of contributions uploaded to VGI initiatives. Statistics[9] and predictive models (Jokar Arsanjani et al., 2015a) for the OSM project confirm an increasing growth in both the number of new contributors and submitted data, while Mooney and Winstanley (2015) have argued that VGI contributions can be considered a form of big data. In turn, the increase in VGI may also increase the heterogeneity of contributions and hence solving quality issues for assessing VGI usability may become harder in the future.

In citizen science projects, especially those in the field of conservation and ecology, protocols and guidelines for data collection are generally well developed and clearly accepted by the contributors. In contrast, by its very same nature, the world of VGI has developed in a much freer, diverse and often uncontrolled fashion. Even OSM, which since its birth has dominated the VGI scene, features a culture of freedom in terms of what is mapped and which tags are provided. Hence, this chapter has investigated the need and opportunity to integrate protocols in order to rule and guide the data collection process in active VGI projects, with the purpose of increasing the quality of volunteer contributions. A general and flexible protocol was introduced and described, which can be exploited to standardise data collection processes in

VGI initiatives. The protocol is suitable for implementation in new as well as existing VGI projects and can serve as a reference tool, not just for the project volunteers, but also for the project managers and developers who need to put in place the best possible system to facilitate collection of high-quality data. The implementation of the proposed protocol was illustrated through two different hypothetical examples.

The first example sees an NMA developing an application for crowdsourced data collection aimed at enriching and improving its topographic buildings theme. Data collection includes improving and updating existing building objects (geometry and thematic information) and capturing new features related to buildings and associated thematic information such as entrances. The implementation of the vector data protocol for this application will ensure that helpful advice and guidance is provided to all users in an attempt to maintain and ensure good quality as citizens are contributing changes and new content. The protocol provides guidance on building scale, building shape, logical consistency of building polygon, geometric consistency of entry points to buildings, thematic quality and the provision of metadata. Crucially, the use of a protocol here will allow the NMA to outline guidance on these issues so that high-quality data can be captured. The workflow of the protocol (initialisation, data collection, self-assessment/quality control, data submission and feedback to the community) provides more structure to the contribution process for all users regardless of their background skills or technical abilities.

The second example, an example of implementing the protocol for the collection of geotagged photographs for LULC mapping, involved the hypothetical development of a customised photograph-sharing application by an NMA. However, it could also be beneficial for existing photograph-sharing sites like Flickr and Panoramio to adopt elements of the proposed data collection protocol, recording and providing access to a minimum set of metadata. First, locational information is a common feature of modern mobile phones and some digital cameras, so storing and providing the location as standard information does not present any additional burden to these providers. Moreover, the positional accuracy of handheld devices continues to increase, and there are early efforts to also expand this increased accuracy to indoor positioning (Mautz, 2009; Kuo et al., 2014), so the locational quality of information will continue to become better in the future. Similarly, it could be beneficial to record other elements, such as camera orientation, tilt, etc. These metadata are not only useful for geomatics applications but are also of interest to other domains. A prime example is that of user-contributed tags. From touristic applications (Majid et al., 2013) to early response systems (Masó et al., 2011), tags are considered a semantically rich source of information that need to be further enhanced. Also, the photograph-sharing repositories themselves can gain valuable insights from more complete and rich contributions, since these can be analysed to improve the repositories' own services and attract more participants.

The recognition of the need for protocols to guide future VGI projects is clearly lacking. Hence this chapter has attempted to provide a generic set of guidelines that can help VGI projects consider what elements are necessary to ensure that a minimum data standard is reached while still motivating and sustaining participation. Within this broader project protocol, a protocol for data collection is needed, where we would argue that technology should be used to seamlessly integrate components of the protocol as much as possible, thereby reducing the burden of compliance by contributors. This work provides fruitful ground for future research. The proposed protocol was conceived in a sufficiently general way so that it can be potentially applied to any VGI project. Based on the multiple recommendations and suggestions provided in this chapter, we feel that detailed, customised versions of the protocol can now be created and applied easily to specific VGI initiatives, and that future VGI projects would benefit greatly from adhering to the protocol when designing the data collection process. Applying the protocol to existing or future projects would also serve as a way to determine the value of the protocol itself and to suggest possible improvements. Finally, exploiting the protocol to revise the way in which VGI is collected in a project would allow for the comparison of the quality of data produced before and after the protocol's introduction and therefore to help assess its effectiveness.

Notes

[1] https://hotosm.org
[2] http://www.openstreetmap.org
[3] http://wiki.openstreetmap.org/wiki/Map_Features
[4] http://navigator.er.usgs.gov/help/vgistructures_userguide.html
[5] https://www.geohistoricaldata.org
[6] http://confluence.org
[7] http://www.fotoquest-europe.com
[8] http://ec.europa.eu/eurostat/web/lucas/overview
[9] http://wiki.openstreetmap.org/wiki/Stats

Reference list

Al-Bakri, M., Fairbairn, D., 2012. Assessing similarity matching for possible integration of feature classifications of geospatial data from official and informal sources. *International Journal of Geographical Information Science* 26, 1437–1456. DOI: https://doi.org/10.1080/13658816.2011.636012

Antoniou, V., Fonte, C., See, L., Estima, J., Arsanjani, J., Lupia, F., Minghini, M., Foody, G., Fritz, S., 2016. Investigating the feasibility of geo-tagged photographs as sources of land cover input data. *ISPRS International Journal of Geo-Information* 5, 64. DOI: https://doi.org/10.3390/ijgi5050064

Antoniou, V and Skopeliti, A. 2017. The Impact of the Contribution Microenvironment on Data Quality: The Case of OSM. In: Foody, G, See, L, Fritz, S, Mooney, P, Olteanu-Raimond, A-M, Fonte, C C and Antoniou, V. (eds.) *Mapping and the Citizen Sensor*. Pp. 165–196. London: Ubiquity Press. DOI: https://doi.org/10.5334/bbf.h.

Batini, C., Cappiello, C., Francalanci, C., Maurino, A., 2009. Methodologies for data quality assessment and improvement. *ACM Computing Surveys* 41, 1–52. DOI: https://doi.org/10.1145/1541880.1541883

Bearden, M.J., 2007. The National Map Corps. Presented at the Specialist Meeting on Volunteered Geographic Information, University of California at Santa Barbara. Available at: http://www.ncgia.ucsb.edu/projects/vgi/docs/position/Bearden_paper.pdf [Last accessed 1 May 2017].

Bonney, R., Cooper, C.B., Dickinson, J., Kelling, S., Phillips, T., Rosenberg, K.V., Shirk, J., 2009. Citizen science: A developing tool for expanding science knowledge and scientific literacy. *BioScience* 59, 977–984. DOI: https://doi.org/10.1525/bio.2009.59.11.9

Bordogna, G., Carrara, P., Criscuolo, L., Pepe, M., Rampini, A., 2015. A user-driven selection of VGI based on minimum acceptable quality levels. *ISPRS Annals of Photogrammetry, Remote Sensing and Spatial Information Sciences II-3/W5*, 277–284. DOI: https://doi.org/10.5194/isprsannals-II-3-W5-277-2015

Brovelli, M.A., Minghini, M., Molinari, M., Mooney, P., 2016a. Towards an automated comparison of OpenStreetMap with authoritative road datasets. *Transactions in GIS* 21, 191–206. DOI: https://doi.org/10.1111/tgis.12182

Brovelli, M.A., Minghini, M., Zamboni, G., 2016b. Public participation in GIS via mobile applications. *ISPRS Journal of Photogrammetry and Remote Sensing* 114, 306–315. DOI: https://doi.org/10.1016/j.isprsjprs.2015.04.002

Cohn, J.P., 2008. Citizen science: Can volunteers do real research? *BioScience* 58, 192–197. DOI: https://doi.org/10.1641/B580303

Dorn, H., Törnros, T., Zipf, A., 2015. Quality evaluation of VGI using authoritative data—A comparison with land use data in Southern Germany. *ISPRS International Journal of Geo-Information* 4, 1657–1671. DOI: https://doi.org/10.3390/ijgi4031657

Du, H., Anand, S., Alechina, N., Morley, J., Hart, G., Leibovici, D., Jackson, M., Ware, M., 2012. Geospatial information integration for authoritative and crowd sourced road vector data. *Transactions in GIS* 16, 455–476. DOI: https://doi.org/10.1111/j.1467-9671.2012.01303.x

Estima, J., Painho, M., 2014. Photo based Volunteered Geographic Information initiatives: A comparative study of their suitability for helping quality control of Corine Land Cover. *International Journal of Agricultural and Environmental Information Systems* 5, 73–89. DOI: https://doi.org/10.4018/ijaeis.2014070105

Fan, H., Yang, B., Zipf, A., Rousell, A., 2016. A polygon-based approach for matching OpenStreetMap road networks with regional transit authority data. *International Journal of Geographical Information Science* 30, 748–764. DOI: https://doi.org/10.1080/13658816.2015.1100732

Fan, H., Zipf, A., Fu, Q., Neis, P., 2014. Quality assessment for building footprints data on OpenStreetMap. *International Journal of Geographical Information Science* 28, 700–719. DOI: https://doi.org/10.1080/13658816.2013.867495

Fonte, C C, Antoniou, V, Bastin, L, Estima, J, Arsanjani, J J, Bayas, J-C L, See, L and Vatseva, R. 2017. Assessing VGI Data Quality. In: Foody, G, See, L, Fritz, S, Mooney, P, Olteanu-Raimond, A-M, Fonte, C C and Antoniou, V. (eds.) *Mapping and the Citizen Sensor*. Pp. 137–163. London: Ubiquity Press. DOI: https://doi.org/10.5334/bbf.g.

Foody, G., Boyd, D., 2012. Using volunteered data in land cover map validation: Mapping tropical forests across West Africa, in: Proceedings of the 2012 IEEE International Geoscience and Remote Sensing Symposium (IGARSS), Munich, Germany, 22–27 July 2012, pp. 2368–2371. DOI: https://doi.org/10.1109/IGARSS.2012.6352675

Fritz, S., McCallum, I., Schill, C., Perger, C., Grillmayer, R., Achard, F., Kraxner, F., Obersteiner, M., 2009. Geo-Wiki.Org: The use of crowdsourcing to improve global land cover. *Remote Sensing* 1, 345–354. DOI: https://doi.org/10.3390/rs1030345

Fritz, S., McCallum, I., Schill, C., Perger, C., See, L., Schepaschenko, D., van der Velde, M., Kraxner, F., Obersteiner, M., 2012. Geo-Wiki: An online platform for improving global land cover. *Environmental Modelling & Software* 31, 110–123. DOI: https://doi.org/10.1016/j.envsoft.2011.11.015

Fritz, S, See, L and Brovelli, M. 2017. Motivating and Sustaining Participation in VGI. In: Foody, G, See, L, Fritz, S, Mooney, P, Olteanu-Raimond, A-M, Fonte, C C and Antoniou, V. (eds.) *Mapping and the Citizen Sensor*. Pp. 93–117. London: Ubiquity Press. DOI: https://doi.org/10.5334/bbf.e.

Gao, S., Li, L., Li, W., Janowicz, K., Zhang, Y., 2014. Constructing gazetteers from volunteered Big Geo-Data based on Hadoop. *Computers, Environment and Urban Systems*. DOI: https://doi.org/10.1016/j.compenvurbsys.2014.02.004

Girres, J.-F., Touya, G., 2010. Quality assessment of the French OpenStreetMap dataset. *Transactions in GIS* 14, 435–459.

Goodchild, M.F., 2007. Citizens as sensors: the world of volunteered geography. *GeoJournal* 69, 211–221. DOI: https://doi.org/10.1007/s10708-007-9111-y

Haklay, M., 2013. Citizen science and volunteered geographic information: Overview and typology of participation, in: Sui, D., Elwood, S., Goodchild, M. (Eds.), *Crowdsourcing Geographic Knowledge*. Springer Netherlands, Dordrecht, Netherlands, pp. 105–122.

Haklay, M., 2010. How good is volunteered geographical information? A comparative study of OpenStreetMap and Ordnance Survey datasets. *Environment and*

Planning B: Planning and Design 37, 682–703. DOI: https://doi.org/10.1068/b35097

Hargittai, E., 2002. Second-level digital divide: Differences in people's online skills. *First Monday* 7.

Harvey, F., 2013. To volunteer or to contribute locational information? Towards truth in labeling for crowdsourced geographic information, in: Sui, D., Elwood, S., Goodchild, M. (Eds.), *Crowdsourcing Geographic Knowledge.* Springer Netherlands, Dordrecht, Netherlands, pp. 31–42.

Johnson, P., Sieber, R., 2013. Situating the adoption of VGI by government, in: Sui, D., Elwood, S., Goodchild, M. (Eds.), *Crowdsourcing Geographic Knowledge.* Springer Netherlands, pp. 65–81.

Jokar Arsanjani, J., Mooney, P., Helbich, M., Zipf, A., 2015a. An exploration of future patterns of the contributions to OpenStreetMap and development of a Contribution Index: Future Patterns of the Contributions to OpenStreetMap. *Transactions in GIS* 19, 896–914. DOI: https://doi.org/10.1111/tgis.12139

Jokar Arsanjani, J., Mooney, P., Zipf, A., Schauss, A., 2015b. Quality assessment of the contributed land use information from OpenStreetMap versus authoritative datasets, in: Jokar Arsanjani, J., Zipf, A., Mooney, P., Helbich, M. (Eds.), *OpenStreetMap in GIScience, Lecture Notes in Geoinformation and Cartography.* Springer International Publishing, Cham, pp. 37–58.

Jokar Arsanjani, J., Zipf, A., Mooney, P., Helbich, M. (Eds.), 2015c. *OpenStreetMap in GIScience, Lecture Notes in Geoinformation and Cartography.* Springer International Publishing, Cham.

Kasperowski, D., Kullenberg, C., 2015. Learning in citizen science projects: types, behaviour and protocols, in: *The Conference on Computer Science Collaborative Learning, CSCLROWD2015: Designing Futures for Learning in the Crowd: New Challenges and Opportunities for CSCL*, Gothenburg, Sweden, 7 June 2015. Available at: https://csclcrowd2015.files.wordpress.com/2015/04/csclcrowd2015_submission_4.pdf [Last access 16 May 2017]

Kerle, N., Hoffman, R.R., 2013. Collaborative damage mapping for emergency response: the role of Cognitive Systems Engineering. *Natural Hazards and Earth System Science* 13, 97–113. DOI: https://doi.org/10.5194/nhess-13-97-2013

Kuo, Y.-S., Pannuto, P., Dutta, P., 2014. Demo: Luxapose: Indoor positioning with mobile phones and visible light, in: Proceedings of the 20th Annual International Conference on Mobile Computing and Networking, MobiCom '14. ACM, New York, NY, USA, pp. 299–302. DOI: https://doi.org/10.1145/2639108.2641747

Leung, D., Newsam, S., 2012. Exploring geotagged images for land-use classification, in: Proceedings of the ACM Multimedia 2012 Workshop on Geotagging and Its Applications in Multimedia (GeoMM '12). Presented at the GeoMM '12, ACM Press, New York, NY, pp. 3–8. DOI: https://doi.org/10.1145/2390790.2390794

Ludwig, I., Voss, A., Krause-Traudes, M., 2011. A comparison of the street networks of Navteq and OSM in Germany, in: Geertman, S., Reinhardt, W., Toppen, F. (Eds.), *Advancing Geoinformation Science for a Changing World*. Springer Berlin Heidelberg, Berlin, Heidelberg, pp. 65–84.

Majid, A., Chen, L., Chen, G., Mirza, H.T., Hussain, I., Woodward, J., 2013. A context-aware personalized travel recommendation system based on geotagged social media data mining. *International Journal of Geographical Information Science* 27, 662–684. DOI: https://doi.org/10.1080/13658816.2012.696649

Masó, J., Diaz, P., Pons, X., 2011. Mapping geo-tagged pictures in a disaster management event, in: *ISPRS Annals of the Photogrammetry, Remote Sensing and Spatial Information Sciences*, Antalya, Turkey, 3–8 May 2011, available at: http://www.isprs.org/proceedings/2011/Gi4DM/PDF/OP24.pdf [Last accessed 16 May 2017].

Mautz, R., 2009. Overview of current indoor positioning systems. *Geodezija ir Kartografija* 35, 18–22. DOI: https://doi.org/10.3846/1392-1541.2009.35.18-22

Michel, F., 2015. How many photos are uploaded to Flickr every day, month, year? Available at https://www.flickr.com/photos/franckmichel/6855169886 [Last accessed 27 June 2016].

Mooney, P and Minghini, M. 2017. A Review of OpenStreetMap Data. In: Foody, G, See, L, Fritz, S, Mooney, P, Olteanu-Raimond, A-M, Fonte, C C and Antoniou, V. (eds.) *Mapping and the Citizen Sensor*. Pp. 37–59. London: Ubiquity Press. DOI: https://doi.org/10.5334/bbf.c.

Mooney, P., Minghini, M., Laakso, M., Antoniou, V., Olteanu-Raimond, A.-M., Skopeliti, A., 2016. Towards a protocol for the collection of VGI vector data. *ISPRS International Journal of Geo-Information* 5(11), 217. DOI: https://doi.org/10.3390/ijgi5110217

Mooney, P., Winstanley, A., 2015. Is VGI Big Data? in: *Proceedings of GISRUK 2015*, extended abstracts, University of Leeds, Leeds, UK, 15–17 April 2015. Available at: http://leeds.gisruk.org/abstracts/GISRUK2015_submission_65.pdf [Last accessed 27 June 2016].

Neis, P., Zielstra, D., Zipf, A., 2011. The street network evolution of crowdsourced maps: OpenStreetMap in Germany 2007–2011. *Future Internet* 4, 1–21. DOI: https://doi.org/10.3390/fi4010001

Olteanu-Raimond, A.-M., Hart, G., Foody, G., Touya, G., Kellenberger, T., Demetriou, D., 2017a. The scale of VGI in map production: A perspective of European National Mapping Agencies. *Transactions in GIS* 21, 74–90. DOI: https://doi.org/10.1111/tgis.12189

Olteanu-Raimond, A-M, Laakso, M, Antoniou, V, Fonte, C C, Fonseca, A, Grus, M, Harding, J, Kellenberger, T, Minghini, M, Skopeliti, A. 2017b. VGI in National Mapping Agencies: Experiences and Recommendations.

In: Foody, G, See, L, Fritz, S, Mooney, P, Olteanu-Raimond, A-M, Fonte, C C and Antoniou, V. (eds.) *Mapping and the Citizen Sensor*. Pp. 299–326. London: Ubiquity Press. DOI: https://doi.org/10.5334/bbf.m.

Panorank, 2016. Global stats. Available at http://www.panorank.com/index.php?lang=en&op=global. [Last accessed 27 June 2016].

Perret, J., Gribaudi, M., Barthelemy, M., 2015. Roads and cities of 18th century France. *Scientific Data* 2, 150048. DOI: https://doi.org/10.1038/sdata.2015.48

Pilz, D., Ballard, H.L., Jones, E.T., 2006. Broadening Participation in Biological Monitoring: Handbook for Scientists *and Managers (General Technical Report No. PNW-GTR-680)*. United States Department of Agriculture, Forest Service, Pacific Northwest Research Station, Portland, OR, USA.

Pocock, M.J.O., Chapman, D.S., Sheppard, L.J., Roy, H.E., 2014. *Choosing and Using Citizen Science: a guide to when and how to use citizen science to monitor biodiversity and the environment*. Centre for Ecology & Hydrology, UK.

Pourabdollah, A., Morley, J., Feldman, S., Jackson, M., 2013. Towards an authoritative OpenStreetMap: Conflating OSM and OS openData national maps' road network. *ISPRS International Journal of Geo-Information* 2, 704–728. DOI: https://doi.org/10.3390/ijgi2030704

Ruas, A., Plumejeaud, C., Nahassia, L., Grosso, E., Olteanu, A.-M., Costes, B., Vouloir, M.-C., Motte, C., 2014. GéoPeuple: The creation and the analysis of topographic and demographic data over 200 Years, in: Buchroithner, M., Prechtel, N., Burghardt, D. (Eds.), *Cartography from Pole to Pole*. Springer Berlin Heidelberg, Berlin, Heidelberg, pp. 3–17.

See, L, Estima, J, Pőd.r, A, Arsanjani, J J, Bayas, J-C L and Vatseva, R. 2017. Sources of VGI for Mapping. In: Foody, G, See, L, Fritz, S, Mooney, P, Olteanu-Raimond, A-M, Fonte, C C and Antoniou, V. (eds.) *Mapping and the Citizen Sensor*. Pp. 13–35. London: Ubiquity Press. DOI: https://doi.org/10.5334/bbf.b

See, L., Mooney, P., Foody, G., Bastin, L., Comber, A., Estima, J., Fritz, S., Kerle, N., Jiang, B., Laakso, M., Liu, H.-Y., Milčinski, G., Nikšič, M., Painho, M., Pődör, A., Olteanu-Raimond, A.-M., Rutzinger, M., 2016. Crowdsourcing, citizen science or Volunteered Geographic Information? The current state of crowdsourced geographic information. *ISPRS International Journal of Geo-Information* 5, 55. DOI: https://doi.org/10.3390/ijgi5050055

Sui, D.Z., 2007. Volunteered Geographic Information: A tetradic analysis using McLuhan's law of the media. Position Paper for the NCGIA and Vespucci Workshop on Volunteered Geographic Information. Santa Barbara, CA, 13–14 Dec 2007. Available at: http://ncgia.ucsb.edu/projects/vgi/docs/position/Sui_paper.pdf [Last accessed 27 June 2016].

Touya, G, Antoniou, V, Christophe, S and Skopeliti, A. 2017. Production of Topographic Maps with VGI: Quality Management and Automation. In: Foody, G, See, L, Fritz, S, Mooney, P, Olteanu-Raimond, A-M, Fonte, C C and Antoniou, V. (eds.) *Mapping and the Citizen Sensor*. Pp. 61–91. London: Ubiquity Press. DOI: https://doi.org/10.5334/bbf.d.

Touya, G., Coupé, A., Jollec, J., Dorie, O., Fuchs, F., 2013. Conflation optimized by least squares to maintain geographic shapes. *ISPRS International Journal of Geo-Information* 2, 621–644. DOI: https://doi.org/10.3390/ijgi2030621

Tweddle, J.C., Robinson, L.D., Pocock, M.J.O., Roy, H., 2012. *Guide to citizen science: developing, implementing and evaluating citizen science to study biodiversity and the environment in the UK*. Natural History Museum and NERC Centre for Ecology & Hydrology.

Wiggins, A., Newman, G., Stevenson, R.D., Crowston, K., 2011. Mechanisms for data quality and validation in citizen science. Presented at the Seventh International Conference on e-Science Workshops, IEEE, pp. 14–19. DOI: https://doi.org/10.1109/eScienceW.2011.27

CHAPTER 11

Data and Metadata Management for Better VGI Reusability

Lucy Bastin[*,†], Sven Schade[*] and Christian Schill[‡]

[*]European Commission, Joint Research Centre, Ispra, Italy,
[†]Aston University, Birmingham UK, lucy.bastin@jrc.ec.europa.eu
[‡]Albert-Ludwig University, Freiburg, Germany

Abstract

The rapid expansion of citizen science projects and crowdsourcing applications is yielding a huge and varied pool of Volunteered Geographic Information (VGI) on a wide variety of themes. This VGI may be of huge value for institutions, individuals and decision-makers, but only if it can be discovered, evaluated for quality and fitness-for-purpose and combined with data from other sources. If VGI data are to be discovered, used and reused to their full potential, they must be actively managed. In this chapter we assess the current state of the art regarding data management practices in VGI, identify some challenges, obstacles and best-practice examples, and review a range of developing and established open source technologies which can underpin robust and sustainable data management for VGI. We conclude that VGI is likely to remain patchy and heterogeneous and that existing standards may not be exploited to their full potential. Nevertheless, automated support for documenting the generation and use of VGI, as well as annotations following the Linked Data paradigm, can help to improve interoperability and reuse. We were able to identify good practices within different existing systems, but more research and development work is needed in order to support their joint application for the

How to cite this book chapter:

Bastin, L, Schade, S and Schill, C. 2017. Data and Metadata Management for Better VGI Reusability. In: Foody, G, See, L, Fritz, S, Mooney, P, Olteanu-Raimond, A-M, Fonte, C C and Antoniou, V. (eds.) *Mapping and the Citizen Sensor.* Pp. 249–272. London: Ubiquity Press. DOI: https://doi.org/10.5334/bbf.k.

benefit of VGI. New data management methodologies can only succeed if their benefits (for example, simplifying administration or lowering the entry barrier to data publication) exceed the implementation costs.

Keywords

Data Management; Quality Assurance; Quality Control; Interoperability; Open Standards

1 Introduction

The visibility and perceived importance of VGI projects and citizen science is continuously increasing, and this book offers insight into many aspects of user-generated content and VGI collections. In this chapter, we summarise some insights on good practice for the storage and dissemination of this type of data.

Data collection and information retrieval in crowdsourcing or VGI projects may happen on very different spatial and temporal scales and diverse thematic areas, and may involve very varied groups of contributors in terms of expertise and interests. VGI campaigns can include, for example, short-term emergency response projects (e.g. after earthquakes and other natural disasters) that exploit volunteered observations along with repurposed information harvested from social media; Citizens' Observatories such as those funded by the European Commission[1], which have structured and strategic goals to foster '... general public engagement in scientific research activities when citizens actively contribute to science either with their intellectual effort or surrounding knowledge or with their tools and resources...'(Socientize, 2013); or well established infrastructures and frameworks such as the Global Biodiversity Information Facility (GBIF), which has collated and registered decades-worth of global species data.

Inherently, such initiatives have quite heterogeneous requirements for data cataloguing, access to data, licensing and long-term availability of data, but they do (or at least they should) share some general 'good practice principles' of data management. These principles include aspects such as how to securely store data; how to grant access and to whom; how to document data so they can be found by humans or machines for specific purposes; and how to develop a common understanding of the meaning of collected information so that data can be understood and used, at the very least within the context of the original project, but potentially also outside that domain.

In 2014, the Joint Research Centre (JRC; the EC's science service) in Ispra, Italy, conducted a 'Citizen Science and Smart Cities Summit' and summarised in a technical report (Craglia and Granell, 2014) that at the time when they wrote '... there [was] little interoperability and reusability of [user-generated] data, apps, and services developed in each project.' A follow-up survey reinforced these conclusions, especially in relation to data management practices in citizen science projects (Schade and Tsinaraki, 2016). Acknowledging these observations, this chapter summarises good practice recommendations in data/metadata management and curation, as well as details on international standards and cross-community interoperability that can potentially overcome the identified shortcomings. Proper application of these principles could permit seamless integration of data sources from different domains into coherent information that can be reused beyond the scope of the original problem – thus leveraging user-contributed content 'to the next level', i.e. making the data discoverable, easier to reuse and thus even more valuable.

2 Data Management Overview

This section first introduces the required background about the topic. It is then devoted to some of the most central aspects of data management. We focus on those items that cut across all types of data and data sources, and highlight the foundational issues that should be addressed in data management and the related planning processes.

2.1 Background

Data appear in many different forms and originate from an ever-increasing number of sources – and VGI is no exception. VGI has huge potential to enrich the data portfolios of the public sector (e.g. environmental measurement stations, earth observing satellites, land surveys and consultations) and of the private/corporate sectors (e.g. mobile phone data, sensor measurements inside vehicles, market studies, etc.). However, the heterogeneous nature of VGI presents challenges for integrating with these 'traditional' data assets, which are generally structured according to the application domains from which they arise, and formatted according to industry standards, which may or may not be open-source. As seen from the concrete examples in this book, VGI can encompass a wide range of measurement and observation types, including GPS tracks, digitised vector graphics, occurrence information, tagged photographs and sound recordings, and observations of individual species over time.

Each of these datasets is generated/collected for an intended purpose (i.e., to deliver some value for a beneficiary), and is dealt with in a particular way. In other words, it is 'managed' in one way or another – independently of the availability of any form of data management plan. The approaches by which data in general, and VGI in particular, are managed diverge greatly, and are highly dependent on the context of generation and use. For example, data collected locally in a field trip to teach a small group of students about digital cartography might be kept on an SD card, be copied to several desktop computers at the university and be deleted as soon as the course ends. By contrast, worldwide observations about species occurrences might be fed into a well networked structure in order to contribute to a global collection effort which will curate those data for generations of scientists and environmental organisations.

Although it might be debatable whether every single collected dataset should be preserved for potential future use, sharing of volunteer-generated data is a part of the unspoken contract with the original contributors that underlies citizen science, and can be crucial in maintaining the commitment of volunteers. Bearden (2007) records how, in the absence of feedback on their mapping efforts, volunteer USGS contributors '... would become alienated when they realized that their meticulous work would not be used in the foreseeable future ...'. In a broader context, if data are likely to be usable for science, then, following recent moves towards reproducibility, they must be made reusable. These requirements for repeatability, transparency and independent evaluation inevitably suggest a need to curate and preserve data collections. With the growing availability of data storage and data sharing capacities, many of the technical needs are well addressed. However, organisational peculiarities and the differences between communities of practice mean that, in reality, multiple different approaches can be applied. While some thematic areas and communities have well established and internally consistent approaches to data handling and sharing, those experiences and practices are rarely exchanged widely across parties with different interests. To give an example: the geospatial community (or, more strictly speaking, the spatial data infrastructure (SDI) community), has developed in-depth knowledge and best-practice recommendations on managing geographic and other spatial information using web services – especially under the ISO Technical Committee on Geographic Information/Geomatics (ISO/TC211) and the Open Geospatial Consortium (OGC). However, interconnections with the biodiversity and nature conservation community have until recently been limited to a few dedicated projects, including, for example, EU BON[2] and COBWEB[3]. However, as citizen science moves into a new era of data aggregation and harmonisation, this situation is changing fast, making a discussion of data management practices especially topical in the domain of VGI. We will re-visit some of the SDI community standards below, in order to indicate reuse potentials.

While each individual collection of VGI is valuable to preserve *per se*, VGI also has reuse potential for purposes that might not have been initially foreseen. These purposes might include longitudinal studies on the use and evolving concept of VGI itself, but could also involve integration with other data sources and interconnection with previously unknown data flows and systems. It is therefore an emerging practice to follow common standards and support interoperability, in order to avoid introducing artificial barriers to such novel and unforeseen usages of VGI. The Group on Earth Observation (GEO) recently published just such a set of data management principles for the Global

Earth Observation System of Systems (GEOSS)[4]. Simultaneously, and along the same lines, the Belmont Forum – a group of the world's major and emerging funders of global environmental change research – released their data principles[5]. The latter principles focus on Findability, Accessibility, Interoperability and Reuse (FAIR) and will be used as a lens through which to assess the state of the art in Section 2.

2.2 Organising Data

One of the very first challenges is the organisation of the data themselves. Before even considering the concrete storage format and structure used, it has to be decided at some point which items are considered data in an 'atomic' form, and how these items might be packaged. As we will see later in the chapter, these early decisions will impact other areas, such as the provision of (persistent) identifiers or the granularity of metadata (data about data). In the context of airborne imagery, the decision could be whether to make accessible as one unit a whole series of images from airborne imagery gathered in a single flight or whether to treat each single scene (image) as a single dataset. Analogously, a species observation could be put into a collection that unites all data relating to a particular day, person, sensor type (e.g. smartphone), administrative region, area of interest (e.g. a natural park), field campaign, etc. The particular choice of grouping will depend on the intended use, which in turn will define the discovery and access needs.

2.3 Persistent Identifiers

Data can only be unambiguously recognised – especially when they are shared with other people – if they can be uniquely and persistently identified. In other words, the data need to be branded in some way that does not change over time. If the data are to be accessible, it must also be possible to resolve that persistent and unique identifier into an appropriate data request.

Without going into too much detail about the meaning of uniqueness and identity, it obviously makes a difference whether a persistent and unique identifier is assigned to every 'atomic' data item or to collections that apply any of the criteria listed above.

The meaning of persistency also has to be challenged: which authorities can guarantee the persistency and uniqueness of identifiers? What if identifiers contain the names of institutions or groups that disappear in real life? Who can guarantee a service that resolves certain identifiers in order to retrieve the actual dataset? Furthermore, it has to be noted that in cases where unique and persistent identifiers are allocated to a data stream, for example one generated by a person or a sensor, the retrieved data will change over time. In practice, the identifier could resolve to the latest data item that has been collected, or to an accumulated collection. Some specific mechanisms for minting and managing persistent identifiers are detailed and described in Section 3.

2.4 Data Documentation

Are we able to use a dataset that we created ourselves? Can we use it again a few years after we collected it? How are others supposed to find that dataset, understand what it really encapsulates (and assess if it might be valuable for their work), access it and provide their experiences and impressions about it? The answer to all of these questions lies in metadata, or, in other words, the appropriate documentation of data – an answer which is more easily given than implemented.

Documentation is required for a wide range of purposes (e.g. discovery, evaluation and use), and therefore possible forms of documentation vary greatly. Here, again, the packaging of VGI is one determining factor, since one might document a range of possible 'entities', for example: a single observation; observations from one person (including also a description of that person); and VGI collected for a particular area (including also documentation about the area). A dataset stored as a collection of individual observations or measurements might include information about the accuracy of each single value; it has to be determined how this accuracy information is then propagated to a collection of measurements in order to achieve an overall quality measure for the dataset. If a user is filtering this dataset for potential use in an analysis and their fitness-for-purpose criteria include accuracy, then, in theory, this aggregate measure of quality should be recalculated for each candidate set of observations – a considerable challenge for the architecture within which the data are being curated and made accessible for discovery. To give another example, in a VGI dataset where observations can be attributed to an individual, the documentation might include the reputation of this individual in the context of a particular activity or community; but how should such values be propagated when talking about a group of people? At the time of writing, accessible and robust tools for this type of aggregation are lacking.

Another important feature of documentation is the semantics used to describe what is actually being measured. Terms and units that are implicit in one domain are often taken for granted, and not necessarily well recorded for communication with potential users in other fields. For example, the choice of code list, (i.e. determined terminologies of a particular community) to constrain keywords about a data collection might hinder others in finding the data collection because they use other words to say the same thing, or might confuse people expecting something completely different because they use the same word to say something else. Only where semantic mappings between code lists are available can these cross-domain discoveries be made possible and reliable.

Such 'cross-walking' initiatives are very valuable, because, by contrast to free text, which is complicated and laborious to parse and mine, code lists and restricted vocabularies are extremely valuable ways to speed up the filtering and fitness-for-purpose assessment of datasets. Natural language processing is powerful and becoming more so, as can be seen from the increasing support for automated systems such as chatbots. However, these systems model primarily social contexts, and are not yet coupled to the kind of semantic matching and inference that are needed to distinguish the correct context in which a word is being used to describe an indicator, unit of measure or phenomenon across different scientific fields. For example, if a user is searching globally for datasets that include numerical estimates of uncertainty or variability, they could search for free text descriptions that include terms such as '*variance*', '*standard deviation*', '*ecart-type*' or '*intervalo de confianza*'. However, the presence of such words does not guarantee that variability is indeed mathematically described within the dataset, since, for example, the word 'variance' can also be used in a qualitative sense. By contrast, a URI[6] identifies, via the vocabulary server of the UK's National Environmental Research Council, a definition of 'variance' that is explicitly mathematical and that can be related to other defined statistical concepts, across spoken languages and scientific domains. A similar clarification of terms such as 'sea level' can be seen at the SeaDataNet vocabulary server[7].

For this reason, many classic metadata elements allow free text only for titles and descriptions but require selection from code lists for everything else. We will consider some examples of this practice below, in the section relating to standards. However, there are times when there is no substitute for human-readable material such as manuals and descriptions of research methods, and so methods for adding or linking these to VGI datasets as annotations must be considered. Such documentation can encourage the dissemination of a dataset and might raise the reputation of those who created it – see, for example, the first publication within the newly established geospatial dataset description section of the *International Journal of Spatial Data Infrastructures Research* [8], or the recently launched *Data in Brief* journal[9]. Such documents can convey organisational priorities that are hard to capture otherwise: they can help others to understand the deeper intentions behind why a dataset has been collected, and the reasons for organisational decisions, thereby contributing to the understanding of the overall purpose and potential reusability of a dataset.

Last but not least, it should be considered whether feedback can be collected on the dataset (at whatever level of granularity the packaging allows). Such feedback might include ratings, written statements and references to cases of reuse, but also more direct indications of potential error, identified needs for updating, etc.

2.5 Sharing - With Whom?

The management and curation of datasets not only is an exercise for those gathering and hosting data, but also benefits the users, whether those are the originally-intended beneficiaries or new user groups that find value in reusing a dataset for their own purposes. Access and use conditions may vary – e.g. depending on privacy and legal issues (see also Chapter 6, Mooney and Minghini, 2017 on privacy, legal issues and ethics), commercial interests, or an organisation's commitment to Open Science. However, VGI can only be exploited to its full potential if these conditions are clearly articulated and, ideally, accompanied by the relevant licences. The decision to integrate or split VGI into collections will have an impact here, since permissions on different elements of a VGI dataset could be different, meaning that different consumers would access different collections of records.

Having persistent identifiers and a minimum set of documentation (including contributors, title and release date) in place also enables proper data citation – an element that should not be underestimated. On the one hand, citable VGI allows clear reuse, since reference can now be made not only to other scientific articles, but also unambiguously to data used within a particular activity. On the other hand, data citation also provides a means of acknowledging the source – thereby contributing to the recognition of the data contributors and owners and providing an incentive for the provision of metadata and curation of VGI. It is likely that new metrics for scientific reputation (altmetrics) will very soon take these achievements into account; the cross-referencing of datasets and the numbers of citations will become essential measures of impact.

3 The Role of Open Standards for VGI Data Management

In the above discussion we have identified a number of crucial practices for ensuring the usability and usefulness of VGI data. A number of tools and protocols exist which can support these practices, and key among these are the various open standards which allow data to be described, structured, exchanged, discovered and documented in ways which best promote interoperability and reuse. In this context, we use the word 'standards' not to denote quality standards, which are addressed in Chapter 7, but agreed schemas, formats and protocols from bodies such as the World Wide Web Consortium (W3C)[10] and OGC[11], which, by virtue of being open for free use, are accessible to a wide range of users across scientific and other domains.

In the following section, the FAIR principles will be used to structure discussion of the tools and approaches that are available. This minimum set of foundational principles originally derives from a 2014 workshop that brought together a wide range of 'academic and private stakeholders all of whom had an interest in overcoming data discovery and reuse obstacles'. The principles have been subsequently developed and refined with the goal of ensuring that 'research objects should be Findable, Accessible, Interoperable and Reusable (FAIR) both for machines and for people' – allowing stakeholders to 'more easily discover, access, appropriately integrate and re-use, and adequately cite, the vast quantities of information being generated by contemporary data-intensive science' (Wilkinson et al., 2016). FAIR is intended to be domain-independent and to be applicable to data archival, management, exploration, discovery and reuse across a range of research fields and scholarly disciplines.

Examples have been chosen from the current practice of the Global Biodiversity Information Facility to illustrate certain sections of FAIR. The reason for this choice is that GBIF is an extremely good example of cross-domain strategic thinking where standards from different fields have been employed, adapted, influenced and developed in order to generate a highly usable, scientifically robust repository of data from hugely varying sources that supports hundreds of high-quality peer-reviewed scientific analyses each year[12].

The FAIR principles are as follows:

F1. (meta)data are assigned a globally unique and persistent identifier
F3. metadata clearly and explicitly include the identifier of the data it describes

As described above, data can only be sensibly shared and reused if the data resource can be identified and reliably retrieved. Persistent identifiers are unique strings of numbers and/or characters that are assigned to a digital resource (e.g. datasets, documents, images) in order to allow long-term, reliable access to that specific item. Persistent identifiers should ideally be managed separately from the physical location of the resource, ensuring the continued accessibility and discoverability of the resource 'no matter how many times the object moves to different servers or property rights owners' (USGS, 2017). *Actionable* persistent identifiers permit access to the resource via a link, which should remain resolvable for the long term. An example that is widely used in the scientific domain is the Digital Object Identifier (DOI; ISO standard 26324:2012)[13], which allows published documents and datasets to be tracked and cited, and which is assigned to journal publications (or prepublications) by CrossRef[14], Figshare[15], Zenodo[16] and other platforms. Recent moves towards data DOIs have been hugely supported by initiatives such as DataCite[17], NOAA's EZID[18], or DryadLab[19], which enable a data producer to mint a DOI and, in some cases, register associated metadata.

An example current practice for VGI is the ability of the GBIF website to produce and maintain a DataCite DOI for a specific user request, guaranteeing that this request can be reliably repeated at a future date. Different query filters (date, type of record, species' scientific name, country, etc.) are collated and stamped with a DOI, which is supplied to the user to ensure future retrieval of records according to the same filters.

A DOI can be allocated at a level of granularity specified by the user, but the maintenance of relationships (e.g. hierarchical 'nestings' of DOIs) is the responsibility of the resource owner, and can be challenging. The ability to discover related datasets in this way is extremely powerful, and can support the Linked Data approach described more fully in the next section. Attention to versioning is also important: a DOI may represent the final version of a resource, approved for release; an extension or annotation of a resource; or a model/algorithm version used in a reproducible workflow (in this context, a github or subversion version ID can be adapted to fulfil at least some of the role of a DOI). However, there are cases where a DOI will always return 'the latest version' of a resource, and, here, scientific reproducibility is not guaranteed. GBIF DOIs are a good example: the data underlying a query are regularly improved and updated, and historical records may be retrospectively added, meaning that the exact same set of records is not guaranteed to be returned when a DOI is used at a later date.

It is possible to embed dataset identifiers within metadata using existing geospatial metadata standards, such as ISO 19115[20], which offers a CI_Citation element that allows an identifier such as a DOI to be supplied in a structured manner and to be associated with a namespace that can help to ensure the uniqueness of the identifier. However, the real-world practice is less consistent, as evidenced when exploring records in the GEOSS Common Infrastructure (GCI): here, metadata and data identifiers are found in a wide variety of locations within catalogued metadata documents, and are sometimes completely absent. This problem is more cultural than technical: because ISO 19115:2003 is not completely clear about the difference between data and metadata identifiers, and lacks a clear recommendation on the use of Unique and Universal Identifiers (UUIDs), profilers have generated a variety of different identifiers (if they have generated them at all in the first place) and have located these identifiers in at least four different locations within metadata documents (Maso, 2013). The US FGDC metadata standard also allows the encoding of a variety of references to data and metadata[21], but also requires some investment of time and effort for proper use. In the next section we discuss the implications of these standards' complexity for VGI initiatives that may be ephemeral and poorly resourced.

I3. (meta)data include qualified references to other (meta)data R1.2. (meta)data are associated with detailed provenance

In the above section, we described potential ways in which the identifier of a dataset can be embedded in a traditional geospatial metadata document. However, an important consideration in the context of VGI is the rather complex and laborious nature of generating such 'traditional' metadata documents, which require a significant investment of time and effort. Geospatial metadata standards such as ISO 19115/19157 and FGDC offer a rich and expressive range of descriptive elements, but the reality is that many VGI initiatives are unlikely to generate such detailed documentation. In the face of this reality, other, more lightweight alternatives are likely to be taken up.

In those cases where metadata that are compliant with the ISO standard are generated, there is a huge opportunity for documenting provenance in a machine-readable way that can, if necessary, encode a full production workflow. The Lineage element of an ISO document, stored as part of the data quality statement, permits the description of any number of processing steps, complete with references to input and output data, descriptions of algorithms of software processing and citations of published reports/articles[22]. Figure 1 shows a single ProcessStep taken from such a lineage statement, rendered in a more human-readable format. It consists of a description of the processing that was carried out, and the three data sources (all of which may be optionally identified with persistent identifiers) that were used in the processing.

The standard and schema implementations of ISO 19115/19157 allow for a series of such ProcessSteps to be combined to generate a highly detailed, and, to some extent, machine-readable description of a dataset's provenance. However, in practice, the rich array of available elements are rarely used as intended, and it is far more common, if a lineage statement is provided at all, to see a single ProcessStep with a long and descriptive text account of the means by which the data were produced. This is in part because of the basic nature of many editing tools for ISO metadata and the lack of best-practice examples, but it is also evidence of the investment required to generate detailed metadata compliant to standards, and of the fact that this investment is not always budgeted into research projects – especially not citizen science projects. The FGDC approach to documenting data provenance is simpler, relying primarily on citations to scientific papers rather than on a fully modular description of the processing, but it is still common to find FGDC-compliant metadata with no real information on data provenance.

An alternative, or potentially a complement, to traditional geospatial metadata is a Linked Data approach (Heath and Bizer, 2011). Here, triples (in the form of subject-predicate-object) are used to describe relationships between entities. This mechanism, further discussed in Section 4.3, extends the potential for resource discovery to off-the-shelf web browsers, rather than just specialised portals and catalogues. Such an encoding, which is, in effect, returning to the roots of Geography Markup Language (GML) – GML version 1.0 came with an encoding in the Resource Description Framework (RDF) – can be adapted to include provenance information on a dataset. This strategy is of particular interest because it could be used to improve or enrich data documentation after data are published, or when they are reused for a different purpose than the original intended use case. For example, user reviews, reports of usage, discovered issues relating to particular observations, spatial regions or observers could be attached, post-hoc, to a published dataset and used in filtering and assessing fitness-for-purpose. Initial research along these lines can be seen in the outputs of the CHARMe project[23], which adapted the proposed OGC Geospatial User Feedback standard (Maso and Bastin, 2015) to permit lightweight annotations to be added to climate data in order to document quality issues, anomalies and user opinions on the value of the data. Another promising approach is the use

ProcessStep

description

Discriminant Analysis (DA) involves a linear combination of the original variables to produce a new set of variables that maximise the statistical difference between the predefined groups. DA acts as a standard classifier (applied to each date) because it enables an unknown pixel to be assigned to one of the predefined classes using discriminant functions obtained from a set of training areas. Training areas were obtained from fieldwork carried out in the Ebro Delta on 29 October 2006 and 17 January 2007. The surface of the training areas collected during fieldwork was 79.7 ha and 40% of them were reserved for an independent test of the results (random sampling).

source

description

Training areas collection (47.8 ha): Several sites representative of each class were visited and georeferenced with the aid of a Global Positioning System (Garmin etreX VistaC, Garmin International, Olathe, KS, USA), the cadastre cartography and the most recently available Landsat image.

source

description

Test areas collection (31.9 ha): Several sites representative of each class were visited and georeferenced with the aid of a Global Positioning System (Garmin etreX VistaC, Garmin International, Olathe, KS, USA), the cadastre cartography and the most recently available Landsat image.

source

description

Each of the 5 previous Landsat-5 images after geometric and radiometric correction and SIGPAC masking

Fig. 1: The content of a ProcessStep in an ISO 19115 metadata document. Namespaces and XML-specific formatting have been removed for clarity.

of the W3C PROV specifications in combination with RDF triples to create queryable databases representing the steps by which a dataset has been generated. A particular advantage of this approach is its amenability to extension when products are derived by some process which needs to be documented. In particular, the documentation of uncertainty introduced by data processing has been explored by Car et al. (2015), who combined UncertML (Williams et al., 2009) – a model and schema for documenting probabilistic uncertainty – with the PROV-O provenance ontology in such a way that quality issues in multi-part datasets can be encoded, and automated uncertainty propagation is made much more feasible.

F4. (meta)data are registered or indexed in a searchable resource
A2. metadata are accessible, even when the data are no longer available

The geospatial community has widely adopted the use of catalogues, which can be harvested, aggregated and searched in order to yield metadata that in turn reference the location of data resources. In many cases, the data referenced in these metadata documents are no longer available at the specified locations – though this is usually an accidental result of poor curation, rather than a demonstration of conscious compliance with principle A2. The prevalent standard underlying geospatial catalogues is the OGC's Catalogue Service standard[24], of which there are many free and open-source implementations, including the Java-based GeoNetwork and the Python implementation pycsw. Acknowledging that the OGC and SDI community to a large extent complements mainstream Internet developments through specific additions and extensions, the provision of metadata in the form of indexing files for common Internet search engines should also be considered.

A1. (meta)data are retrievable by their identifier using a standardized communications protocol
A1.1 the protocol is open, free, and universally implementable
A1.2 the protocol allows for an authentication and authorization procedure, where necessary

As described above, a variety of free and open standards exist for the search and retrieval of metadata from catalogues through an identifier. In terms of data service protocols, a powerful and widely adopted set of standards has been agreed to and maintained by the OGC: namely, the Web Map Service (for images), Web Feature Service (for data about geospatial objects) and Web Coverage Service (for data about geospatial fields). These standards are widely used, and implemented in a variety of languages and off-the-shelf toolkits such as GeoServer, MapServer, THREDDS and GeoNode, which are free to install and require relatively little configuration effort on the part of a user. When accessing data or imagery via OGC services, a simple HTTP request is parameterised with various user-specified options such as the area of interest and the projection in which the data should be returned. However, it is not specifically the identifier of the data that is used to identify the resource of interest; more commonly, one or more URLs are embedded in the metadata document, incorporating the layer name and namespace and enabling the retrieval of the resource from the service in question, which may not incorporate that unique identifier at all. For example, a typical WFS request contains a parameter with a namespace and layername defining the data to be retrieved (e.g. '*typeName=lrm:wdpa_latest*'), but there is no requirement to use a persistent identifier for the layer name.

Authorisation and authentication are possible with some implementations of these standards, for example GeoServer[25].

I1. (meta)data use a formal, accessible, shared, and broadly applicable language for knowledge representation
I2. (meta)data use vocabularies that follow FAIR principles
R1.3. (meta)data meet domain-relevant community standards
F2. data are described with rich metadata
R1. meta(data) are richly described with a plurality of accurate and relevant attributes

In order to represent the knowledge of data producers, some clear and well structured approaches have been developed. These identify core sets of vital information which **must** be provided, and supplement these cores with optional descriptive elements that can enrich the metadata and assist in assessment of fitness-for-purpose. For example, both ISO and FGDC standards have a subset of compulsory elements without which the metadata are invalid, and a wide array of optional descriptors that can be extremely detailed – for example, reports on quality, representativity, licensing and data provenance. Thus these standards support the generation of rich and informative metadata. In order to make these metadata more easily machine-readable and avoid large amounts of text mining, many elements can be populated with strings selected from code lists, which map to defined meanings in vocabularies and may be further maps to terms in other vocabularies. A good example of this is the 'occurrence issue' vocabulary used by GBIF to describe potential problems with a record, ranging from swapped coordinates to incorrectly inferred country origin for a record. Using values constrained by this list, extremely detailed information about quality assurance can be recorded in a very systematic way, which enables easy filtering and querying of records based on the nature of their errors, and avoids confusion where different assessors might describe an issue using different technical terms[26].

Similar vocabularies have been devised for ISO standards[27] and for taxonomic terms that allow the FDGC standard to be extended to cover biological data[28]. This last point is another strength of these agreed standards: they can be profiled to produce domain-relevant standards, while core elements remain consistent and interoperable with metadata produced using the base standard. In the context of GBIF, the Darwin Core standard, which is fundamental for structuring and harmonising species occurrence data, has been recently extended with new elements that permit the representation of sample data reporting species abundance information[29].

4 Representative Examples of Cross-Community Interoperability Approaches

Following the considerations so far, GBIF has already been considered as a good example to learn from. In addition to some of the highlights of the underlying approach, we see additional value in including two more examples in order to cover a wider spectrum of existing (or emerging) good practices in VGI data management.

4.1 The GBIF Data Publishing Framework

GBIF[30] was founded in 2001 upon a recommendation of the Biodiversity Informatics Subgroup of the Megascience Forum and a subsequent endorsement by the OECD science ministers, to 'enable users to navigate and put to use vast quantities of biodiversity information, advancing scientific research … serving the economic and quality-of-life interests of society, and providing a basis from which our knowledge of the natural world can grow rapidly and in a manner that avoids duplication of effort and expenditure.'[31]

Since then, GBIF has established a renowned cross-community data and metadata infrastructure to function as a single point of access to hundreds of institutions and services offering biodiversity data, based upon a data publishing framework as advised by the GBIF Data Publishing Framework Task Group with the central recommendation that 'all data relevant to the understanding of biodiversity and to biodiversity conservation should be made freely, openly and effectively available' (Moritz et al., 2011). GBIF facilitates responsible use and sharing of data by emphasising the need for proper publishing and citation, and by citing contributing nodes as data curators. It claims to offer data about more than 1.6 million species, collected in 300 years of exploration, from volunteers, researchers and monitoring programmes (see the organisation's 'what is GBIF' website section[32] and the GBIF Data Policy[33]).

As a mature and open infrastructure, the GBIF architecture supports several standards, the most important ones being Darwin Core, Ecological Metadata Language (EML[34]), Access to Biological Collections Data (ABCD[35]) for metadata and also access protocols like TDWG Access Protocol for Information Retrieval (TAPIR[36]) and Distributed Generic Information Retrieval (DiGIR[37]), in order to register and connect hundreds of different data holders and service providers within the GBIF portal. Most of the 'biodiversity standards' are being developed in the context of the Taxonomic Databases Working Group (TDWG)[38].

The principal workflow within the GBIF (2011) infrastructure is described as follows:

1. Digitization: The initial capturing of information in electronic form, through imaging, databasing, maintaining spreadsheets etc.

2. Publishing: The act of making data sources available in a well known format (standard) and with appropriate metadata for access on the internet.
3. Integration: The process of aggregating published datasets, applying consistent quality control routines and normalizing formats.
4. Discovery and access: By building network wide indexes, discovery services are offered for users through portals and for machines by extensive web service APIs (GBIF, 2011).[39]

In order to collect standardised information from contributing nodes, GBIF offers its community several tools, the most prominent one being the Integrated Publishing Toolkit (IPT):

> The IPT's two primary functions are to
> 1) encode existing species occurrence datasets and checklists, such as records from natural history collections or observations, in the Darwin Core standard to enhance interoperability of data, and
> 2) publish and archive data and metadata for broad use in a Darwin Core Archive, a set of files following a standard format (Robertson et al., 2014).

A further functionality is the possibility to convert metadata into 'data papers' that may be published as peer-reviewed scholarly articles in a journal. This is a direct incentive for publishing, as data can then be cited, raising the profile of the researcher or institution[40]. It also encourages the user to directly choose a public domain licence for the data (which is in line with GBIF's data policy and also leads to easier reuse of the data; see FAIR principles in previous section).

The Integrated Publishing Toolkit is one prominent example of how GBIF tries to lower the barriers for new data publishers and to promote this community's standards.

4.2 The OGC Interoperability Program, Cross Community Interoperability

VGI data often lack a common understanding associated to the meaning of the data or are user-contributed without any specific purpose, via social media platforms such as Twitter and Flickr. Nonetheless, often these data contain geographic reference and are *tagged* with other useful and queryable information, and the social media platforms offer application programming interfaces (APIs) to harvest from their services. In photo-community platforms, for example, the position of the published image may be (sometimes unintentionally) recorded in the GPS tags of EXIF metadata. This is likely to increase with the widespread use of smartphones equipped with capable GPS sensors. These sensors may eventually provide even more sophisticated information – for example, orientation and tilt angle of the camera. Such ancillary information is useful in a wide variety of use cases: for example as additional 'ground truth data' in the validation of global land cover products, or as one source among others in realtime crisis management. Several authors (Goodchild, 2007; Jürrens et al., 2009; Schade et al., 2011) have suggested viewing citizens [or humans] as *sensors* and using the OGC Sensor Web Enablement (SWE) as a reference framework to describe these *sensors* and their readings (or *observations*). In short, this framework aims at making sensor readings of all kinds discoverable and accessible via the net as near real-time streams in a standardised way, thus allowing for e.g. additional information streams beyond authoritative data from satellite images (in the case of crisis response for example). The SWE consists of a set of relevant standards, for example:

- *O&M* – Observations and Measurements: This standard describes the general data model and specifies XML encodings on how to represent data.
- *SOS* – Sensor Observation Service: The standard description of the service offering sensor descriptions and their observations.
- *SensorML* – Sensor Model Language: The standard models and XML Schema for describing the processes within sensor and observation processing systems.

(See the OGC website's Sensor Web Enablement description[41] for details.)

The data model of O&M is generic in the sense that its core element, an observation event, can be mapped against all kinds of physical properties:

'An observation is an act associated with a discrete time instant or period through which a number, term, or other symbol is assigned to a phenomenon. It involves application of a specified procedure, such as a sensor, instrument, algorithm, or process chain. The procedure may be applied in situ, remotely, or ex situ with respect to sampling location. The result of an observation is an estimate of the value of a property of some feature' (Cox, 2013).

In a series of so-called testbeds, the OGC Interoperability Program (IP) addresses fundamental questions regarding testing, prototyping and early adoption of OGC standards. These testbeds consist of several threads in specific application domains, such as aviation. In one of these threads – on Cross-Community-Interoperability (CCI) – the OGC has taken up the idea of mapping VGI information against the O&M data model (see *testbed 10 CCI VGI Engineering report* (OGC, 2014)). By transforming social media content into the O&M data model, the data can further be served by OGC service components in a standardised way, as observations made by the human observer, by using the Sensor Observation Service (SOS). The testbed report also states some real-world problems – since the prototype was tested against several clients, some of which could not deal with the SOS interface (at the time of writing SOS is not yet as widespread as the Web Feature Service (WFS) interface), the data were also encoded as features for usage within a WFS. In this scenario, the social media content was harvested by using the REST interface of the service (Flickr in their example) and uploaded as observations to the SOS after being transformed into the O&M model. This development was taken up as 'SWE for Citizen Science' as part of the discussions that led to the proposal of a new OGC Domain Working Group on Citizen Science (that was adopted at the OGC Technical Committee Meeting in September 2016).

4.3 The Provision of OpenStreetMap (OSM) as Linked Data

An interesting case builds on one of the most prominent VGI initiatives so far: OpenStreetMap (OSM). In the provision of OSM as Linked Data (Stadler et al., 2012), the traditional OSM dataset gets translated into a model that implements the Linked Data paradigm using RDF. Technically, the OSM data are periodically extracted from the official web page (openstreetmap.org), transformed into an RDF representation and loaded into a publicly available triple store that is essentially an RDF database. This processing is enabled by the open licensing model of OSM.

Apart from changing the data model (i.e. data formats and structures that are used to encode the points, lines, polygons, etc. that are used within OSM), the transition to a Linked Data approach also provides a step change in respect to (semantic) interoperability. While OSM defined its own structures and map elements (features) that are at most known to its own community, RDF is a recognised standard of the W3C and thereby well known to web developers around the globe, i.e. far beyond the original OSM contributors and the geospatial community. As such, datasets that are translated to so-called RDF triples (subject-predicate-object) can be easily connected to other triples by adding standard or self-defined relationships. In this way, datasets from multiple providers become interconnected and can be cross-navigated within the Linked Data Cloud[42].

In addition to introducing a standard way of modelling and related encodings, RDF also provides the possibility to reuse existing vocabularies so that the expressions used to represent subjects, predicates and objects are understood by many different communities (and not only by those that are familiar with a particular VGI dataset, such as, in this case, OSM). Considering geospatial data, for example, one might use the Location Core Vocabulary[43] for describing any place in terms of its name, address or geometry. In a similar manner vocabularies exist to describe persons and their social network[44] or even relationships between terms in two different vocabularies[45]. The most important point here is that the use of RDF is a well established step to breaking down the silos between closed communities, such as the SDI or the VGI community (see also Schade and Smits, 2012). Compared to many current OGC standards, which mostly evolve in parallel worlds, RDF provides common grounds for all sorts of different communities. This is because RDF builds on the (semantic) web as the common denominator and enables the specification of community-specific vocabularies, together with shared terms and well defined mappings. The mechanisms of vocabulary reuse and matching avoid the need for additional architectural approaches to join information from separately operating communities, such as wrappers, brokers or proxies.

While the above holds for all data models, it particularly also holds for models of data quality. Returning to the concrete example of OSM, the overall quality assurance and data management mechanisms remain core business within the traditional platform that underlies OSM (available from openstreetmap.org). The architecturally loosely coupled Linked Data representation adds, for example, the possibility to apply W3C vocabularies related to data quality – most notably the W3C Data on the Web Best Practices: Dataset Quality Vocabulary (W3C, 2016a) and Data Usage Vocabulary (W3C, 2016b). Whereas DQV provides the means to describe 'the quality of a dataset …, whether by the dataset publisher or by a broader community of users' (W3C, 2016a), DUV specifies 'a number of foundational concepts used to collect dataset consumer feedback, experiences, and cite references associated with a dataset' (W3C, 2016b). Together, both vocabularies could also be used for VGI, in order to support providers to express quality parameters of their offerings, but also to enable users to add their experiences and feedback to these parameters.

Yet, at the time of writing, both of these best practices are only available in draft versions and so far (to our knowledge) we still lack tangible access to using this concrete approach in a VGI context. We consider it as an extremely exciting area that is worth exploring (and comparing to dedicated OGC-centric approaches) in respect to VGI data management. The example of OSM as Linked Data may be the most straightforward use case for testing these possibilities.

5 Conclusion

In this chapter, we have looked into some generic – and not only VGI projects-specific – principles and good practices of data management, with the central paradigm being the FAIR principle: data should be findable, accessible, interoperable and reusable. To be reusable, it is vital that (meta)data are released with a clear and accessible data usage licence (see Chapter 6, Mooney and Minghini, 2017). Furthermore, we have summarised standards that support these principles, both from the Open Geospatial Consortium and from ISO TC/211, as well as from W3C, and we have investigated three examples where these principles and standards are utilised to maximise cross-discipline interoperability.

A key conclusion from this review into the current state of the art is that metadata for VGI are, and are likely to remain, patchy and extremely heterogeneous. 'Traditional' standards aimed at complete documentation of a one-off production workflow, such as ISO 19115/19157, are rich in descriptive elements that, if used properly, can enable the provenance and quality of geospatial data to be documented in very useful and machine-readable ways that support uncertainty propagation and fitness-for-use assessment. However, an investigation of open geospatial catalogues quickly shows that these standards are not being exploited to their full potential, even by large institutional data producers – partly because of the resource-intensive nature of metadata generation, and partly because of an ongoing shortage of tools and examples to simplify the process. For VGI, where even a single 'dataset' can contain observations produced by a wide variety of observers, instruments and methods, such monolithic standards may only be of use for periodic review and documentation of aggregated and quality-controlled data. In addition, the nature of VGI is such that observations may be accessed and used in a variety of different combinations and groupings. With such a fluid granularity, tools and APIs that allow annotation and documentation of individual records or groups of records are likely to be more useful, as are any tools and processing methods that permit the collection and storage of metadata automatically at the point of observation. Ongoing developments in RDF and Linked Data appear very promising for supporting data annotation, but are still too immature to be easily usable within most VGI initiatives. However, this is a key angle of research that should be developed, not least because the annotation/commentary approach to metadata permits information and quality reports to be attached to data after their production, so that VGI can be mobilised and made more usable and reusable.

We have not looked into software solutions of how to access, store and back up data, for example which database management solution to use, such as PostgreSQL (with its language extension PostGIS), MySQL or the lightweight SpatiaLite, to name a few. We have also only touched the surface of the topic of software suites like GeoServer, deegree or GeoNetwork, all of which offer substantial building blocks for Spatial Data Infrastructures. We encourage the use of Open Source software like these, as well as open and freely accessible standards.

In this text we have not addressed Environmental Sensor Networks (ESNs) that may comprise a backbone in data assessment from distributed heterogeneous sensors. We expect that the Sensor Web Enablement, as an OGC reference framework, will play an important role in citizen sensing. For further reading, the FP7 funded Citizen Observatory 'COBWEB' has defined a 'Generic Infrastructure Platform to facilitate the collection of Citizen Science data for Environmental Monitoring'(Higgins et al., 2016).

In terms of actual formulation of Data Management Plans, substantial resources are available; see for example DataOne's 'Data Management Guide for Public Participation in Scientific Research'[46] or COBWEB's 'Generic Data Management Plan Check' in their 'deliverable 7.1 on Data Management Guidelines.'[47]

Data management methodologies can only succeed if their benefits overcome their implementation costs; i.e. existing solutions and best practices will have to be tailored to the needs and capabilities of individual projects, and feasibility needs to be assessed on a case by case basis. However, it is imperative to recognise that a precise knowledge of the provenance and meaning of data is a most precious asset that should be highly valued.

Notes

1 https://ec.europa.eu/programmes/horizon2020/en/news/citizens%E2%80%99-observatories-empowering-european-society-open-conference
2 http://www.eubon.eu/
3 https://cobwebproject.eu/
4 https://www.earthobservations.org/documents/dswg/201504_data_management_principles_long_final.pdf
5 http://www.bfe-inf.org/info/data-principles
6 E.g. http://vocab.nerc.ac.uk/collection/P15/current/CFCM0010/
7 E.g. http://seadatanet.maris2.nl/v_bodc_vocab_v2/vocab_relations.asp?lib=P02
8 http://ijsdir.jrc.ec.europa.eu/index.php/ijsdir/article/view/389
9 https://www.journals.elsevier.com/data-in-brief
10 https://www.w3.org/
11 http://www.opengeospatial.org/
12 http://www.gbif.org/mendeley
13 http://www.iso.org/iso/catalogue_detail?csnumber=43506
14 https://www.crossref.org/
15 https://figshare.com/
16 https://zenodo.org/
17 https://www.datacite.org/
18 http://ezid.cdlib.org/
19 http://datadryad.org/
20 http://www.iso.org/iso/catalogue_detail.htm?csnumber=53798
21 http://www.ngdc.noaa.gov/wiki/index.php/Data_Set_Identifiers_and_other_Unique_IDs
22 https://geo-ide.noaa.gov/wiki/index.php?title=File:LI_Lineage-2.png
23 http://charme.org.uk/
24 http://www.opengeospatial.org/standards/cat
25 http://docs.geoserver.org/stable/en/user/security/service.html
26 http://gbif.github.io/gbif-api/apidocs/org/gbif/api/vocabulary/OccurrenceIssue.html
27 https://geo-ide.noaa.gov/wiki/index.php?title=ISO_19115_and_19115-2_CodeList_Dictionaries

[28] http://www.fgdc.gov/standards/projects/FGDC-standards-projects/metadata/biometadata/biodatap.pdf
[29] http://www.gbif.org/sites/default/files/gbif_IPT-sample-data-primer_en.pdf
[30] http://www.gbif.org
[31] http://www.gbif.org/what-is-gbif#background
[32] http://www.gbif.org/what-is-gbif
[33] http://www.gbif.org/resource/80527
[34] https://knb.ecoinformatics.org/#external//emlparser/docs/index.html
[35] http://www.tdwg.org/activities/abcd/
[36] http://www.tdwg.org/activities/tapir/
[37] http://digir.sourceforge.net/
[38] http://www.tdwg.org/standards/
[39] http://www.gbif.org/infrastructure/summary
[40] http://www.gbif.org/publishing-data/data-papers
[41] http://www.opengeospatial.org/ogc/markets-technologies/swe
[42] http://lod-cloud.net/
[43] https://www.w3.org/ns/locn
[44] http://www.foaf-project.org/
[45] https://www.w3.org/2004/02/skos/
[46] https://www.dataone.org/sites/all/documents/DataONE-PPSR-DataManagementGuide.pdf
[47] https://cobwebproject.eu/sites/default/files/COBWEB%20D7.1%20Data%20Management%20Guidelines%20v1_0.pdf

Reference list

Bearden, M.J., 2007. The National Map Corps. Presented at the Specialist Meeting on Volunteered Geographic Information, University of California at Santa Barbara. Available at http://www.ncgia.ucsb.edu/projects/vgi/docs/position/Bearden_paper.pdf [Last accessed 1 May 2017]

Car, N., Cox, S., Fitch, P., 2015. *Associating uncertainty with datasets using Linked Data and allowing propagation via provenance chains.* EGU General Assembly Conference Abstracts, p. 4392. Available at http://adsabs.harvard.edu/abs/2015EGUGA..17.4392C [Last accessed 1 May 2017].

Global Biodiversity Information Facility (GBIF), 2011. GBIF Position Paper on Data Hosting Infrastructure for Primary Biodiversity Data. Version 1.0. (Authored by Goddard, A., Wilson, N., Cryer, P., & Yamashita, G.) Available at http://www.gbif.org/resource/80733 [Last accessed 1 May 2017]

Cox, S., 2013. OGC and ISO 19156:2011(E) OGC Abstract Specification Geographic information — Observations and measurements. Available at http://portal.opengeospatial.org/files/?artifact_id=41579 [Last accessed 1 May 2017].

Craglia, M., Granell, C., 2014. *Citizen Science and Smart Cities*. Publications Office of the European Union, Luxembourg. DOI: https://doi.org/10.2788/80461

Goodchild, M.F., 2007. Citizens as sensors: the world of volunteered geography. *GeoJournal* 69, 211–221. DOI: https://doi.org/10.1007/s10708-007-9111-y

Heath, T., Bizer, C., 2011. Linked Data: Evolving the Web into a Global Data Space. *Synthesis Lectures on the Semantic Web: Theory and Technology* 1, 1–136. DOI: https://doi.org/10.2200/S00334ED1V01Y201102WBE001

Higgins, C.I., Williams, J., Leibovici, D.G., Simonis, I., Davis, M.J., Muldoon, C., van Genuchten, P., O'Hare, G., Wiemann, S., 2016. Citizen OBservatory WEB (COBWEB): A Generic Infrastructure Platform to Facilitate the Collection of Citizen Science data for Environmental Monitoring. *International Journal of Spatial Data Infrastructures Research* 11, 20–48. DOI: https://doi.org/10.2902/1725-0463.2016.11.art3

Jürrens, E.H., Bröring, A., Jirka, S., 2009. A Human Sensor Web for Water Availability Monitoring, in *Proceedings of OneSpace 2009 – 2nd International Workshop on Blending Physical and Digital Spaces on the Internet*, Berlin, Germany, 1 Sep 2009. Available at: http://onespace.ace.ed.ac.uk/2009/docs/onespace2009_submission_2.pdf [Last accessed 16 May 2017].

Maso, J., 2013. Unique Identifiers within Systems of Systems, Slide 8: Id's in metadata records. Metadata and data id's. GEOSS Future Products Workshop, NOAA, Silver Springs, USA, 27 March 2013. Available at: https://portal.opengeospatial.org/files/?artifact_id=53340 [Last accessed 1 May 2017].

Maso, J., Bastin, L., 2015. OGC Geospatial User Feedback Standard. Conceptual Model and XML Encoding Extension. Available at: http://www.opengeospatial.org/standards/requests/144 [Last accessed 1 May 2017].

Mooney, P and Minghini, M. 2017. A Review of OpenStreetMap Data. In: Foody, G, See, L, Fritz, S, Mooney, P, Olteanu-Raimond, A-M, Fonte, C C and Antoniou, V. (eds.) *Mapping and the Citizen Sensor*. Pp. 37–59. London: Ubiquity Press. DOI: https://doi.org/10.5334/bbf.c.

Moritz, T., Krishnan, S., Roberts, D., Ingwersen, P., Agosti, D., Penev, L., Cockerill, M., Chavan, V., 2011. Towards mainstreaming of biodiversity data publishing: recommendations of the GBIF Data Publishing Framework Task Group. *BMC Bioinformatics* 12, 1–10. DOI: https://doi.org/10.1186/1471-2105-12-S15-S1

OGC, 2014. OGC Testbed-10 CCI VGI Engineering Report. Edited by A. Broring, S. Jirka, M. Rieke, B. Pross. Available at: https://portal.opengeospatial.org/files/?artifact_id=58925 [Last accessed 1 May 2017].

Robertson, T., Döring, M., Guralnick, R., Bloom, D., Wieczorek, J., Braak, K., Otegui, J., Russell, L., Desmet, P., 2014. The GBIF Integrated Publishing Toolkit: Facilitating the Efficient Publishing of Biodiversity Data on the Internet. *PLOS ONE* 9, e102623. DOI: https://doi.org/10.1371/journal.pone.0102623

Schade, S., Díaz, L., Ostermann, F., Spinsanti, L., Luraschi, G., Cox, S., Nuñez, M., Longueville, B.D., 2011. Citizen-based sensing of crisis events: sensor web enablement for volunteered geographic information. *Applied Geomatics* 5, 3–18. DOI: https://doi.org/10.1007/s12518-011-0056-y

Schade, S., Smits, P., 2012. Why linked data should not lead to next generation SDI, in: 2012 IEEE International Geoscience and Remote Sensing Symposium. Presented at the 2012 IEEE International Geoscience and Remote Sensing Symposium, pp. 2894–2897. DOI: https://doi.org/10.1109/IGARSS.2012.6350721

Schade, S., Tsinaraki, C., 2016. *Survey report: data management in Citizen Science projects.* JRC Technical Report; EUR 27920 EN, Luxembourg (Luxembourg), Publications Office of the European Union. DOI: https://doi.org/10.2788/539115.

Socientize, 2013. White Paper on Citizen Science in Europe. Available at http://www.socientize.eu/?q=eu/content/download-socientize-white-paper [Last accessed 1 May 2017].

Stadler, C., Lehmann, J., Höffner, K., Auer, S., 2012. LinkedGeoData: A core for a web of spatial open data. *Semantic Web* 3, 333–354. DOI: https://doi.org/10.3233/SW-2011-0052

USGS, 2017. Persistent Identifiers. USGS Data Management. 17 February 2017. Available at https://www2.usgs.gov/datamanagement/preserve/persistentIDs.php [Last accessed 4 April 2017].

W3C, 2016a. Data on the Web Best Practices: Data Quality Vocabulary. Edited by A. Albertoni, A. Isaac. Available at: https://www.w3.org/TR/vocab-dqv/ [Last accessed 1 May 2017].

W3C, 2016b. Data on the Web Best Practices: Data Usage Vocabulary. Edited by B Farais-Losco, E. Stephan, S. Purohit. Available at: https://www.w3.org/TR/vocab-duv/ [Last accessed 1 May 2017].

Wilkinson, M.D., Dumontier, M., Aalbersberg, Ij.J., Appleton, G., Axton, M., Baak, A., Blomberg, N., Boiten, J.-W., Santos, L.B. da S., Bourne, P.E., Bouwman, J., Brookes, A.J., Clark, T., Crosas, M., Dillo, I., Dumon, O., Edmunds, S., Evelo, C.T., Finkers, R., Gonzalez-Beltran, A., Gray, A.J.G., Groth, P., Goble, C., Grethe, J.S., Heringa, J., Hoen, P.A.C. 't, Hooft, R., Kuhn, T., Kok, R., Kok, J., Lusher, S.J., Martone, M.E., Mons, A., Packer, A.L., Persson, B., Rocca-Serra, P., Roos, M., Schaik, R. van, Sansone, S.-A., Schultes, E., Sengstag, T., Slater, T., Strawn, G., Swertz, M.A., Thompson, M., Lei, J. van der, Mulligen, E. van, Velterop, J., Waagmeester, A., Wittenburg, P., Wolstencroft, K., Zhao, J., Mons, B., 2016. The FAIR Guiding Principles for scientific data management and stewardship. *Scientific Data* 3, 160018. DOI: https://doi.org/10.1038/sdata.2016.18

Williams, M., Cornford, D., Bastin, L., Pebesma, E. 2009. Uncertainty Markup Language (UncertML). OGC Discussion Paper, Document Number: 08–122r1. Available at: http://www.opengeospatial.org/docs/discussion-papers [Last accessed 16 May 2017].

CHAPTER 12

Integrating Spatial Data Infrastructures (SDIs) with Volunteered Geographic Information (VGI) for creating a Global GIS platform

Demetris Demetriou[*,†], Michele Campagna[‡], Ivana Racetin[§], Milan Konecny[¶]

[*]Public Works Department, 165 Strovolos Avenue, 2048 Nicosia, Cyprus
[†]School of Geography, University of Leeds, LS2 9JT, Leeds, UK, demdeme@cytanet.com.cy
[‡]DICAAR, Università Di Cagliari, Via Marengo 2, 09123 Cagliari, Italy
[§]Faculty of Civil Engineering, Architecture and Geodesy, University of Split, Matice hrvatske 15, 21000 Split, Croatia
[¶]Department of Geography, Faculty of Science, Masaryk University, Kotlarska 2, 61137 Brno, Czech Republic

Abstract

Spatial Data Infrastructures (SDIs) are a special category of data hubs that involve technological and human resources and follow well defined legal and technical procedures to collect, store, manage and distribute spatial data. INSPIRE is the EU's authoritative SDI in which each Member State provides access to their spatial data across a wide spectrum of data themes to support policy-making. In contrast, Volunteered Geographic Information (VGI) is one type of user-generated geographic information (GI) where volunteers use the

How to cite this book chapter:

Demetriou, D, Campagna, M, Racetin, I, Konecny, M. 2017. Integrating Spatial Data Infrastructures (SDIs) with Volunteered Geographic Information (VGI) for creating a Global GIS platform. In: Foody, G, See, L, Fritz, S, Mooney, P, Olteanu-Raimond, A-M, Fonte, C C and Antoniou, V. (eds.) *Mapping and the Citizen Sensor.* Pp. 273–297. London: Ubiquity Press. DOI: https://doi.org/10.5334/bbf.l.

web and mobile devices to create, assemble and disseminate spatial information. There are similarities and differences between SDIs and VGI, as well as advantages and disadvantages to both. Thus, the integration of these two data sources will enhance what is offered to end users to facilitate decision-making. This idea of integration is in its early stages, because several key issues need to be considered and resolved first. Therefore, this chapter discusses the challenges of integrating VGI with INSPIRE and outlines a generic framework for a global integrated GIS platform, similar in concept to Digital Earth and Virtual Geographic Environments (VGEs), as a realistic scenario for advancements in the short term.

Keywords

SDIs, INSPIRE, VGI, Global Integrated GIS platform

1 Introduction

Data hubs have arisen through the evolution of information technology, and aim to provide a centralised, unified data source that can be easily accessed by certain groups of users, or more widely by the public, to support a diversity of professional and/or other needs (Mangano, 2013). A special category of data hub is that of Spatial Data Infrastructures (SDIs; Williamson et al., 2003), which emerged during the mid-1990s (Delaney and Pettit, 2014). SDIs involve technological and human resources that follow well defined legal and technical procedures to collect, store, manage and distribute spatial data. On 14 March 2007, the European Parliament and Council adopted a Directive establishing the Infrastructure for Spatial Information in the European Community (INSPIRE) European SDI (European Commission, 2007). Following the INSPIRE Directive, Public Authorities (PAs) in each Member State should provide access to their SDI across a wide spectrum of data themes through a community geoportal, aiming thus to support policy-making and activities aimed at, but not limited to, the protection of the environment.

Whilst INSPIRE tries to unite and standardise existing Authoritative Geographic Information (AGI) made available by PAs in EU Member States, technologies that enable User-Generated Content (UGC) have also appeared (Moens et al., 2014) in web-based platforms (e.g. blogs, wikis, discussion forums, posts, chats, tweets), mobile computing and GPS devices. Hence, users have started to create and share data and information. Volunteered Geographic Information (VGI) is one type of user generated GI (Goodchild, 2007), where volunteers use the web and mobile devices to create, assemble and disseminate spatial information. Among the most well known VGI platforms are OpenStreetMap (OSM; Demetriou, 2016) and Wikimapia, but there are many others, covering a range of fields such as conservation, planning, and crisis management. Thus, there is a potential for VGI to become an important source of information that could benefit INSPIRE and similar projects and efforts; on the other hand, VGI could also benefit from INSPIRE through integration with official and reliable data and the need to adopt more strict specifications.

Although INSPIRE[1] is a well organised, official and reliable platform that is based on strict standards, it provides data that are mainly used by experts and involves static information (with a limited level of detail in some cases) that is not updated very regularly because of the high costs involved. VGI, on the other hand, is captured unofficially by volunteers, often using cheap devices, e.g. a handheld GPS or smartphones; hence the data quality is usually limited and the data collection is not based on strict standards. However, real-time data can be collected anywhere by anybody, opening up concrete possibilities for data to be updated very regularly at little or no cost. Therefore, the integration of both types of data (Craglia, 2007; Budhathoki et al., 2008; Craglia et al., 2008; McDougall, 2009; Parker et al., 2012; Massa and Campagna, 2016) could potentially enhance what is delivered to end users, supporting the full spectrum of related needs, both professional, e.g. planning and spatial decision-making, and of the daily activities of citizens.

The idea of integration of VGI and authoritative data has arisen recently and been emphasised by several researchers (Budhathoki et al., 2008; Craglia et al., 2008; McDougall, 2009; Parker et al., 2012). In addition, the benefits of integration refer to both the organisations involved, i.e. National Mapping Agencies (NMAs; Olteanu-Raimond et al., 2017) that operate national INSPIRE geoportals, and those who run VGI initiatives, as well the end users. Although some efforts towards this integration have already been made (Craglia, 2007; Wiemann and Bernard, 2014), the literature suggests that this endeavour is in its early stages because several critical issues need to be considered and resolved. As a result, the available literature is limited and focuses on specific projects or technical issues (Botshelo, 2009) without attempting to investigate the broader picture of integration or setting out a conceptual framework. Further to this integration, the vision is the development of a global integrated GIS platform, which extends the capabilities of a typical data hub and the benefits of integration of SDIs with VGI by embedding on-line geospatial tools, to deliver both static and dynamic outputs to support planning and decision-making. Such visionary and/or applied advanced geospatial tools and frameworks moving in this direction are the GeoWeb (Dangermond, 2005), Digital Earth (Craglia et al., 2008) and Virtual Geographic Environments (VGEs; Lin et al., 2013).

Based on the above, this chapter aims to discuss the challenges of integrating VGI with INSPIRE, and to outline a generic framework for a global integrated GIS platform, similar in concept to Digital Earth and VGEs, as a realistic scenario for advancements in the future. The remainder of this chapter is organised as follows: Section 2 provides an overview of SDIs and VGI, contrasting these two sources of data. This is followed by a discussion about critical issues that arise in INSPIRE and VGI integration (Section 3). In Section 4, the prospects of integration are examined, with some examples. Section 5 then presents an outline of a conceptual framework for an ideal global integrated GIS platform, while conclusions are summarised in Section 6.

2 Spatial Data Infrastructures (SDIs) and Volunteered Geographic Information (VGI)

Before discussing the various issues of integration between SDIs and VGI, an overview of each infrastructure and a comparison are presented, providing the necessary background.

2.1 Spatial Data Infrastructures (SDIs)

Data hubs are defined as community-run catalogues of useful, online datasets, which store a copy of the data or host them in a database and provide some basic visualisation tools (Open Knowledge Foundation, 2013). A typical data hub consists of four basic elements, as shown in Figure 1: Data, a Facilitator, a Custodian and End Users, which together form a dynamic communication cycle (Delaney and Pettit, 2014).

In particular, the Facilitator should provide a connection between the Custodian, i.e. the data hub's administrator, and the End Users; negotiate with the

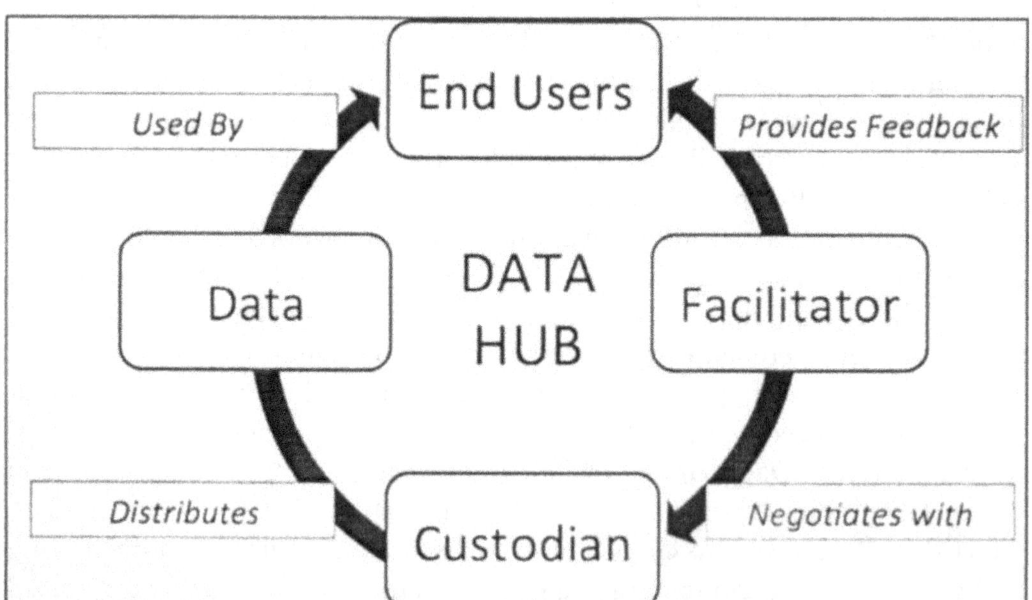

Fig. 1: Data hub conceptual communication – feedback cycle (adapted from Delaney and Pettit, 2014).

Custodian in terms of the needs or problems; and provide feedback to end users. The role of the Custodian is to provide and distribute data, which will be used by the End Users. It is to be noted that the terms 'end users' and 'users', as used in this chapter, have a slightly different meaning: namely, while 'end users' utilise the data provided by the hub, they do not necessarily contribute to the development of the hub voluntarily, i.e. by delivering new data, updating existing data or sharing data – tasks that are carried out by 'users'. Obviously, 'users' can also be 'end users'; that is, they can have a double role.

Access to data hubs can be free and/or licensed. A data hub allows users to access, search and use a variety of data with associated metadata provided as a discrete set of formats. The data hub concept has been realised in many locations and contexts globally. Many scientific fields have collaborated to create research-specific data hubs to store and discover data and to distribute them to other researchers (Delaney and Pettit, 2014).

SDIs are a special category of data hubs (Williamson et al., 2003) that involve a framework of interacting elements, aiming to acquire, store, preserve, process, distribute, use and maintain data with 'a direct or indirect reference to a specific location or geographical area' (European Commission, 2007). The main elements of this framework are: spatial datasets and their metadata; networks services and technologies; standards that define the quality of the data; policies for distributing and managing the data; human resources; and a mechanism for coordinating and monitoring the whole infrastructure (European Commission, 2007; Iliffe, 2012). An SDI may be developed by national public bodies to support all of the spatially relevant activities in a country. Each national, regional or local SDI, as a node of INSPIRE, recognises the significance of metadata by ensuring all contributed data align to a minimum standard and aims to deliver up-to-date data and information to other government agencies and the general public (Steven, 2005) to support effective decision-making. Several SDIs have been developed (Craglia, 2007), e.g. the National Spatial Data Infrastructure (NSDI) in the United States in 1994 and INSPIRE in Europe.

2.2 Volunteered Geographic Information (VGI)

UGC is divided into two main types: non-georeferenced and georeferenced, as illustrated in Figure 2. The most popular forms of the former type include text messaging, social media interactions, photos, videos, blog entries, etc. Georeferenced UGC involves various forms of location-based technologies, such as location-based services (LBSs), location-based social networks (LBSNs), social network location sharing (SNLS), location-based games (LBGs) and location-based social network games (LBSNGs; Odobašić et al., 2013). In particular, the LBS industry has profited from UGC primarily because ubiquitous and affordable smartphones equipped with multiple sensors foster geographic data collection. Similarly, LBSN leverage the power and high adoption rate of modern

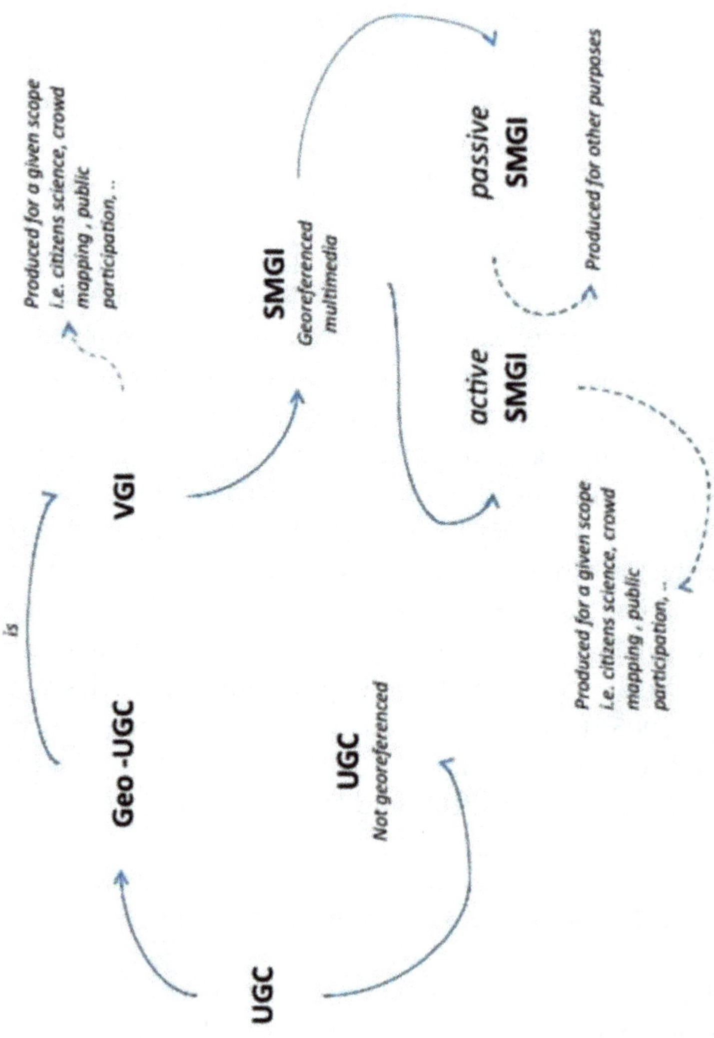

Fig. 2: Types of user-generated content.

mobile devices to provide applications and services that allow users to share and discuss the real-world places they visit, as a part of their virtual interactions (Furey et al., 2013). In terms of social networks, location sharing has changed from a purpose-driven to a social-driven activity. Users traditionally shared their location with one other person (one-to-one) or with a small group (one-to-few); social networks, depending on the privacy/user settings, enable users to share their location with a large group (one-to-many) or with everyone (one-to-all; Tang et al., 2010). LBGs are games in which the game play somehow evolves and progresses based on a player's location. Thus, LBGs almost always support some kind of georeferencing technology, for example by using, WiFi, Near Field Communication, Bluetooth and satellite positioning such as GPS. The blend of LBGs and LBSNs creates LBSNGs, which are exemplified by a service like Foursquare.

Among the most popular geo-UGC-based technologies is VGI (Goodchild, 2007), or crowdsourced GI, which has arisen since 2007. VGI involves harnessing tools to create, assemble and disseminate geographic data provided voluntarily by individuals, and it can be generated through geobrowsers or smartphone apps, making use of georeferencing or geocoding tools and techniques. Two widely popular VGI platforms are OSM (Haklay, 2010) and Wikimapia (Wikimapia, 2015), but there are many others, covering many kinds of fields, such as conservation, planning, and crisis management. A special class of VGI is Social Media Geographic Information (SMGI), which can generally be divided into active and passive type (Figure 2). The former type is produced for a given scope, e.g. citizen science, crowd mapping or public participation, and users (i.e. volunteer contributors) are fully aware of this, such as in the case of OSM or Wikimapia. In contrast, the latter is produced for other purposes (i.e. users share passively or share unvolunteered information for undefined purposes, such as in the case of social network interaction) and may be accessed independently at a later stage for reuse by third parties for a variety of disparate aims.

2.3 A Comparison of SDIs and VGI

There are similarities and differences between SDIs and VGI (Castelein et al., 2010) regarding data, as well as advantages and disadvantages, and these are outlined in Figure 3. In particular, data provided by SDIs are captured by well trained specialists who are employed by formal public or private organisations, and through well defined workflows, using state-of-the-art technology (Castelein et al., 2010); hence the SDI approach is an official, top-down approach involving high costs. On the other hand, VGI is captured unofficially by volunteer-citizens (classified by Coleman et al. (2009) into five categories), through smart phones/devices that provide GPS and Internet access or using other simple aids to take measurements; it is a bottom-up process with limited or no operational costs. Whilst the former data are generally free of charge or can be licensed through a fee, the latter are always provided for free. Moreover, SDIs have a data-centric scope as they mainly provide data used by experts through GIS portals, while VGI delivers information to a broader audience of mainly non-experts through user-friendly GI platforms.

In addition, SDIs involve static information provided periodically and in some cases with a limited level of detail, while VGI has both static and dynamic (real-time) information, since it can process real-time, spatiotemporal information, and can provide a much greater level of detail in some cases. This suggests that VGI could be a potentially complementary source to SDI in providing relevant real-time data related to physical catastrophes, crisis management situations or humanitarian missions. Furthermore, SDI provides certified data based on strict and professional international standards and specifications such as that provided by the Open Geospatial Consortium (OGC) and International Standardisation Organisation (ISO), while VGI is based on essential data standards that vary from platform to platform; most importantly, the quality of their data is unknown.

The above comparison also reveals two weaknesses of SDIs: the lack of capacity for real-time data to be collected anywhere by anybody and the lack of the flexibility of very regular data updates at low or no cost. Thus, a combination of both technologies will enhance what is offered to end users to facilitate decision-making, and the idea of integration has been discussed by several researchers (Budhathoki et al., 2008; Craglia et al., 2008; McDougall, 2009; Parker et al., 2012). However, this challenge will not be an easy one, because the institutional framework of the integration will be complex due to the different requirements and scope underlying each technology.

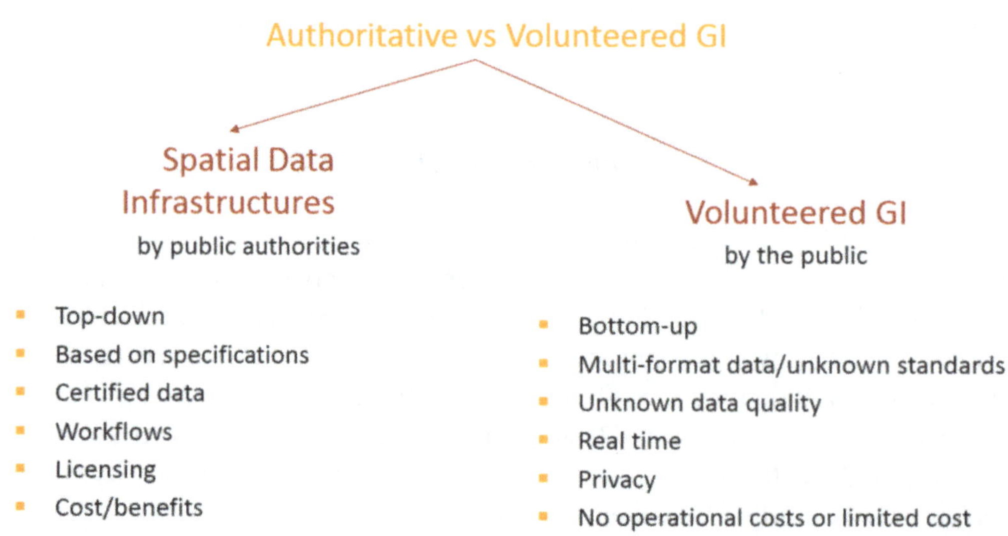

Fig. 3: The differences between SDIs and VGI.

3 Integrating VGI to INSPIRE

The dominant European SDI is INSPIRE, and its integration with VGI is a difficult task because of several critical issues regarding the common implementing rules, which are discussed below. An overview of INSPIRE is first provided.

3.1 The INSPIRE Directive

INSPIRE, which has been defined by EU Directive 2007/2/EC (European Commission, 2007) and was adopted in 2007, establishes the requirement that each Member State should provide access to their SDI through a community geoportal operated by the European Commission or any other access point they wish to operate. The INSPIRE implementation provides a large-scale application of the open geoportal environment and is a big step forward in the development of an SDI in Europe. INSPIRE will overcome existing weaknesses and gaps in the interoperability of information resources across Europe by integrating them into a common framework (Craglia, 2007). The aim of INSPIRE is to assist policy-making and activities related to the environment and beyond; hence it involves data regarding a broad spectrum of fields, which are reflected in 34 spatial-data themes. The INSPIRE implementation represents a significant investment from all Member States, and has resulted in close to 300,000 spatial datasets being made available to the community through a standardised data-discovery site. The main INSPIRE portal allows users to search for datasets from across the EU from a single interface, and allows advanced search filters to be used to narrow down searches by geography, format or spatial theme. The INSPIRE portal only displays metadata for each dataset; it does not allow users to directly access any of the datasets, either manually or programmatically. However, each metadata resource contains a link to the data source, which may be a file, service or web application.

It should be noted that INSPIRE involves some general rules: it is based on existing SDI of Member States, and hence does not require the collection of new data, but demands the transformation of existing data to comply with its specification structure; and it does not affect intellectual property rights. In particular, the Directive also requires that common implementing rules be adopted in four main specific areas: **metadata**, **data specifications**, **network services**, and **data and service sharing**. These areas face critical integration issues, as discussed below.

3.2 Critical Issues for Integration

Following the INSPIRE Directive, Member States should provide **metadata** for spatial datasets/data series and/or for spatial data services. The metadata consist of 27 elements of information regarding the data resources, elements of information which are grouped into 10 categories: identification; classification; keywords; geographic location; temporal reference; quality and validity; conformity with the interoperability implementing rules; constraints related to access and use; organisation responsible for the resource; and metadata for metadata (European Commission, 2007). Clearly, populating all of these elements of metadata for VGI data will have a consequential time and cost. Furthermore, these elements cannot be gathered comprehensively by volunteers given current VGI practices. An issue is therefore who will be responsible for inputting all of these metadata and validating their reliability. Therefore, VGI metadata can be limited to only the basic information among the 27 elements provided by INSPIRE that can be input by the contributor, by the VGI system administrator or automatically by the system.

Similarly to metadata, the employment of common **data specifications** is a vital aspect of integration. Specifically, in order to ensure the interoperability of spatial information in INSPIRE, common international standards (those defined by ISO), technical specifications (e.g. regarding data types, code lists and enumerations, encoding, updating, the life cycle of spatial objects, reference temporal systems, and metadata) and minimum performance criteria for download services and transformation services have been defined (for each of the 34 related themes mentioned earlier). The issue of how to accommodate the diversified, dynamic and easy-to-access VGI data types to SDI is not a serious problem in technical terms; the problem is to define and apply minimum data requirements for VGI that are reasonable and achievable in order to satisfy data quality requirements (Wiemann and Bernard, 2014). Aspects of data quality such as positional accuracy, classification correctness and accuracy of the time measurement may follow the ISO 19157 standard (ISO, 2013; see Chapter 7 by Fonte et al. (2017) for more information on quality); a legally binding aspect is that of the topological consistency of the network data. VGI data quality and credibility vary from contributor to contributor (Flanagin and Metzger, 2008; Goodchild and Li, 2012; Foody et al., 2013); thus it is only up to a data provider whether they will respect data quality recommendations and whether they will report on recommendations in the metadata. Although some case studies on popular VGI platforms such as OSM have shown good and acceptable outcomes (Haklay, 2010), NMAs should evaluate the risks and problems that arise from the adoption of this new production system (Coleman et al., 2009; Bégin, 2012). Users should always be aware of how can they assess the credibility of data (Flanagin and Metzger, 2008) and contributors should be aware of the quality of the data used (Dassonville et al., 2003) and of whether they are fit for purpose. It is essential to develop tools that enable this evaluation. In addition, data quality can be improved by providing training on the needs of SDIs and on their protocols, and incentives can be awarded to contributors providing good work (see Chapter 5 by Fritz et al. (2017) for a discussion of incentives for volunteers).

The interoperability of **network services** is also crucial for the joint operation of the systems. In particular, INSPIRE network services utilise one standard communication-protocol and binding technology for all service types to avoid mixing technologies: the Simple Object Access Protocol (SOAP), which ensures streamlined integration and implementation, as well as getting a maximum benefit from the offered services. SOAP is a protocol specification for exchanging structured information in the implementation of web services in computer networks. It uses the XML Information Set for its message format, and relies on other application layer protocols, most notably Hypertext Transfer Protocol (HTTP) or Simple Mail Transfer Protocol (SMTP), for message negotiation and transmission. In contrast to INSPIRE, it is reasonable that the various VGI platforms should use different communication-protocols and binding technologies through the platform owner's Application Programming Interfaces (APIs). However, VGI may reuse the two types of services provided by INSPIRE, i.e. viewing and downloading. The former operation is typically based on OGC Web Map Services (WMSs) or OGC Web Map Tile Services (WMTSs), which are easy to integrate into a VGI application from the technical as well as the legal point of view; the VGI application acts like a client application to a server, publishing data under the INSPIRE Directive. Most of the INSPIRE view services are provided free of charge, but there may be conditions that prevent their reuse for commercial purposes (European Commission, 2007). The latter type of service, download, is based on OGC Web Feature Services (WFSs), OGC Web Coverage Services (WCSs) and OGC Sensor Observation Services (SOSs), among others, which are also easy to integrate from a technological point of view. Data published through INSPIRE download services may also have associated fees, but these charges should not exceed the cost of collection, production, reproduction and dissemination, together with a reasonable return on investment (European Commission, 2007).

Once the aforementioned technical issues are resolved, an integrated **data and service sharing** policy should be defined. Currently, INSPIRE requires Member States to provide the institutions and bodies of the community with access to spatial datasets and data services in accordance with harmonised conditions based on a minimum set of conditions to be respected. Member States are permitted exceptions to data sharing, and can even completely restrict access to certain data or can set security measures for obtaining access to these datasets and data services; for example, public-data access that may threaten individual privacy or national security can be restricted. While SDI data are under the full control of each Member State and several data are provided free of charge, VGI data are generally freely accessible, even though in some cases access is limited through restrictions. However, inherently, VGI platforms encourage registration of new users not only in terms of access, but also in terms of inputting new data and editing existing data. As a result, some critical security aspects may arise for society. For instance, how can a criminal VGI contributor be identified if they try to promote illegal activities and fraudulent information? (Legal issues of VGI are discussed in Chapter 6 by Mooney et al., 2017.) The above discussion indicates that VGI cannot be ruled through a strict framework such as that applied for INSPIRE, because it involves volunteered pieces of many GI infrastructures without an authoritative structure and scope. Therefore, the focus should be on the minimum aspects that will ensure interoperability, credibility and security of services and data.

4 The Prospects of Integration

4.1 Integration for Supporting Conventional Spatial Tasks

The combination of INSPIRE and VGI provides great potential for creating a comprehensive information platform by linking the advantages of authoritative information, i.e. quality assurance and normative status, with VGI advantages, i.e. rapid, up-to-date and dynamic information (Wiemann and Bernard, 2014). As a result, this integration can benefit NMAs, administrators of VGI projects and end users, with consequent socio-economic impacts (Campagna and Craglia, 2012). In particular, NMAs may have a real opportunity to use crowdsourced data to update some of their databases when the update is not done by them regularly due to the high costs involved or to add new data that are not available to them (Coleman et al., 2009). They can also use crowdsourced data to detect changes or vernacular place names (Olteanu-Raimond et al., 2017). On the other hand, INSPIRE can serve as a basis for validating VGI information (Wiemann and Bernard, 2014). Furthermore, end users may use this mix of official and spatio-temporal data for any relevant purpose, i.e. for leisure (to walk in unexplored natural tracks), for receiving notifications about a fact (e.g. the impacts of an earthquake), for travelling (i.e. which travel route to follow) and for professional/authoritative decision-making (e.g. how to manage a physical catastrophe or a crisis; Craglia, 2007; Wiemann and Bernard, 2014).

Some efforts towards VGI/SDI integration for the aforementioned purposes have already occurred (Craglia, 2007), e.g. the Linked Map project, which links GI from different sources, in particular SDI and VGI, through the paradigm of Linked Data (Lopez-Pellicer and Barrera, 2014). Linked Data connects related data through Web technologies. The Linked Map project has converted government datasets provided by the Spanish National Geographic Institute to Linked Data into Resource Description Framework (RDF) data, so that these datasets can be linked to VGI sources (OSM, DBpedia, etc.) and can be integrated using RDF links. RDF is a standard model for data interchange on the Web; RDF links enable Linked Data browsers and crawlers to navigate between data sources and to discover additional data. Another successful example is the case of the Ordnance Survey, which has linked an administrative geography dataset to other datasets on the Web, demonstrating the advantages of explicitly encoding topological relations between geographic entities over traditional spatial queries (Goodwin et al., 2008).

4.2 Integration with Social Media

Both active and passive Social Media Geographic Information (SMGI) can be integrated with SDIs in a GIS environment to perform qualitative and quantitative spatial, or more complex, multidimensional, analyses (Jankowski et al., 2010; Bugs, 2014; Campagna et al., 2015; Longley and Adnan, 2016). In particular, the integration of INSPIRE and VGI may generate a higher level of knowledge than INSPIRE alone, especially in those domains where the social component of data plays a relevant role, such as in politics, geo-marketing, tourism or spatial planning. The INSPIRE model may be extended through integration with SMGI, where multimedia data (i.e. texts, images, videos or audio) and user evaluations of the portrayed objects or phenomena are given with a time-stamp, enabling various kinds of new analysis, such as the spatial, temporal and statistical analysis of user interests and preferences; multimedia analyses; behavioural analyses; or combinations of these analyses, among others. Regarding the spatial analysis of user interests, the high number of georeferenced posts on social media platforms such as Twitter, Instagram, YouTube, Panoramio and Flickr can be used to investigate the patterns of user interests in space using density (Campagna, 2014) and clustering functions (Massa and Campagna, 2014). Data from such platforms can be accessed through APIs, georeferenced and saved as spatial data layers. Using SDI services such as WMSs or WFSs, GIS software can easily access the social media platform through the API, enabling the seamless integration of AGI and geo-UGC, as demonstrated by Massa and Campagna (2014). The overlay of spatial data layers with topographic SDIs such as administrative boundaries may offer useful hints to public authorities in understanding not only which places are important to the community and how they are perceived (Campagna, 2014), but also the composition of a community, e.g. local people, commuters, tourists or others.

Similarly, the temporal reference is often an available attribute in SMGI, which enables the study of when given places or infrastructures and services are used at different points in time. In addition, spatial statistics of user preferences, i.e. the collecting of posts by location, enables planners to analyse patterns in user interests at different scales. An example is given in Floris and Campagna (2014), where hotspot analysis has been used at the regional level to study tourist preferences by profile, before further analysing single hotspots with a tool embedded in ArcGIS called the Spatio-Temporal Textual analysis (Spatext-STTx) suite and with geographically weighted regression to explore, at the local level, what physical and locational factors may affect those preferences. Furthermore, multimedia analysis is well developed in the case of text analytics. However, it is currently more difficult to automatically extract useful information from images, video or audio. In the case of text, many software packages can be used to apply simple (i.e. calculating word frequency, or tag clouds) to more advanced (e.g. sentiment analysis) text analysis techniques. These techniques can be easily applied to subsets of SMGI obtained by spatial, temporal or user query. Moreover, user behavioural analysis, i.e. querying SMGI by a user, enables the study of user behaviour in space and time. This information can be used to analyse, for example, whether a public space is visited by local people or by outside visitors. This information may also be useful for profiling: for the users visiting a certain place or service, user spatiotemporal footprints can be defined to identify people who mainly move locally, regionally or internationally, and where they come from.

An additional application of the Spatext (STTx) suite is that made in a case study for the cyclone Cleopatra in Sardinia (Italy) to extract all relevant data and information (e.g. perceptions, opinions and needs from the local communities) from social media, i.e. Twitter, YouTube, Wikimapia and Instagram. These data were then integrated with the latest official datasets for further analysis and relevant action by decision-makers. Another related web application called 'Place, I care' was employed to support urban and regional planning processes. In particular, the aim was to collect information from concerned citizens about the physical, environmental and socio-cultural space to support collaborative and participatory planning. Although they have not been verified yet through a systematic analysis, there have been several case studies on the application of STTx in the same areas with different SMGI sources, where different types of users returned similar results, suggesting further research should be devoted to better understanding the issue of representativeness.

The above novel analytics may result not only in increasing the real-time monitoring capability of geo-UGC in representing the state of territorial systems, but also in supporting public participation and dialogue among digitally-enabled communities, which increasingly represent a substantial share of the total population in most countries. Other similar examples can be found in several domains. For example, the US Geological Survey (USGS) uses social networking to collect real-time, earthquake-related messages and early information to accelerate the delivery risk and response. Other related initiatives aim at (spatial) data collection, e.g. Project Noah[2], which is a citizen science web/mobile tool developed to explore and document wildlife around the globe. Similarly, the ZmapujTo.cz mobile application[3] was developed in 2012 in the context of an ecological project to combat illegal dumping grounds in the Czech Republic and contribute to solving this problem with the involvement of citizens and relevant authorities. At the time of creation, there was only a database of old ecological burdens, which covered the illegal dumps only marginally. In order to cover the largest possible area and utilise the potential of crowdsourced data, a platform was founded for information-gathering from citizens. The modern, efficient and widely-accepted platform was chosen for mapping while the mobile application and interactive web form were used for reporting. More than 2 500 illegal dumps were reported, and more than 40 municipalities and towns took part during the lifetime of the first version. In March 2014, the second version of ZmapujTo.cz was launched. This version introduced several new features. The most important change was the ability to report not only illegal dumping, but also a variety of other problems that one can encounter both in town and in the countryside. The entire website was redesigned, including an interactive map for efficient, fast and intuitive work. Further to the aforementioned applications, many other initiatives are aimed at supporting pluralism and public participation in decision-making, such as in the case of the SoftGIS approach (Kahila and Kyttä, 2009) adopted in the design of the Maptionnaire web platform (Kahila-Tani et al., 2016).

While early experiences in SDI/VGI integration and analyses may still be limited to expert research laboratories or to the fortresses of the social media corporations, institutional initiatives such as MYGEOSS may trigger further development in this domain. MYGEOSS is an ongoing project (2015–16) of the European Commission to develop smart Internet applications based on the Global Earth Observation System of Systems (GEOSS) to inform European citizens about the changes affecting their local environment. Specifically, within this project, a number of interactive apps were developed that reuse official spatial data to offer interactive services to the end users. For example, an application called 'Know Your City!', developed by UbikGS, presents social, economic and environmental indicators on a map-based quiz. Similarly, 'Loss of the Night', created by Interactive Scape GmBH & GFZ, is an application enabling citizen scientists all over the world to collect quantitative information on the changing nighttime environment, and MYGEOSS Phenology App Response was produced by the Friedrich-Schiller University to support vegetation phenology analysis using satellite data and data collected by citizens[4].

Despite the aforementioned efforts, Lopez-Pellicer and Barrera (2014) note that the integration of INSPIRE with VGI has not gained the expected attention yet, and this especially from large producers of GI, because of the technical disadvantages of the current Linked Data mechanism (Schade et al., 2010). Similarly, Wiemann and Bernard (2014) state that this integration effort is in its early stages, because several critical issues, which have been discussed earlier, need to be considered. Therefore, it seems that there is still a long way ahead for a full integration and operation of a global GIS platform, which is a concept set out in the next section.

5 Towards a Global Integrated GIS Platform

During the last few decades, the world has evolved rapidly because of the continuous increase in the urban population, new needs, modern lifestyles and technological advancements, creating millions of individual activities with environmental, economic and social impacts at different levels. As a result, sustainability at various levels and contexts has been introduced as one of the core aims of society, sustainability which will be better met if we understand the complexity of interactions and interrelations between the parameters involved. This suggests the need for dynamic information systems that provide reliable, accurate and real-time data to support intelligent planning and management in order to reach optimum decisions. Visionary and/or applied advanced geospatial tools and frameworks that move in this direction, such as the GeoWeb (Dangermond, 2005), Digital Earth (Craglia et al., 2008) and VGEs (Lin et al., 2013), have been proposed.

The GeoWeb is a computer network providing the ability to integrate and share geospatial information locally or globally via the Internet. Through the GeoWeb, the ideal system would be a wide network of distributed GIS services constructed and implemented by various inter-organisational collaborative agreements so that individual systems and communities might use each other's services, splitting the world into geographic components and allowing the dynamic integration of knowledge. The communities involved may range from simple users to governments, business enterprises and professionals focusing on improving their decision-making. Gradually, these communities may expand, interoperate more and become increasingly synergistic; hence the system might be driven by the thousands to millions of participants currently using websites such as Google Earth and OSM. Eventually, these services could provide a global network of open-access geographic knowledge about the planet and online applications (open access and licence-based) for processing this information to produce the outputs for decision-making. These functionalities may support a whole range of applications and purposes, supporting regional, national and even global applications, solving issues ranging from routine, static and structured problems to problems that are complex and unstructured (including those demanding real-time responses) and that depend on cross-organisation and cross-discipline collaboration. Both GIS professionals and citizens sensors have a role in this system. The former have the skills, knowledge and experience of authoritative system development and operation, while the latter represent the 'VGI-soldiers' across space and time who voluntarily collect and share valuable static or real-time information not available to SDIs (Dangermond, 2015).

Similarly, the vision of Digital Earth as defined by Craglia et al. (2008), which refers to a virtual globe system, would provide access to vast amounts of spatiotemporal multi-geoinformation for various levels of users – including modelling tools to facilitate decision-making. Digital Earth has eight key characteristics: it has multiple connected globes/infrastructures addressing the needs of different audiences; it is problem-oriented, i.e. focused on various key application themes such as the environment, health and societal issues; it enables space-temporal search in real-time from both sensors and humans; it allows spatial-based queries and advanced spatial analysis; it provides access to models as well as to 'what if' scenarios and forecasts; it supports the visualisation of abstract concepts and data types regarding global social issues, e.g. low income and poor health; it is based on open access and public participation across multiple technological platforms and media; and it is engaging, to enhance interactive and exploratory learning for multidisciplinary education and science. Five use cases that would comprise the vision of Digital Earth involving a unique platform have been provided by Goodchild (2012). These use cases involve Digitial Earth as a geoportal, a visualisation service, a platform for simulation and prediction, a source of unprecedented spatial and temporal resolution, and a technology fully integrated into human activities.

In a similar vein, VGEs involve a new generation of Web-based virtual geographic analysis platforms to facilitate the advanced exploration of physical, environmental, socio-economic and other phenomena to solve related problems at a deeper level by combining state-of-the-art geotechnology and knowledge. Such a VGE system would consist of four basic components: (i) the data component for the integration, organisation and management of geographic information; (ii) the modelling and simulation component for the dynamic analysis of geographic phenomena by providing experts from various disciplines with an open access platform to develop and disseminate distributed advanced models in an easy and collaborative way; (iii) the interactive component between the system and users that includes external and internal data collection tools; and (iv) the collaborative component that enables group decision-making for significant societal problems through public participation in the processes carried out by experts.

Although the concept of Digital Earth, the existing technology of the GeoWeb and the use cases for VGEs have a common aim and functions, i.e. to provide advanced geodata hubs and sophisticated spatial analysis tools on the Web, they have some differences in terms of their focus. In particular, Digital Earth and VGEs involve extended capabilities beyond sharing knowledge and geoinformation such as the GeoWeb's, by providing advanced virtual reality, processing, simulation and analysis models for solving a wide range of complex spatial problems. In addition, VGEs involve more problem-oriented geotechnology tools that inherently have some of the features of planning and decision support systems, while the Digital Earth concept aims to provide more abstract tools for investigating the spatial interactions of certain domains.

Based on the aforementioned visions, we try to shift from a conceptual context for creating a new-generation geographic tool, to a more practical and tangible framework for developing a global integrated GIS platform, as illustrated in Figure 4. This framework extends the capabilities of a typical data hub and the benefits of integration of SDIs with VGI. In particular, the system consists of three main components: integrated data infrastructures, integrated online applications and a system for providing outputs (both static and dynamic) that could lead to decision-making and actions. As an alternative to providing wide access to a single source of data, the Integrated Data Infrastructures component can provide distributed data mashups by integrating vast stores of information (from many sources in the public and private sector as well as from citizens) and of many different types of data, along with geospatial services that can interact and be used to create new information. The data sources can be SDIs such as INSPIRE or NSDI in the United States, VGI platforms created through various projects (e.g. OSM and Wikipedia), social media (e.g. Facebook or Twitter) and other media such as emails, mobile phones, Instant messenger, etc. Existing services can be combined to make new services, and Geocommunities, which are currently fragmented, may be consolidated in a loosely coupled environment and create new synergies (Esri, 2006).

The integration of online applications could provide functionalities from simple publishing and mapping/visualisation to advanced GeoComputation modelling (Abrahart and See, 2014). In particular, the current Web-GIS services can be extended to provide not only easy map publishing and viewing through VREs, but also basic GIS functions, such as querying, buffering, overlays, etc., through Open Access (or licence-based) online GIS software. In addition, focused GIS applications, in the form of different thematic modules (i.e. for planning, transport, the environment, etc.) embedded in the online GIS, may be offered through distributed geo-services based on Web, GIS server-technology and service-oriented architecture (SOA) that is open, interoperable, and dynamic, based on common data and service standards and specifications. Using the SOA model with GIS services, users can integrate their desktop and departmental solutions into implementations that connect many departments and organisations (Dangermond, 2008). The Web Services architecture allows users to both federate their distributed systems and integrate GIS and spatial processing with other IT business systems, such as Enterprise resource planning (ERP), Customer relationship management (CRM) and Supervisory control and data acquisition (SCADA). While this has been possible for some time, the advent of SOA and simple technologies to integrate these services has made it much easier and promises to greatly expand the GIS market. Ideally, in this context, easy-to-build ad-hoc advanced spatial models for GeoComputation that employ artificial intelligence techniques, for example, for solving complicated problems might be the biggest achievement of this system.

The results of the system could take the form of Dynamic Outputs. Outputs, which result from the processing of static or real-time information, can have any form, i.e. they can take the form of maps, reports and messages, and mass notification alerts. In particular, maps and reports in text or tabular form are the custom outputs of a GIS and can be used by users for decision-making and appropriate actions. Messages, e.g. through phone calls, emails, SMS, Viber etc., refer to real-time reporting to administrations and organisations. Similarly, mass notification alerts refer to broad notifications, or alerts, sent to people in a specific geographic region in emergency or crisis management situations. The tremendous high-speed evolution of the Web and Geospatial technologies suggests that this 'super' global Geo-system is not far away.

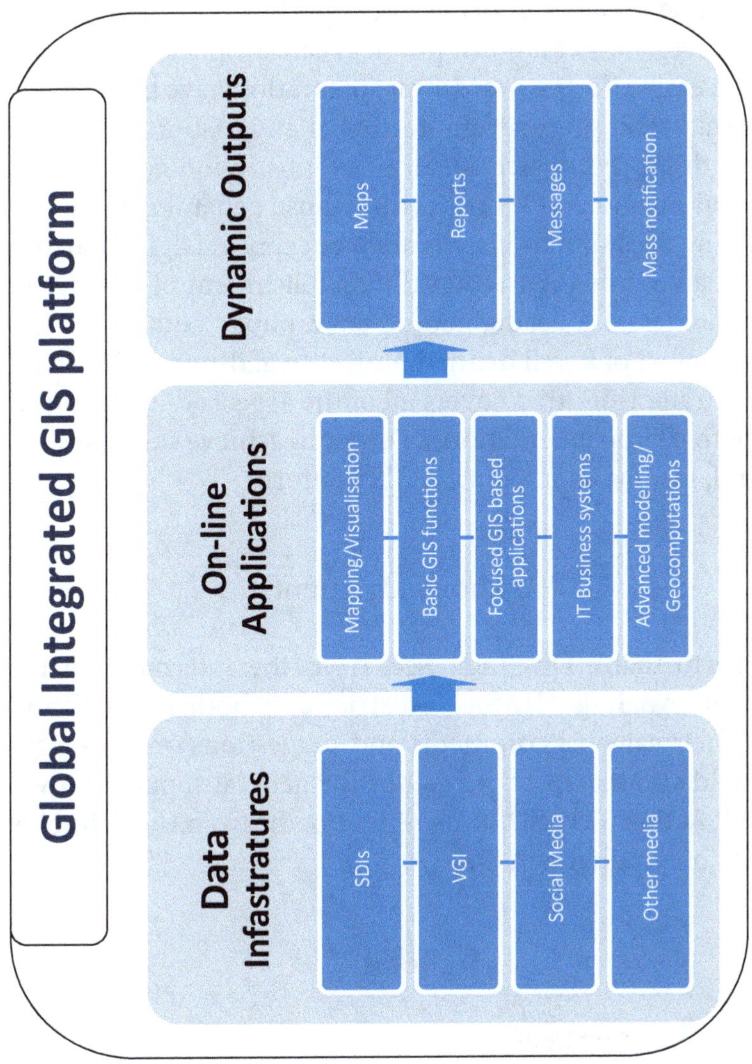

Fig. 4: The framework of an ideal global integrated GIS platform.

6 Conclusions

The integration of SDIs, and in particular INSPIRE, with VGI may potentially provide considerable benefits for all stakeholders involved, i.e. public and private organisations, professionals and citizens, because each technology may complement the other. In particular, benefits may include benefits for specific professional groups dealing with spatial problems; for planning and decision-making; and for the wider community, which may enable the dissemination and uptake of real-time updated information regarding daily activities (e.g. traffic incidents) or emergency situations, physical catastrophes or unknown threats. Although some early efforts towards this integration have been made, this project is not an easy task, since several technical and institutional issues need to be resolved, as discussed earlier. Ideally, the integration could be extended to creating a global integrated GIS platform, whose general framework has been presented and involves similar visions and concepts to Digital Earth and VGEs. The next steps should be focused on the establishment of a wider network of involved stakeholders, i.e. academia, industry, public authorities, citizens and NGOs, in the context of a well defined project (e.g. through a COST Action) to set up a robust framework that covers all of the aspects of the project, from the initial concept to its implementation, in order to achieve successful examples of integration and, ideally, an integrated GIS platform.

Acknowledgements

We would like to thank Dr Linda See, from the International Institute for Applied Systems Analysis (IIASA), Austria, and both internal and external reviewers for their valuable comments and suggestions on the original manuscript. We would also like to acknowledge the general support provided by the Chair of COST Action TD1202, Prof. Giles Foody, from the School of Geography, University of Nottingham, UK.

Notes

1 http://inspire.ec.europa.eu/
2 http://www.projectnoah.org/
3 http://www.ZmapujTo.cz
4 http://digitalearthlab.jrc.ec.europa.eu/mygeoss/results.cfm

Reference list

Abrahart, R.J., See, L. (Eds.), 2014. *GeoComputation*, 2nd ed. CRC Press, Boca Raton, FL.

Bégin, D., 2012. Towards integrating VGI and national mapping agency operations - A Canadian case study, in: Proceedings of the GIScience Conference, Columbus, OH, USA, 18–21 September 2012. Available at: http://web.ornl.gov/sci/gist/workshops/2012/documents/Begin,%20Daniel%20-%20Paper.pdf [Last accessed 16 May 2017].

Botshelo, S., 2009. Assessing Alternative Technologies for Use of Volunteered Geographic Information in Authoritative Databases. M.Sc.E. Thesis Department of Geodesy and Geomatics Engineering Technical Report No. 269. University of New Brunswick.

Budhathoki, N.R., Bruce, B. (Chip), Nedovic-Budic, Z., 2008. Reconceptualizing the role of the user of spatial data infrastructure. *GeoJournal* 72, 149–160. DOI: https://doi.org/10.1007/s10708-008-9189-x

Bugs, G., 2014. Tecnologias da informação e comunicação, sistemas de informação geográfica e a participação pública no planejamento urbano. Tesi de dotourado. Universidade Federal do Rio Grande du Sul.

Campagna, M., 2014. The geographic turn in social media: Opportunities for spatial planning and Geodesign, in: Murgante, B., Misra, S., Rocha, A.M.A.C., Torre, C., Rocha, J.G., Falcão, M.I., Taniar, D., Apduhan, B.O., Gervasi, O. (Eds.), *Computational Science and Its Applications - ICCSA 2014*. Springer International Publishing, Cham, pp. 598–610.

Campagna, M., Craglia, M., 2012. The socioeconomic impact of the Spatial Data Infrastructure of Lombardy. *Environment and Planning B: Planning and Design* 39, 1069–1083. DOI: https://doi.org/10.1068/b38006

Campagna, M., Floris, R., Massa, P., Girsheva, A., Ivanov, K., 2015. The role of Social Media Geographic Information (SMGI) in spatial planning, in: Geertman, S., Ferreira, Jr, J.F., Goodspeed, R., Stillwell, J. (Eds.), *Planning Support Systems and Smart Cities, Lecture Notes in Geoinformation and Cartography*. Springer International Publishing, pp. 41–60.

Castelein, W., Grus, L., Crompvoets, J., Bregt, A.K., 2010. A characterization of VGI, in: Proceedings of the 13th AGILE International Conference on Geographic Information Science. Guimarães, Portugal, 10–14 June 2010. Available at: https://agile-online.org/conference_paper/cds/agile_2010/shortpapers_pdf/106_doc.pdf [Last accessed 16 May 2017].

Coleman, D.J., Georgiadou, Y., Labonte, J., 2009. Volunteered geographic information: The nature and motivation of produsers. *International Journal of Spatial Data Infrastructures Research* 4, 332–358.

Craglia, M., 2007. Volunteered Geographic Information and Spatial Data Infrastructures: when do parallel lines converge? Position Paper for the NCGIA and Vespucci Workshop on Volunteered Geographic Information. Santa Barbara, CA, 13–14 Dec 2007. Available at: http://ncgia.ucsb.edu/projects/vgi/docs/position/Craglia_paper.pdf [Last accessed 16 May 2017].

Craglia, M., Goodchild, M.F., Annoni, A., Camara, G., Gould, M., Kuhn, W., Mark, D., Masser, I., Maguire, D., Liang, S., Parsons, E., 2008. Next-Generation Digital Earth: A position paper from the Vespucci Initiative for the Advancement of Geographic Information Science. *International Journal of Spatial Data Infrastructures Research* 3, 146–167. DOI: https://doi.org/10.2902/

Dangermond, J., 2015. Geospatial Technology and the Future of the City, ArcNews Online. Available at http://www.esri.com/esri-news/arcnews/winter1415articles/geospatial-technology-and-the-future-of-the-city [Last accessed 1 May 2017].

Dangermond, J., 2008. GIS and the GeoWeb, ArcNews Online. Available at http://www.esri.com/news/arcnews/summer08articles/gis-and-geoweb.html [Last accessed 1 May 2017].

Dangermond, J., 2005. GIS Helping Manage our World, ArcNews Online. Available at http://www.esri.com/news/arcnews/fall05articles/gis-helping-manage1of2.html [Last accessed 1 May 2017].

Dassonville, L., Vauglin, F., Jakobsson, A., Luzet, C., 2003. Quality management, data quality and users, metadata for geographical information, in: Shi, W., Fisher, P., Goodchild, M.F. (Eds.), *Spatial Data Quality*. CRC Press, pp. 214–227.

Delaney, P., Pettit, C., 2014. Urban Data Hubs Supporting Smart Cities, in: Winter, S., Winter, S., Rizos, C. (Eds.), *Research@Locate 14*. Canberra, Australia.

Demetriou, D., 2016. Uncertainty of OpenStreetMap (OSM) data for the road network in Cyprus, in: Proceedings of SPIE 9688, Fourth International Conference on Remote Sensing and Geoinformation of Environment (RSCy2016), 968806, Paphos, Cyprus, 4–8 April 2016. DOI: https://doi.org/10.1117/12.2239612

Environmental Systems Research Institute (Esri), 2006. The GeoWeb: A Vision for Supporting Collaboration, ArcNews Online. Available at http://www.esri.com/news/arcuser/0206/geoweb.html [Last accessed 1 May 2017].

European Commission, 2007. DIRECTIVE 2007/2/EC OF THE EUROPEAN PARLIAMENT AND OF THE COUNCIL: Establishing an Infrastructure for Spatial Information in the European Community (INSPIRE). Official Journal of the European Union.

Flanagin, A., Metzger, M., 2008. The credibility of volunteered geographic information. *GeoJournal* 72, 137–148.

Floris, R., Campagna, M., 2014. Social Media Geographic Information in tourism planning. Tema. *Journal of Land Use, Mobility and Environment* 0.

Fonte, C C, Antoniou, V, Bastin, L, Estima, J, Arsanjani, J J, Bayas, J-C L, See, L and Vatseva, R. 2017. Assessing VGI Data Quality. In: Foody, G, See, L, Fritz, S, Mooney, P, Olteanu-Raimond, A-M, Fonte, C C and Antoniou, V. (eds.) *Mapping and the Citizen Sensor*. Pp. 137–163. London: Ubiquity Press. DOI: https://doi.org/10.5334/bbf.g.

Foody, G., See, L., Fritz, S., Van der Velde, M., Perger, C., Schill, C., Boyd, D.S., 2013. Assessing the accuracy of volunteered geographic information arising from multiple contributors to an internet based collaborative project. *Transactions in GIS* 17, 847–860. DOI: https://doi.org/10.1111/tgis.12033

Fritz, S, See, L and Brovelli, M. 2017. Motivating and Sustaining Participation in VGI. In: Foody, G, See, L, Fritz, S, Mooney, P, Olteanu-Raimond, A-M, Fonte, C C and Antoniou, V. (eds.) *Mapping and the Citizen Sensor*. Pp. 93–117. London: Ubiquity Press. DOI: https://doi.org/10.5334/bbf.e.

Furey, E., Curran, K., Kevitt, P.M., 2013. Probabilistic indoor human movement modeling to aid first responders. *Journal of Ambient Intelligence and Humanized Computing* 4, 559–569. DOI: https://doi.org/10.1007/s12652-012-0112-4

Goodchild, M.F., 2012. The future of Digital Earth. *Annals of GIS* 18, 93–98. DOI: https://doi.org/10.1080/19475683.2012.668561

Goodchild, M.F., 2007. Citizens as sensors: the world of volunteered geography. *GeoJournal* 69, 211–221. DOI: https://doi.org/10.1007/s10708-007-9111-y

Goodchild, M.F., Li, L., 2012. Assuring the quality of volunteered geographic information. *Spatial Statistics* 1, 110–120. DOI: https://doi.org/10.1016/j.spasta.2012.03.002

Goodwin, J., Dolbear, C., Hart, G., 2008. Geographical linked data: The administrative geography of Great Britain on the semantic web. *Transactions in GIS* 12, 19–30. DOI: https://doi.org/10.1111/j.1467-9671.2008.01133.x

Haklay, M., 2010. How good is volunteered geographical information? A comparative study of OpenStreetMap and Ordnance Survey datasets. *Environment and Planning B: Planning and Design* 37, 682–703. DOI: https://doi.org/10.1068/b35097

Iliffe, M., 2012. Towards understanding the factors of integrating Spatial Data Infrastructures and crowdsourced geographic data, in: Proceedings of the 1st AGILE PhD School. Wernigerode, Germany, 13–14 March 2012, pp. 71–76. Available at: https://agile-online.org/conference_paper/images/phddocs/proceedings_agile_phd_school_2012.pdf [Last accessed 16 May 2017].

ISO, 2013. *ISO 19157:2013–Geographic information–Data Quality*. International Organization for Standardization, Geneva, Switzerland.

Jankowski, P., Andrienko, N., Andrienko, G., Kisilevich, S., 2010. Discovering landmark preferences and movement patterns from photo postings. *Transactions in GIS* 14, 833–852. DOI: https://doi.org/10.1111/j.1467-9671.2010.01235.x

Kahila, M., Kyttä, M., 2009. SoftGIS as a bridge-builder in collaborative urban planning, in: Geertman, S., Stillwell, J. (Eds.), *Planning Support Systems Best Practice and New Methods*. Springer Netherlands, Dordrecht, pp. 389–411.

Kahila-Tani, M., Broberg, A., Kyttä, M., Tyger, T., 2016. Let the citizens map—Public participation GIS as a planning support system in the Helsinki Master Plan process. *Planning Practice & Research* 31, 195–214. DOI: https://doi.org/10.1080/02697459.2015.1104203

Lin, H., Chen, M., Lu, G., Zhu, Q., Gong, J., You, X., Wen, Y., Xu, B., Hu, M., 2013. Virtual Geographic Environments (VGEs): A new generation of geographic analysis tool. *Earth-Science Reviews* 126, 74–84. DOI: https://doi.org/10.1016/j.earscirev.2013.08.001

Longley, P.A., Adnan, M., 2016. Geo-temporal Twitter demographics. International *Journal of Geographical Information Science* 30, 369–389. DOI: https://doi.org/10.1080/13658816.2015.1089441

Lopez-Pellicer, F., Barrera, J., 2014. Linked Map requirements definition and conceptual architecture, Deliverable D15.1, Available at http://www.planet-data.eu/sites/default/files/PD%20D15.1.pdf [Last accessed 1 May 2017].

Mangano, D., 2013. *The Integrated Data Hub: The Next Generation Data Warehouse*. Createspace, S.l.

Massa, P., Campagna, M., 2016. Integrating authoritative and Volunteered Geographic Information for spatial planning, in: Capineri, C., Haklay, M., Huang, H., Antoniou, V., Kettunen, J., Oostermann, F., Purves, R. (Eds.), *European Handbook of Crowdsourced Geographic Information*. Ubiquity Press, London, UK, pp. 401–418. DOI: https://doi.org/10.5334/bax

Massa, P., Campagna, M., 2014. Social Media Geographic Information: Current developments and opportunities in urban and regional planning, in: Proceedings of the 19th International Conference on Urban Planning and Regional Development in the Information Society. Presented at the GeoMultimedia 2014. DOI: https://doi.org/10.6092/1970-9870/2500

McDougall, K., 2009. Volunteered Geographic Information for building SDI, in: Ostendorf, B., Baldock, P., Bruce, D., Burdett, M., Corcoran, P. (Eds.), Proceedings of the Surveying and Spatial Sciences Institute: Biennial International Conference. Adelaide, Australia, 28 Sep-2 Oct 2009, Surveying and Spatial Sciences Institute, pp. 645–653.

Moens, M.-F., Li, J., Chua, T.-S. (Eds.), 2014. *Mining User Generated Content, Social Media and Social Computing*. CRC Press, Boca Raton, FL.

Mooney, P, Olteanu-Raimond, A-M, Touya, G, Juul, N, Alvanides, S and Kerle, N. 2017. Considerations of Privacy, Ethics and Legal Issues in Volunteered Geographic Information. In: Foody, G, See, L, Fritz, S, Mooney, P, Olteanu-Raimond, A-M, Fonte, C C and Antoniou, V. (eds.) *Mapping and the Citizen Sensor*. Pp. 119–135. London: Ubiquity Press. DOI: https://doi.org/10.5334/bbf.f.

Odobašić, D., Medak, D., Miler, M., 2013. Gamification of geographic data collection, in: Thomas, J., Car, A., Strobl, J., Griesebner, G. (Eds.), *GI_Forum: Creating the GISociety*. Wichman-Verlag, Berlin, pp. 328–337.

Olteanu-Raimond, A.-M., Hart, G., Foody, G., Touya, G., Kellenberger, T., Demetriou, D., 2017. The scale of VGI in map production: A perspective of European National Mapping Agencies. *Transactions in GIS* 21, 74–90. DOI: https://doi.org/10.1111/tgis.12189

Open Knowledge Foundation, 2013. The Data Hub - The easy way to get, use and share data. Available at http://datahub.io/sq/about [Last accessed 1 May 2017].

Parker, C.J., May, A., Mitchell, V., 2012. Understanding design with VGI using an information relevance framework. *Transactions in GIS* 16, 545–560. DOI: https://doi.org/10.1111/j.1467-9671.2012.01302.x

Schade, S., Granell, C., Diaz, L., 2010. Augmenting SDI with Linked Data. Available from: http://ceur-ws.org/Vol-691/paper3.pdf.

Steven, A.R., 2005. The US National Spatial Data Infrastructure: What is new? Presented at the ISPRS Workshop on Service and Application of Spatial Data Infrastructure, Hangzhou, China, 14–16 October 2005. Available at: http://www.isprs.org/proceedings/XXXVI/4-W6/papers/313-318FGDC-PaprChina14Oct05.pdf [Last accessed 16 May 2017].

Tang, K.P., Lin, J., Hong, J.I., Siewiorek, D.P., Sadeh, N., 2010. Rethinking location sharing: exploring the implications of social-driven vs. purpose-driven location sharing, in: Proceedings of the 12th ACM International Conference on Ubiquitous Computing. ACM Press, New York, NY, USA, pp. 85–94. DOI: https://doi.org/10.1145/1864349.1864363

Wiemann, S., Bernard, L., 2014. Linking crowdsourced observations with INSPIRE, in: Huerta, J., Schade, S., Granell, G. (eds.), Proceedings of the 17th AGILE International Conference on Geographic Information Science: Connecting a Digital Europe through Location and Place, Castellón, Spain, 3–6 June 2014, Available at: https://agile-online.org/conference_paper/cds/agile_2014/agile2014_80.pdf [Last accessed 16 May 2017].

Wikimapia, 2015. Wikimapia. Available at http://old.wikimapia.org [Last accessed 1 May 2017].

Williamson, I.P., Rajabifard, A., Feeney, M.-E.F. (Eds.), 2003. *Developing spatial data infrastructures: from concept to reality*. Taylor & Francis, London; New York.

CHAPTER 13

VGI in National Mapping Agencies: Experiences and Recommendations

Ana-Maria Olteanu-Raimond*, Mari Laakso†, Vyron Antoniou‡, Cidália Costa Fonte§, Alexandra Fonseca¶, Magdalena Grus‖, Jenny Harding**, Tobias Kellenberger††, Marco Minghini‡‡, Andriani Skopeliti§§

*Univ. Paris-Est, LASTIG COGIT, IGN, ENSG, F-94160 Saint-Mande, France, ana-maria.raimond@ign.fr

†Finnish Geospatial Research Institute, Kirkkonummi 02430, Finland

‡Hellenic Army General Staff, Geographic Directorate, PAPAGOU Camp, Mesogeion 227-231, Cholargos, 15561, Greece

§Department of Mathematics, University of Coimbra, 3001-501 Coimbra, Portugal / INESC Coimbra, Rua Sílvio Lima, Pólo II, 3030-290 Coimbra, Portugal

¶Portuguese Territorial Development Agency (DGT), Center for Environmental and Sustainability Research (CENSE), Portugal

‖Kadaster, Hofstraat 110, 7311 KZ Apeldoorn, the Netherlands

**Ordnance Survey, Adanac Drive, Southampton SO16 0AS, UK

††Federal Office of Topography swisstopo, Seftigenstrasse 264, 3084 Wabern, Switzerland

‡‡Department of Civil and Environmental Engineering, Politecnico di Milano, Piazza Leonardo da Vinci 32, 20133 Milano, Italy

§§School of Rural and Surveying Engineering, National Technical University of Athens, 9 H. Polytechniou, Zografou, 15780, Greece

How to cite this book chapter:

Olteanu-Raimond, A-M, Laakso, M, Antoniou, V, Fonte, C C, Fonseca, A, Grus, M, Harding, J, Kellenberger, T, Minghini, M, Skopeliti, A. 2017. VGI in National Mapping Agencies: Experiences and Recommendations. In: Foody, G, See, L, Fritz, S, Mooney, P, Olteanu-Raimond, A-M, Fonte, C C and Antoniou, V. (eds.) *Mapping and the Citizen Sensor*. Pp. 299–326. London: Ubiquity Press. DOI: https://doi.org/10.5334/bbf.m.

Abstract

Despite the considerable growth in Volunteered Geographic Information (VGI) activities in citizen sensing and the evident opportunities for VGI use in map revision and updating, few European National Mapping Agencies (NMAs) or other types of government bodies have engaged significantly with VGI. Moreover, the level of engagement of NMAs with the VGI community varies greatly, and most of them have proposed their own tools for encouraging citizens and public partners to collect feedback or new data. There are numerous barriers limiting the participation of citizens and public partners in NMA data collection, including data quality issues, the motivation of the contributors and legal issues. The aim of this chapter is to give an overview of the experiences of some European NMAs in engaging with VGI. Guidelines and recommendations to support wider engagement with the VGI community are also proposed to help NMAs and interested government bodies exploit the potential of VGI for authoritative mapping.

Keywords

VGI, authoritative mapping, VGI platform, data collection, data quality

1 Introduction

Volunteered Geographic Information (VGI) initiatives have seen considerable growth in citizen sensing (Goodchild, 2007). Different terms are used in the literature to describe this volunteered activity, such as crowdsourcing and neogeography (Turner, 2006) or user generated spatial content (Antoniou et al., 2009). See et al. (2016) give a complete review on the current terminologies used and the distinctions between them. In this chapter, the focus is on VGI in the context of European National Mapping Agencies (NMAs).

With the adoption of the Open Data Policy[1] which encourages to freely release data that can be used and republished by any user, many government datasets are now freely available to the public, including spatial data from some European NMAs (Brovelli et al., 2016). Some NMAs, such as those of Finland and the Netherlands, have released their datasets under open access licences; these authoritative data have been integrated into OpenStreetMap (OSM), which has improved the OSM database. More studies are necessary to determine if this integration may also have benefits for NMAs. The Open Data Policy can be an opportunity for both NMAs and geographic data end users. Indeed, releasing data under open access licences through a platform can increase the usability of authoritative data, because end users such as citizens can freely download and use data for different purposes. In addition, the motivation for citizens and partners to contribute by adding new information, giving feedback and providing alerts on errors and updates can also increase.

Although local governments had already started during the last ten years to use VGI as a participation platform to engage in a dialogue with citizens rather than as a way to simply gain or share information (Johnson and Sieber, 2013), there has been a noticeable change. Indeed, more recently, different initiatives have been proposed by local governments to collect data for different purposes (such as in urban planning, in order to advertise new regulations) where citizens have been considered both as sensors and as potential partners (Karimipour and Azari, 2015; Sedano, 2016).

Traditionally, almost all mapping agencies have some experience in collecting information from their data users by receiving alerts regarding mapping errors or updates. However, it is important to differentiate between passive processes and more active processes in which the mapping agencies actively engage with the VGI community by proposing platforms to collect and disseminate data (See et al., 2016).

Olteanu-Raimond et al. (2017) have recently undertaken a detailed review of the engagement of European NMAs with VGI. A survey was undertaken to elicit experiences with VGI, which revealed that few European NMAs are currently engaged with VGI and that those have developed their own VGI collection processes, mostly for change detection and the reporting of alerts, with less frequent examples of the reporting of new content, vernacular place names and photo interpretation (see Figure 1). In most cases the information gathered was

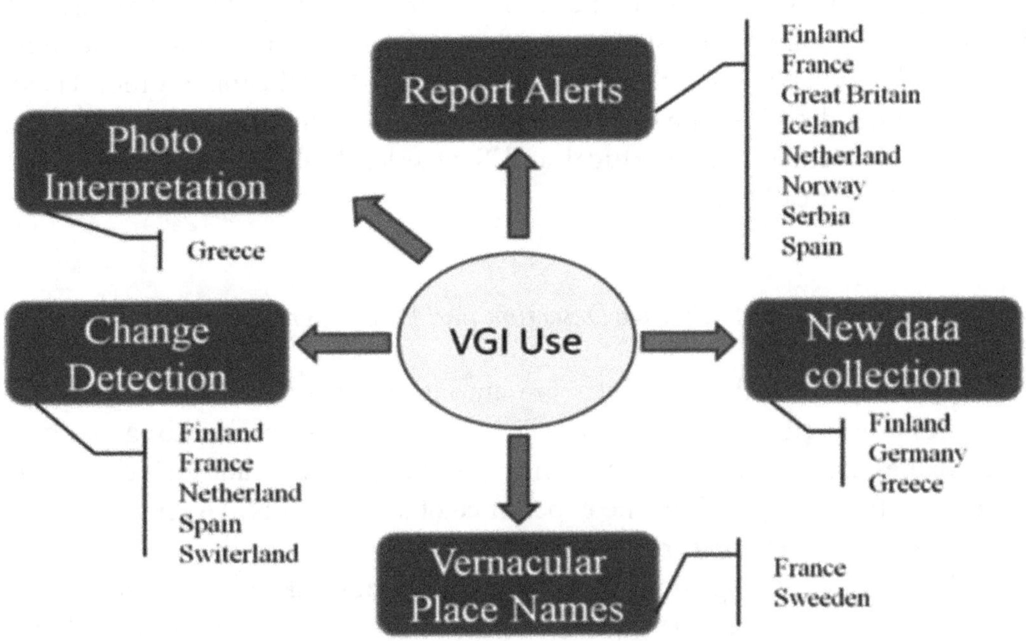

Fig. 1: Use of VGI by European NMAs. Source: Olteanu-Raimond et al. (2017).

on traditional features included in standard topographic maps, such as roads, buildings and names. Very few mapping agencies have harvested and used the data collected by OSM or GeoNames.

The low involvement of European NMAs with VGI is related to five major barriers, which have been discussed in detail in Olteanu-Raimond et al. (2017); these are issues of data quality and validation; legal issues; issues related to the nature and motivation of the crowd; sustainability issues; and employment fears.

This chapter further develops the work of Olteanu-Raimond et al. (2017) by proposing a typical VGI collection workflow, which was considered by many NMAs such a good practice. This type of VGI platform is based on the main idea of a volunteered activity where contributors contribute directly to the platform by adding new features or attributes, correcting existing features, etc. It is important to mention that the integration of data coming from other crowdsourced activities, such as GPS traces from sports activities, are out of the scope of this chapter. The chapter is organised as follows: Section 2 focuses on the experiences of European NMAs with VGI by presenting some specific examples. Section 3 presents some recommendations for NMAs as a response to some of the five major barriers identified in the use of VGI. Finally, conclusions and future research directions are outlined in Section 4.

2 Experiences with VGI

As mentioned previously, most of the NMAs that engage with VGI have developed their own tools to collect data from citizens or from public partners. The aim of this section is to present an overview of some of these tools that completes and provides an update to the review reported in Olteanu-Raimond et al. (2017), which describes the experiences of NMAs in Finland, France, Greece, the UK, the Netherlands, Portugal and Switzerland, all of which responded positively to our call to contribute.

2.1 Change Detection and Error Alerts

Change detection and error alerts are among the most well developed VGI activities proposed by NMAs. Generally, alerts (e.g. to a new building or a new road name) are used as triggers to improve the quality of authoritative databases. The following outlines the experience of a series of NMAs in using VGI for change detection and error alerting.

At IGN France, change detection is generally undertaken by land surveyors who analyse a range of alert types and then contact local governments. Since 2008, IGN France has developed various applications that aim to report alerts concerning errors, change detection or vernacular toponyms (Viglino, 2009).

These applications, deployed on different platforms and via different technologies (e.g. the Web, Android mobile phones and GIS) are mostly community-sourcing systems where professional partners, such as fire services and post offices, make reports on IGN data. A web application, accessible through the French Geoportal, was also developed for citizens, allowing them to make reports. These pioneering applications and their encouraging results have led the IGN to propose a unique community and citizen sourcing portal[2] , on which citizens can complete a form and provide location information, using GPS tracks, photographs or drawings, on an IGN basemap. A new version of the application is being tested that allows partners to access, add and modify features in an up-to-date copy of the topographic database. Contributions are first checked and validated by the surveyors with respect to data specifications, and quality expectations are checked by using quality indicators, visual checking and comparison with different data sources (e.g. construction permits issued by municipalities). Depending on the types of contributions, the VGI can be directly integrated into authoritative databases or used as a trigger for field work to improve the geometric precision of features.

With regard to future engagement with VGI, some research projects are currently under consideration. For example, Ivanovic et al. (2016) are studying the possibility of automatically inferring changes from additional sources found on the Web, including GPS tracks from hiking websites. The EU-funded Horizon 2020 LandSense project (2016–2020) will study the feasibility of updating Land Use/Land Cover (LULC) maps using Sentinel and *in-situ* citizen-derived data. Methods to aid quality assessment and conflict management in order to validate and integrate citizen-derived data into the authoritative database will also be explored (Leibovici et al., 2015).

In the Netherlands, Kadaster is running successful VGI activities, including 'terugmelding BRT' (alert on the Dutch Topographic Registry) and 'terugmelding BGT' (alert on the 'large scale' Topographic Registry), to report new changes and errors. Kadaster works as an open and transparent organisation, and contributors can easily see what has been done with their alerts. To stimulate and effectively motivate contributors, the staff working in the topographic department promptly validates all reported alerts. By directly updating the topographic maps when an error report is accepted, Kadaster shows its appreciation to the contributors and stimulates the further participation of citizens. In addition to the traditional data-updating by means of aerial and panoramic photographs, there is a growing tendency to use thematic data from external sources. The latter sources include governmental organisations, companies and also citizen contributors. In this context, Kadaster has proposed a second pilot (also known as the Sonneveld index) to collect data on religious buildings such as churches, mosques, synagogues, temples, monasteries and chapels; more than 1,000 addresses were collected by a group of enthusiastic contributors. As a result, Kadaster was able to enrich its topographic maps.

Another VGI project was run to collect information on national border markers. On 30 October 1980, the Netherlands and Germany signed an agreement about the maintenance of the markers that define the borders between them; every three years, the national border markers must be inspected and, where necessary, maintained. In 2012, hikers were deployed to gather information about the situation of national border markers by using an *ad hoc* mobile application that also allows sending a picture. As a result, the Kadaster was able to make a decision as to whether it had to maintain a particular marker or not. The border markers application has recently completed its pilot phase, and a continuation of the project is being developed.

Finally, the forest paths project was a recent pilot based on VGI activities. In the Netherlands, the National Dutch Forest Organization (Staatsbosbeheer) is responsible for data on forests. The aim of the forest paths project was to use VGI to update the organisation's datasets. Kadaster provided raw material to forest rangers and asked them to verify and complete the map based on their field work. Kadaster has successfully completed pilot projects in Horsterwold and Flevopolder. The local forest rangers have updated their digital files on forest paths in their region. Kadaster is researching how to implement this method in the rest of the forested area in the Netherlands.

Ordnance Survey (OS), the NMA for Great Britain, has long engaged with customers and the general public for alerts about real-world change or errors reported in its paper or digital map products. While much contact is directly via telephone or written correspondence, a web map-based tool has been successfully trialled with public sector customers for reporting errors or omissions in a range of OS products. Using the 'Tell OS' interface, customers can locate, describe and submit their feedback for the product concerned. Their alerts are acknowledged and the information is fed into product management processes.

Sharing of volunteered information is also enabled for route-based information through the OS Maps application. Aimed at outdoor activities, the application enables the recording and sharing with other users of route information as part of its map display, search and navigation functionality.

2.2 New-Feature Collection

VGI provides the potential to capture new features or new information regarding existing features not previously collected by NMAs as it might not be within their mission priorities or it may be excluded for political or economic reasons.

In the Netherlands, Kadaster is running pilot projects to collect new features. One of these is the 'Crowdsourcing at school!' project, which is part of Kadaster's education programme. The aim of this initiative is to allow children to become familiar with VGI and with advancing society, but also to introduce them to the Kadaster organisation and its products and services. Children get a geographic orientation of the world in a playful way, and they also learn about their position within society. In this pilot, children collect data on emergency services such as police, ambulance and fire services. This project can also be used for data collection for other organisations or public services. The curriculum for this project is in a pilot phase and the first results have highlighted that VGI activities are not only for adults.

Linked to the large-scale renewal of Finland's National Topographic Database, a research project was launched by National Land Survey of Finland (NLS) at the beginning of 2016 to investigate the possibilities that VGI can offer in authoritative data collection. The project will build a concept to define the so-called 'Citizen's layer' to the authoritative topographic data, that is, a platform for data collection where they will be able to import or draw points, lines and polygons representing topographic objects in the real world. The concept will cover principles and tools for VGI data collection, e.g. for building up the service and the user interface as well as developing protocols and methods for engaging with citizens (Mooney et al., 2016). The quality and the best practices for using VGI will be identified in a pilot phase. The project seeks to validate data quality and usability and to investigate the possibilities of integrating VGI collected in the pilot to the authoritative database. As part of another research project, a hyper-local geosocial networking application (hylo.mygeotrust.org) was introduced for school children aged 14- to 15-years-old. With the mobile application, pupils were asked to map different kinds of objects in their neighbourhood to share their knowledge and observations. The initial results are encouraging. Children are interested in their local environments and have volunteered to map and share their knowledge on a map service. Based on these experiences, it seems beneficial to introduce the concept of a 'Citizen's layer' in schools as well.

Greek mapping authorities have been using VGI as a starting point to update or create new mapping outputs. The crowdsourced data are treated as an initial input layer that is compared against imagery backdrops (satellite or aerial). The VGI datasets are corrected, completed and re-assigned to the local nomenclature and then follow the normal processes for internally collected data.

Direção Geral do Território (DGT) is the NMA in Portugal; it coordinates Portugal's National Spatial Data Infrastructure (NSDI), SNIG, and develops research on geographic information. Presently, research on VGI at the DGT is focused on investigating how to use VGI in the production of official topographic data[3]. The general idea is to use case studies to demonstrate the potential benefits of including VGI as part of the authoritative database implementation strategy, benefits which include filling gaps in official data, enlarging the spatio-temporal coverage or addressing the aims of specific communities of interest. These benefits are in line with the more collaborative and participative approach presently adopted for SNIG development.. To identify and analyse the integration of VGI to NSDI, the environment and planning domain will be used as a target. Case studies will be designed to identify required modifications to the NSDI, such as changes to the metadata catalogue to accommodate

VGI types or the interoperability and validation requirements for incorporating VGI within the NSDI. Moreover, a prototype based on a web service may become available through the NSDI geoportal, allowing any registered citizen to edit LULC polygons through the identification of geometry and/or classification changes. This will enable the analysis of thematic and positional inconsistencies reported by the users and define a strategy for including VGI in the production of official mapping.

Looking to future uses of VGI, the research interests of most of European NMAs range from motivational factors of volunteer engagement in VGI to change detection, data capture, and validation and management, all the way through to data or service delivery and associated quality and trust. In addition to VGI involving citizens, community groups and expert groups, exploring how VGI approaches might draw on the local knowledge of internal NMA employees is also of interest.

2.3 Promoting the Usability of Authoritative Data

In the past, within a research context, the Centro Nacional de Informação Geográfica (CNIG), which was then integrated in the Portuguese NMA (Instituto Geográfico Português (IGP), presently named Direção-Geral do Território (DGT) has been involved in the GEOCID (Hipólito et al., 2000) and Senses@watch (Gouveia et al., 2004; Gouveia and Fonseca, 2008) projects, which represented early attempts to promote the involvement of citizens in the use or production of geographic information, and which shared some of the issues associated with the topic of VGI and its integration in an NSDI. The GEOCID project aimed to promote the use of SDI by citizens and represented a first effort to target citizens as users of these infrastructures, although it used a top-down approach (Fonseca and Gouveia, 2005). Senses@watch was a research project centred on the definition and evaluation of strategies to promote the use of environmental spatial information, such as water quality and noise, collected through citizens' senses (e.g. vision, hearing, taste and smell). A prototype of a Web-based collaborative site was developed, including an interface for mobile phones.

The results of these initial projects were successful, with a considerable level of citizen participation, but did not have the intended follow-up in the NMA services and workflows. Nevertheless, lessons could be learned from these experiences that can enrich present approaches to VGI. These projects enabled the confirmation of the SDI data user's increasing role and of the importance of providing participation at multiple levels where VGI can be seen as a resource for SDI. The assessment of the pragmatic implications of using ICT to support citizen participation in environmental monitoring or the identification of the major benefits of involving volunteer contributors (e.g. the promotion of public awareness on environmental issues; the cost-effectiveness of the method to maintain data collection activities; or the facilitation of the creation of early warning systems), as well as the corresponding drawbacks (e.g. the lack of data credibility), are just some of the insights about VGI provided by these activities.

3 Recommendations for NMAs regarding VGI Use

Starting with good/best practices identified in NMA experiences and research work, the goal of this section is to define recommendations for NMAs in organising a platform to collect and manage VGI. Compiling a list of expectations from both crowd or community sourcing and NMAs will ensure a fruitful relationship between both parties, as discussed by Olteanu-Raimond et al. (2017). From the NMA point of view, issues such as motivation, stability, consistency and minimisation of false entries are of concern, while feedback, the citizen layer and transparency for the crowd and community sourcing, among others, are some of the crowd's concerns. Here, we focus on six elements that are either barriers to the use of VGI or key elements that allow for the construction of a successful VGI platform for citizens, public and private partners and governments. The six elements are as follows: the data model and objects; the interface; motivation; identification; licensing; and quality control.

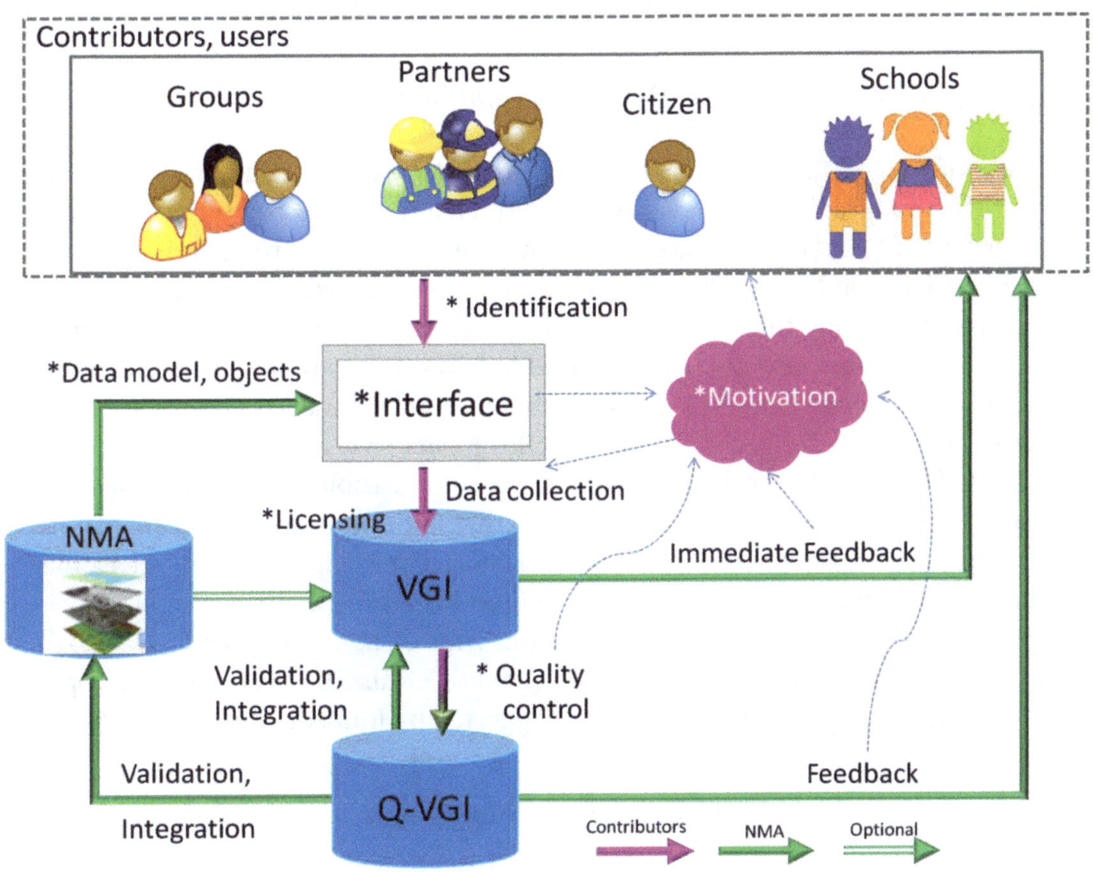

Fig. 2: A typical VGI collection workflow.

A general workflow for VGI data collection is illustrated in Figure 2, where these six elements are marked with an asterisk (*). In Figure 2, green and pink arrows represent NMA and contributor tasks, respectively.

A successful platform should be dedicated to both contributors and users, and should engage with citizens, specific groups of citizens sharing the same interest (e.g. hiking), partners (e.g. governments, emergency services) and the education community. Contributions should be made via user-friendly interfaces that implement an adaptive data model as proposed by NMAs, via secured identification, and via easy-to-use tools to contribute, manage, visualise and download VGI and/or authoritative data, depending on each NMA's data licence. A real added value from NMAs is the quality control of volunteered data, which can be corrected, validated and integrated into the VGI platform (Q-VGI to VGI). Depending on the data specification, some validated VGI can be integrated into the authoritative data (Q-VGI to NMA), in this way improving the accuracy and quality of the NMA's data. The quality control could be performed by contributors in a continuous way through the sharing of opinions on contributions, and step-by-step by the NMAs.

Table 1 summarises the recommendations described in the sections that follow and provides a list of opportunities and threats that can arise from such an NMA-VGI collection system. Opportunities and threats are described with respect to different elements identified in NMA data collection systems.

3.1 Data Model and Objects

Generally, NMAs are in charge of producing topographic databases by mapping the topography of the real world by focusing on specific types of objects described by few thematic features (e.g. number of lanes of a road, building type; Olteanu-Raimond et al., 2017). This implies that the existing features can be enriched by adding thematic information (e.g. number of floors in a building), but also that some new objects can be added. These new objects may be feature classes that are currently lacking in quality in official databases due to frequent real-world change (e.g. POIs, shops etc.), or data that can be most efficiently collected by contributors because collection is not feasible with remote mapping (e.g. hiking trails obscured by trees). Data that are of special interest to citizens or public services such as emergency services and municipalities, e.g. vernacular place names and traditional names of neighbourhoods (Castellote et al., 2013); obstacles, to help the navigation of people with disabilities (Rice et al., 2013); or paths, to improve pedestrian maps (Laakso et al., 2011) could be mapped by citizens having local knowledge, as suggested by Johnson and Sieber (2013).

3.1.1 Citizen and Partner Layer

Two of the identified barriers in using VGI are data quality and legal aspects (Olteanu-Raimond et al., 2017); Johnson and Sieber (2013) have reported the

Table 1: Opportunities, threats and recommendations for different elements that comprise an NMA-VGI data collection system.

Elements in NMA data collection system	Opportunity when integrating VGI	Threat when integrating VGI	Recommendation
Data objects	• New objects • Accuracy improvements • Attribute enrichment • NMAs can provide leadership in the development of standards for contributor data collection	• Volumes of data too large to handle • Thematically too heterogeneous • Incomplete (in coverage and thematically)	• Citizen layer • Protocols for mapping different types of data for existing and new objects • Adaptive data models for new object classes • Forum and online help • Feedback to contributors • User-friendly tools to import and download the data • Clearly stamp the contributed data already quality-controlled by the NMA
Interface		• Low number of contributors and groups of contributors • Lose motivation of contributors	• User-friendly • Easy to use • Rapid • Step-by-step guide to provide the VGI in a form that is as complete as possible following the protocols set out
Motivation	• Recruit and retain a large number of contributors • Improve the quality of VGI	• Contributors get disappointed and do not come back • Interruptions in service (from pilot to production) • Feeling of injustice if not a win-win situation	• Launching punctual campaigns • Gamification • Engaging school children; materials prepared for school curricula • Engaging with different partners, which can also create motivation

Continued.

Table 1: (Continued).

Elements in NMA data collection system	Opportunity when integrating VGI	Threat when integrating VGI	Recommendation
Registration	• Contributors can be contacted and given feedback • Possibility to analyse the group of contributors	• Privacy issues • False or incorrect data provided • Asking for full identity is too onerous for continuous use and contributors can lose motivation • Not everybody wants to register	• Asking for at least a valid email • Possibility to give only feedback (and nothing else) to anonymous contributors
Licensing		• Inconsistent licenses if contributed data is coming from another source of information with different license	• NMAs should prepare clear terms and conditions and they should be signed by contributors • Contributors should give the NMA full rights to the data • Inform contributors that by contributing, they are providing geographic data to the official national basemap
Quality Control	• With automatic control, the time for manual checking can be reduced • Use contributors for validation	• Too much information is required • Number of contributors does not necessarily guarantee the quality of data	Quality Control at different stages: Level 0: Real-time control procedures Level 1: Apply automatic quality methods to the contributed data Level 2: Crowdsourcing revision Level 3: Final validation with respect to a typology of quality assurance

same finding regarding the use of VGI by governments and argued that a more formalised VGI collection process may prove beneficial. A solution to break down these barriers can be a participatory citizen and partner layer proposed by NMAs. In this way, the NMAs will first have the opportunity to add new content, but also to increase the usability of traditional topographic data, and, as a direct consequence, to improve the accuracy of the data and enrich the thematic information (e.g. 'the building is a private school with three entrances'). Then, NMAs can propose a formalised framework and standards, which would be expected by governments, to collect VGI with a focus on data validation, data quality assessment and integration methods, allowing the topographic data to be used to support other types of specific data (e.g. water pump locations for firefighters, billboard locations for local municipalities). We would make the following recommendations:

- Authoritative basemaps should be used to make contributions (e.g. topographic data, orthophotographs, satellite images, DEMs);
- NMAs should assess both the internal quality and the fitness-for-use of the volunteered data and should correct errors by using authoritative data as a topographic support; this could be a big added-value that would encourage different users, such as public authorities, security and emergency services, NGOs and citizens, to both contribute and use VGI;
- Volunteered data should be clearly stamped with quality-control stamps to easily distinguish between quality-controlled contributions and those not yet quality-controlled;
- Campaigns for specific data collection purposes should be organised;
- User-friendly tools to import and download data should be proposed to allow for a detailed selection of suitable data regarding spatiotemporal and social criteria and should be served to end users in different formats.

As mentioned in Section 2, the NLS of Finland has already decided to experiment with this new citizen layer concept through a research project.

3.1.2 Adaptive Data Models for Object Collection

NMAs should propose an adaptive data model to collect and monitor geographic objects. This data model should allow contributors to:

- report updates or errors;
- add new attributes and objects to existing class objects;
- add new class objects identified by NMAs as out of the specifications of the authoritative topographic databases at present, but important for different applications;
- add new class objects if it fits end user needs;

- assess, as a mandatory requirement, the membership of an object to a class object type;
- import data collected *in-situ*; and
- ensure interoperability with the existing topographic data model in order to integrate significant and validated contributions directly into authoritative databases or to generate updates.

Nevertheless, proposing a quite open citizen and partner layer may introduce some threats, such as the possibility of obtaining large volumes of thematically heterogeneous data or data that are characterised by spatial and thematic incompleteness.

3.1.3 Protocols for Object Collection

The lack of protocols and the potential problems that this may entail, as well as recommendations for data collection, are discussed in more detail in Chapter 10 (Minghini et al., 2017). However, our additional recommendations regarding protocols are as follows:

- define a protocol for mapping different types of data for existing or new objects that balances the need to collect a minimum set of attributes and metadata with the desire for completeness in the data collection process;
- update and enrich the protocol regularly by taking into account end user experiences;
- propose a forum and online help facility to share experiences with and between contributors and assist contributors when needed; discussion forums have been proven to contribute, in some cases, to the creation of more reliable data (Haklay, 2010; Perger et al., 2012), and they are also a valuable tool for community building.

3.1.4 Instant Feedback to Contributors

From the different NMA experiences outlined above, it has been shown that engaging with contributors using transparent communication is crucial for a successful and sustainable platform. Good communication can be ensured by:

- one uniform feedback platform for the different products offered by the mapping agency: topographic data, citizen and partner layer, updates and error updates, etc.;
- contributor involvement in the feedback system building process;
- e-mails with updates concerning the status of contributions; and
- the display of contributor data immediately or in near real-time.

3.2 *Interface*

Two kinds of user interface tools can be distinguished (Sabou et al., 2014): acquisition interfaces designed for and used by contributors to carry out crowdsourcing tasks, and management interfaces, which are required by the managers of the VGI project to monitor progress, assess quality and manage contributors. In this section we focus on the acquisition interface used by contributors for data collection, designated here as the contributor interface.

Switzerland's geoportal[4] was recently awarded the '2015 eGovernment special prize'[5] at the ninth national eGovernment Symposium, which was held on 24 November 2015 in Bern involving representatives from the worlds of business, administration, politics and academia. Consistent use of open source software, open standards and cloud computing were the reasons for winning the prize. This geoportal features many properties that an NMA contributor interface should provide (e.g. an intuitive and contributor-friendly interface, a VGI component with the recently renewed revision service with immediate customer feedback, and a smooth, dynamic and interactive map navigation) a report option for customer alerts and the use of open standards. Our recommendations for the contributor-interface are as follows:

- offer a contributor-friendly interface that guides the contributor to supply all the information required by the protocols (e.g. metadata, attributes);
- define contributor-friendly interfaces that incorporate the NMA protocols and best practices without negatively impacting upon contributor enthusiasm or hampering the flow of data;
- provide tools to support the training of contributors and/or groups of contributors according to the goal;
- through the interfaces for collaboration, address the need to educate contributors and easily integrate the non-traditional types of data that they might collect;
- whenever possible, make use of standard, interoperable data formats and services in order to further extend the interface and/or integrate new services and applications;
- use up-to-date basemaps and do not overload the platform with multiple themes;
- adequately accommodate all the VGI types, which are of a diverse nature, dynamic, and sometimes produced in real-time;
- implement the full editing of objects rather than just hints attached loosely to existing data objects.

Additionally to what has been mentioned above, an intuitive contributor-interface can also play another very important role in the field of input data quality. While the basic principles of human-computer interaction (HCI) should be intact and meticulously followed, the contributor-interface could be the vehicle for implementing a number of elements in the protocol regarding the input of high-quality data (more information protocols for data capture can be found in Chapter 10 by Minghini et al., 2017). It is common for NMAs to have protocols in place that must be followed in order to achieve maximum homogeneity in the datasets produced. Volunteered content should also follow similar rules. Thus, the contributor-interface, which in this case serves as the data capturing layer, should be equipped with as many protocol elements as possible, balancing between high data integrity (and thus quality) and adequate freedom for the contributor.

3.3 Motivation

An important part in the success of using VGI is engaging people. Interested readers will find a detailed discussion on user motivation and engagement in Chapter 5 (Fritz et al., 2017). NMA experiences have shown that citizens are often not really interested in getting paid or in being presented with awards or prizes: having the possibility to contribute geographic information from their personal surroundings with a direct impact on publicly visible maps and getting feedback from NMAs are the main positive reasons to contribute. Nevertheless, in order to increase the number of contributors and ensure sustainability, NMAs should first promote, advertise and permeate the crowds, and secondly motivate, activate and reward contributions. However, when implementing reward systems, these rewards should not encourage contributors to favour quantity over quality in their contributions.

3.3.1 Gamification Techniques

Undoubtedly, a contributor-interface that enhances contributor experience can help to engage contributors; however, this factor alone is not enough to create the drivers and support the motivation that need to be achieved to attract a large pool of contributors to an initiative. There are a number of research efforts around the use of gamification techniques (Antoniou and Schlieder, 2014; Yanenko and Schlieder, 2014) to achieve these levels of motivation. Gamification, loosely defined, is the implementation of gaming practices in a non-game context. In essence, gamification, through game mechanics and game design, can have an impact and influence on participant behaviour. The aim of gamification is to make the participant achieve certain goals by enhancing engagement, improving performance and multiplying participation efforts towards a goal. Thus, NMAs can considerably enhance citizen motivation by implementing gamification processes for data capturing or change detection.

3.3.2 Giving Feedback

Feedback to contributors, given by sending updates concerning the status of their individual or group contributions, is an important motivation for contributors. Organisations need to assess the likelihood of such motivations being strong enough in a prospective contributor community to ensure the sustainability of their proposed VGI initiative (Hickling Arthurs Low Corporation, 2012). To help sustain contributions over time, some recommendations are listed here:

- all contributions should be welcome (e.g. attributes such as 'gravel road now paved' can be as valuable as topographic data);
- contributors want to receive acknowledgement for their contributions and to get rapid evidence that these contributions have been used;
- the process of making contributions should be as easy and streamlined as possible, as contributors may not be strongly motivated to contribute to extensive feature classification and the metadata requirements of public mapping programs; and
- different contributor-interfaces may be required for first-time or occasional contributors than for internal production staff or external power contributors; for example, tools that allow inappropriate content to be reported through a link allow contributors some control over data quality
- (Coleman, 2010; Esri, 2010; Hickling Arthurs Low Corporation, 2012).

3.3.3 Engage with Groups of Users

A number of advertising activities can be used to attract contributors to a VGI project. In a study on the impact of contributors to VGI projects (Schmidt et al., 2012), it is proposed to attract diverse groups of contributors with project-related mapping, to make mapping easy for beginners and to keep contributors mapping with social mapping events, as typically happens with OSM (Mooney et al., 2015). Launching campaigns will attract a number of users for a time period, whereas connecting relevant user groups (e.g. land owners having an interest in maintaining boundaries) will create more devoted contributors. In general, people who use the data will feel more attached to the project and will be more willing to contribute. In addition, it will be easier for them to find possible errors and report/make corrections if the procedures are made as easy as possible. Campaigns should target groups such as landowners, school children, cyclists, joggers, scouts, orienteering enthusiasts, hunters, hikers and geocachers, among others, who may be more willing to contribute due to their special interests and because they will take personal advantage of the addition of the VGI to the NMA database.

3.3.4 Engage with Public Partners

Some strategic public partners may be very important for the collection of certain types of data. For example, municipalities can easily engage with citizens for urban planning purposes and security-related partners that manage emergencies, such as civil protection authorities or firefighters, who are very often in the field and have specific needs such as fire hydrants, obstacles, building entrances, etc.

3.3.5 Engage with Schools

Introducing the work of the NMA and the idea and principles of topographic data collection to school pupils may be a good way to disseminate knowledge and could shape a large number of future contributors. To put this idea into practice, the following recommendations can be given:

- the collection of data must be integrated within an education programme (i.e. teaching by collecting data);
- close cooperation with teachers is crucial: teachers are busy with their everyday work, so they need to have some ready, easy-to-adapt teaching materials;
- school pupils are a very motivated group, but an application for data collection by pupils must work perfectly and rapidly;
- before data collection starts, it is important to explain to the pupils what crowdsourcing is and how it works; pupils need to understand that they are an important part of society and that they can deliver valuable data for others;
- data collection projects for pupils need to have some fun aspects that reward pupils who deliver good-quality data, which can even further support their motivation – gamification is one possible approach that may fit well with the needs of this particular group; and
- different stakeholder groups (pupils, parents, teachers, etc.) should be invited to refine/improve the curriculum.

Although unrelated to VGI data collection for NMAs, two successful examples of the engagement of pupils that have taken the above recommendations into account are described in Brovelli et al. (2016) and Ebrahim et al. (2016).

3.4 Registration

Data contributors may be anonymous, but this may permit vandalism (e.g. mapping fake features or deleting features that exist) and the contribution of fraudulent data or spam. It is still not entirely possible to distinguish between a credible VGI contributor on the one hand and an incompetent one, a mischief-maker or an outright vandal on the other hand (Coleman, 2010), although research is ongoing in this area: for example, Ciepłuch et al. (2010) have studied the history and the profiling of contributors; Van Exel et al. (2010) have proposed the experience, recognition and local knowledge of the individual as an indicator of quality input; and D'Antonio et al. (2014) have proposed an evaluation model for the contributor's reputation and data trustworthiness. However, based on the NMAs' experiences, very few bad contributions have been spotted, and in general more than 80–90% of the citizen contributions are useful (Olteanu-Raimond et al., 2017). Registered contributors are expected to have a more consistent contribution, since participating in the registration process proves their motivation and their intention to be identified. Apart from the contributor identification, registration has additional advantages (e.g. the contribution can be saved and finalised later by, for instance, tagging the position in the field and submitting the contribution later by using a computer at home). Three different types of profiles (see Table 2) could be made available when contributors register depending on the type of organisational model for data collection used and the validation process applied by the NMA afterwards; these include:

- Strong registration: this may create more powerful contributors in terms of permitted activities with the data. Full identity should be required in sensitive cases such as cadastral information for property owners. Another category may be a contract contributor (other authorities, for example) with specific permissions but also contribution obligations.
- Light registration: this type of identification allows the organisation collecting the data to contact the contributor if needed and to learn more about contributors, e.g. to determine potentially useful information such as their field of expertise.
- Weak registration: this only requires a valid email and password for registration to create a user account.

Table 2: Types of contributor registration profiles.

Strong registration	Light registration	Weak registration
Full identity, e.g. from e-government authentication systems, such as: full name, full address or postcode, profession and institution, phone number, passport/ID number	Valid email Profession Age / age group Gender	Valid email Password

3.5 *Quality Control*

In Chapter 7 (Fonte et al., 2017), an overview of the quality indicators that can be used to assess VGI is presented. Traditional spatial data quality assessment measures can be used. These can be applied using reference data, such as control data provided by experts, or through the comparison of data coming from several sources, which may even be VGI, enabling the assessment of logical consistency. Additionally, other indicators can be used to assess the reliability of the data, such as metadata on the data acquisition procedure, indicators about the contributor, socio-economic indicators or the consistency of corresponding data with different origins.

The quality control could be carried out at different levels that aim to facilitate the final validation by the NMA (which is mandatory), as outlined in the following subsections.

3.5.1 Level 0: Real-time Control Procedures

This initial level of quality control ensures that the minimum required information specified in the data collection protocol is provided and that no inconsistencies are introduced. It aims to assist the contributor in mapping valid information and is performed during the collection phase. Note that the absence of inconsistencies does not imply that the data are accurate and reliable. It controls:

- Required metadata and attributes. If the minimum information required by the protocols is not provided, alerts to the contributors asking for additional information should be sent. The submission of a new contribution can be approved once this control check is successful, i.e. the contributor can then submit the data. Care should be taken to minimise the mandatory information needed to avoid negatively impacting contributor motivation.
- Logical consistency. Automatic rules implementing some basic topological consistency checks (e.g. a polygon must be closed, roads should be topologically related, etc.) should be applied in real-time when a contribution is submitted. The system should recognise the presence of some types of inconsistencies in contributions and then not allow the submission of these contributions, such as in the case of clear topological mistakes; in other cases, the system should not prohibit submission but generate warning messages to the contributors suggesting corrections or additional checks, as contributors may in fact be providing important information about significant changes in the terrain.

3.5.2 Level 1: Applying Automatic Quality Control Methods to the Volunteered Data

The goal of Level 1 checks concerns data quality assessment through applying automatic methods. Three approaches are recommended:

- Data reliability. Automatic procedures can be used to perform an initial check of data reliability. Several sources of VGI can be used to assess data agreement: this refers to comparison of corresponding positional and attribute information, such as the position of roads in different data sources. The logical consistency of contributions can also be used to assess their reliability; for example, if a building is positioned inside a lake, the information may be considered to have low reliability.
- Contributor-based data reliability. If a prior assessment of contributor reliability is performed, for example by maintaining historical data on the contributors, it is possible to associate a degree of reliability to the data that is related to the reliability of the contributor.
- Specification-based reliability. Reliability of contributions can also be assessed by considering NMA specifications associated with the object being contributed. For example, if a building with an area lower than the NMA specification for the minimum size of buildings is mapped, this contribution can be automatically tagged as not fit-for-purpose.

3.5.3 Level 2: Crowdsourcing Revision

Crowdsourcing revision consists of:

- *In-situ* campaigns. Mapping agencies can organise *in-situ* campaigns asking contributors to assist in the validation of some highlighted complex cases where the NMA has insufficient information to perform a final validation. An example of this can be the assignment of a land cover class to particular locations when no field visits have been made.
- Peer validation between contributors. The VGI platform should provide additional capabilities to enable contributors to vote on ('thumbs up' or 'down') or to comment and discuss contributions. Discussions and/or comments can generate new insights into the main difficulties, and eventually some reliability indicators can be associated with specific types of contributions or to some areas, e.g. the classification of certain land cover types may be difficult for contributors, or contributions from a particular area can be found to have different interpretations. These indicators can be useful for assessing data reliability.

3.5.4 Level 3: Final Validation with Respect to a Typology of Quality Assurance

Methods to visualise quality, discussed in Chapter 9 (Skopeliti et al., 2017), can considerably enhance this step in the process. Some recommendations include:

- Define a typology of quality and associate the indicators previously assessed to either qualitative or quantitative rankings; these rankings can be based on probabilistic, fuzzy or possibilistic approaches.
- To optimise the use of resources and procedures, NMAs may perform validation only on volunteered data that are considered to be worth validating depending on the indicators obtained in the previous step that are considered relevant for each dataset.
- Final decisions are taken on the quality of the contributions and their usefulness, depending on the quality values obtained with the adopted quality typology.
- As the final validation assesses how good the contributor input was, this information may be used to rank contributor performance.

3.6 Licensing

With VGI, an important issue arises regarding the intellectual property of the data, which should be handled through licensing and consent. Contributors should give the NMA full rights to the data so that the NMA can take full advantage of the contributed data; this consent can be obtained either during the registration phase or after the first contribution is made. The contributors should be informed that by contributing, they are providing geographic data to the official national basemap. A well defined licence for the NMA's sharing and use of geographic data should be provided to and agreed upon by the contributors.

Some other legal aspects, such as liability and privacy, can differ from country to country or from product to product. These aspects are discussed in more detail in Chapter 6 (Mooney et al., 2017).

4 Conclusion

In this chapter, a review of the different VGI experiences of a few European NMAs was presented, and guidelines and recommendations were presented to help mapping agencies better exploit the opportunities offered by VGI through volunteered activities made by contributors.

Due to its nature and characteristics, VGI is still seen by NMAs, and more generally by government bodies, as having low quality and as a source of unreliable data. Therefore, few NMAs are engaged with VGI. When they are engaged, they have generally proposed their own tools to collect reports, and only rarely has VGI been used to collect data on features beyond the standard set mapped by NMAs.

Even though this type of data needs the development of new and different procedures for collection (see Chapter 10 by Minghini et al., 2017) or quality assessment (see Chapter 7 by Fonte et al., 2017) to become of major interest, VGI is nevertheless a valuable source of data, as it may help NMAs to provide data that are more up-to-date as well as to collect new, additional data that better address user needs. New features usually not collected by NMAs, either due to cost restrictions or because they represent non-traditional topographic data, could be of value to citizens and to various public services and government agencies.

To engage with the VGI community, the main recommendation for an NMA is to build a VGI platform that allows users to make reports but also to collect new, additional features that are not traditionally collected, to create a citizen and partner layer. An increasing number of VGI projects to collect data have been proposed during the last decade. As noted in the review by See et al. (2016), there is considerable variability in both the sustainability and the goal of the VGI projects. Some of them have been successfully operating for a long time while others have a finite life, being linked to some specific events, or are no longer active or available online. Moreover, few governments and municipalities have proposed platforms to collect data from citizens for purposes such as urban planning. Other public services, such as medical emergency departments or fire services, use their own resources to collect specific spatial data (e.g. water pumps, obstacles, building entrances), which need to be matched to spatial reference data.

Being aware of these current practices and initiatives, the question of why an NMA should also propose a VGI platform is a relevant one. We believe that NMAs, as public bodies, on the one hand are officially responsible for providing accurate and reliable information through SDIs to all potential users and, on the other hand, have the necessary expertise to manage and integrate spatial data. Moreover, all of the public initiatives mentioned earlier could not be implemented without important financial and human resources for deploying the GIS systems to collect, manage and maintain data and to train agents to deal with spatial information. We believe that a stronger collaboration between NMAs and governments through a VGI platform could result in a public-cost reduction and a better service to citizens, where these could be more involved in decision-making or in supporting security issues that affect their lives in a positive or negative way. Thus, for a successful VGI platform, one of the most important recommendations is to engage with citizens in general, specific groups of citizens having the same interests, and groups of public and governmental bodies, including the educational system. Engaging with different public bodies and with the educational sector will increase citizen involvement since these bodies are close to citizens and may invest in the future by educating and raising the awareness of younger generations regarding the relevance of spatial data and their quality. These engagements could create motivation, increase sustainability and promote good-quality data for both NMAs and the contributors.

Another important aspect, more oriented towards citizens, that can increase motivation is gamification. However, when implementing reward systems (of gamified or real-life rewards), attention should be paid to the fact that data quality is much more important than quantity, and this should be clearly explained to the contributor. Thus, a good practice for gamification is to avoid giving rewards or prizes based on (or only on) the number of contributions made.

We feel that a platform based on the recommendations discussed in this chapter is feasible and can be carried out in a step-by-step manner through the development of pilots and research projects, as exemplified by the ongoing initiative of the Finnish mapping agency, which is defining and preparing to test the concept of a citizen layer.

Due to the importance of and increasing trend in VGI, we believe that NMAs should develop national VGI platforms for both data collection and data dissemination, even if it is difficult to predict if, or when, these initiatives will really become a 'standard practice' for all NMAs.

Acknowledgements

The authors would like to thank swisstopo for hosting the WG3 meeting that initiated the discussion on recommendations for NMAs involved in activities with VGI. We would also like to thank the different participants who contributed to the discussions and are not authors of this chapter: Bianka Fohgrub, Jasper Hogerwerf, Oana Popescu, Jean-Christophe Guélat, André Streilein and Peter Mooney. Finally, we are grateful to the referees for their helpful comments on the original version of this chapter.

Previous publication

Figure 1 has been previously published in Olteanu-Raimond, A.-M., Hart, G., Foody, G., Touya, G., Kellenberger, T., Demetriou, D., 2017. The scale of VGI in map production: A perspective of European National Mapping Agencies. *Transactions in GIS* 21, 74–90. DOI: https://doi.org/10.1111/tgis.12189 and is being reproduced here with permission from John Wiley & Sons Ltd.

Notes

1 https://www.gov.uk/government/publications/open-data-charter/g8-open-data-charter-and-technical-annex, published in June 2013

2 https://espacecollaboratif.ign.fr/

[3] http://snig.dgterritorio.pt/portal/
[4] http://www.geo.admin.ch
[5] https://www.news.admin.ch/message/index.html?lang=en&msg-id=59628

Reference list

Antoniou, V., Morley, J., Haklay, M., 2009. The role of user generated spatial content in mapping agencies, in: Proceedings of GIS Research UK 19th Annual conference, Durham, UK, 1–3 April 2009, pp. 251–255.

Antoniou, V., Schlieder, C., 2014. Participation patterns, VGI and gamification, in: Proceedings of AGILE 2014. Presented at the AGILE 2014, Castellón, Spain, pp. 3–6.

Brovelli, M.A., Minghini, M., Molinari, M., Molteni, M., 2016. Do open geodata actually have the quality they declare? The case study of Milan, Italy, in: *International Archives of the Photogrammetry, Remote Sensing and Spatial Information Sciences*. Prague, Czech Republic.

Castellote, J., Huerta, J., Pescador, J., Brown, M., 2013. Towns conquer: a gamified application to collect geographical names (vernacular names/ toponyms). Presented at the AGILE 2013, Leuven, Belgium, 14–17 May 2013. Available from: https://agile-online.org/Conference_Paper/CDs/agile_2013/Short_Papers/SP_S2.3_Castellote.pdf [Last accessed 16 May 2017].

Ciepłuch, B., Jacob, R., Mooney, P., Winstanley, A., 2010. Comparison of the accuracy of OpenStreetMap for Ireland with Google Maps and Bing Maps, in: Proceedings of the Ninth International Symposium on Spatial Accuracy Assessment in Natural Resources and Environmental Sciences, Leicester, UK, 20–23 July 2010, pp. 337–340. Available from: http://www.spatial-accuracy.org/system/files/img-X07133419_0.pdf [Last accessed 16 May 2017].

Coleman, D., 2010. Volunteered geographic information in spatial data infrastructure: An early look at opportunities and constraints, in: Rajabifard, A., Crompvoets, J., Kanantari, M., Kok, B. (Eds.), *Spatially Enabling Society: Research, Emerging Trends and Critical Assessment*. Leuven University Press, Leuven, Belgium, pp. 1–18.

D'Antonio, F., Fogliaroni, P., Kauppinen, T., 2014. VGI edit history reveals data trustworthiness and user reputation, in: Proceedings of AGILE 2014, Huerta, J., Schade, S., Granell, C. (Eds.), Castellón, Spain, 3–6 June 2014. Available from: https://agile-online.org/Conference_Paper/cds/agile_2014/agile2014_140.pdf [Last accessed 16 May 2017].

Ebrahim, M., Minghini, M., Molinari, M., Torrebruno, A., 2016. Minimapathon: mapping the world at 10 years old, in: Proceedings of the 8th Annual International Conference on Education and New Learning Technologies (EDULEARN 2016). Presented at the 8th Annual International Conference on Education and New Learning Technologies (EDULEARN 2016), Barcelona,

Spain, 4–6 July 2016, pp. 4200–4208. DOI: https://doi.org/10.21125/edulearn.2016.2018

Esri, 2010. Simplifying Citizen Reporting. *ArcUser Online*. Available at http://www.esri.com/news/arcuser/0111/citysourced.html [Last accessed 1 May 2017].

Fonseca, A., Gouveia, C., 2005. Spatial multimedia for environmental planning and management, in: Campagna, M. (Ed.), *GIS for Sustainable Development*. CRC Press, pp. 143–165.

Fonte, C C, Antoniou, V, Bastin, L, Estima, J, Arsanjani, J J, Bayas, J-C L, See, L and Vatseva, R. 2017. Assessing VGI Data Quality. In: Foody, G, See, L, Fritz, S, Mooney, P, Olteanu-Raimond, A-M, Fonte, C C and Antoniou, V. (eds.) *Mapping and the Citizen Sensor*. Pp. 137–163. London: Ubiquity Press. DOI: https://doi.org/10.5334/bbf.g.

Fritz, S, See, L and Brovelli, M. 2017. Motivating and Sustaining Participation in VGI. In: Foody, G, See, L, Fritz, S, Mooney, P, Olteanu-Raimond, A-M, Fonte, C C and Antoniou, V. (eds.) *Mapping and the Citizen Sensor*. Pp. 93–117. London: Ubiquity Press. DOI: https://doi.org/10.5334/bbf.e.

Goodchild, M.F., 2007. Citizens as sensors: the world of volunteered geography. *GeoJournal* 69, 211–221. DOI: https://doi.org/10.1007/s10708-007-9111-y

Gouveia, C., Fonseca, A., 2008. New approaches to environmental monitoring: the use of ICT to explore volunteered geographic information. *GeoJournal* 72, 185–197. DOI: https://doi.org/10.1007/s10708-008-9183-3

Gouveia, C., Fonseca, A., Câmara, A., Ferreira, F., 2004. Promoting the use of environmental data collected by concerned citizens through information and communication technologies. *Journal of Environmental Management* 71, 135–154. DOI: https://doi.org/10.1016/j.jenvman.2004.01.009

Haklay, M., 2010. How good is volunteered geographical information? A comparative study of OpenStreetMap and Ordnance Survey datasets. *Environment and Planning B: Planning and Design* 37, 682–703. DOI: https://doi.org/10.1068/b35097

Hickling Arthurs Low Corporation, 2012. *Volunteered Geographic Information (VGI) Primer*. Natural Resources Canada. Available from: http://ftp.maps.canada.ca/pub/nrcan_rncan/publications/ess_sst/291/291948/cgdi_ip_21e.pdf [Last accessed 16 May 2017].

Hipólito, J., Fonseca, A., Gouveia, C., Fava, S., Henriques, R.G., Neves, J.N., 2000. GEOCID: Implementation of a National Geographic Information Infrastructure for the Citizen, in: The 6th EC and GIS Workshop – The Spatial Information Society - Shaping the Future, Lyon, France, 28–30 June 2000. Available from: http://www.ec-gis.org/Workshops/6ec-gis/papers/hipolito-geocid.doc [Last accessed 16 May 2017].

Ivanovic, S., Olteanu-Raimond, A.-M., Mustière, S., Devogele, T., 2016. Detection of outliers in crowdsourced GPS traces, in: Bailly, J.-S., Griffith, D., Josselin, D. (eds.), Proceedings of Spatial Accuracy 2016, Montpellier, France, 5–8 July 2016, pp.130–135. Available from: http://www.spatial-accuracy.org/system/files/Detection%20of%20outliers%20in%20crowdsourced%20GPS%20traces.pdf [Last accessed 16 May 2017].

Johnson, P., Sieber, R., 2013. Situating the adoption of VGI by government, in: Sui, D., Elwood, S., Goodchild, M. (Eds.), *Crowdsourcing Geographic Knowledge*. Springer Netherlands, pp. 65–81.

Karimipour, F., Azari, O., 2015. Citizens as expert sensors: One step up on the VGI ladder, in: Gartner, G., Huang, H. (Eds.), *Progress in Location-Based Services 2014*. Springer International Publishing, Cham, pp. 213–222.

Laakso, M., Sarjakoski, T., Sarjakoski, L.T., 2011. Improving accessibility information in pedestrian maps and databases. *Cartographica: The International Journal for Geographic Information and Geovisualization* 46, 101–108. DOI: https://doi.org/10.3138/carto.46.2.101

Leibovici, D.G., Evans, B., Hodges, C., Wiemann, S., Meek, S., Rosser, J., Jackson, M., 2015. On data quality assurance and conflation entanglement in crowdsourcing for environmental studies, in: ISPRS Annals of the Photogrammetry, Remote Sensing and Spatial Information Sciences, Volume II-3/W5, La Grande Motte, France, 28 September to 3 October 2015, pp. 195–202. Available from: http://www.isprs-ann-photogramm-remote-sens-spatial-inf-sci.net/II-3-W5/195/2015/isprsannals-II-3-W5-195-2015.pdf. DOI: https://doi.org/10.5194/isprsannals-II-3-W5-195-2015 [Last accessed 16 May 2017].

Minghini, M, Antoniou, V, Fonte, C C, Estima, J, Olteanu-Raimond, A-M, See, L, Laakso, M, Skopeliti, A, Mooney, P, Arsanjani, J J, Lupia, F. 2017. The Relevance of Protocols for VGI Collection. In: Foody, G, See, L, Fritz, S, Mooney, P, Olteanu-Raimond, A-M, Fonte, C C and Antoniou, V. (eds.) *Mapping and the Citizen Sensor*. Pp. 223–247. London: Ubiquity Press. DOI: https://doi.org/10.5334/bbf.j.

Mooney, P., Minghini, M., Laakso, M., Antoniou, V., Olteanu-Raimond, A.-M., Skopeliti, A., 2016. Towards a protocol for the collection of VGI vector data. *International Journal of Geographic Information* 5 (11), 217. DOI: https://doi.org/10.3390/ijgi5110217

Mooney, P., Minghini, M., Stanley-Jones, F., 2015. Observations on an OpenStreetMap mapping party organised as a social event during an open source GIS conference. *International Journal of Spatial Data Infrastructures Research* 10, 138–150. DOI: https://doi.org/10.2902/ijsdir.v10i0.395

Mooney, P, Olteanu-Raimond, A-M, Touya, G, Juul, N, Alvanides, S and Kerle, N. 2017. Considerations of Privacy, Ethics and Legal Issues in Volunteered Geographic Information. In: Foody, G, See, L, Fritz, S, Mooney, P, Olteanu-Raimond, A-M, Fonte, C C and Antoniou, V. (eds.) *Mapping and the Citizen Sensor*. Pp. 119–135. London: Ubiquity Press. DOI: https://doi.org/10.5334/bbf.f.

Olteanu-Raimond, A.-M., Hart, G., Foody, G., Touya, G., Kellenberger, T., Demetriou, D., 2017. The scale of VGI in map production: A perspective of European National Mapping Agencies. *Transactions in GIS* 21, 74–90. DOI: https://doi.org/10.1111/tgis.12189

Perger, C., Fritz, S., See, L., Schill, C., van der Velde, M., McCallum, I., Obersteiner, M., 2012. A campaign to collect volunteered geographic information on land cover and human impact, in: Jekel, T., Car, A., Strobl, J., Griesebner, G. (Eds.), *GI_Forum 2012: Geovisualisation, Society and Learning*. Herbert Wichmann Verlag, Berlin / Offenbach, pp. 83–91.

Rice, M.T., Jacobson, R.D., Caldwell, D.R., McDermott, S.D., Paez, F.I., Aburizaiza, A.O., Curtin, K.M., Stefanidis, A., Qin, H., 2013. Crowdsourcing techniques for augmenting traditional accessibility maps with transitory obstacle information. *Cartography and Geographic Information Science* 40, 210–219. DOI: https://doi.org/10.1080/15230406.2013.799737

Sabou, R., Bontcheva, K., Derczynski, L., Scharl, A., 2014. Corpus annotation through crowdsourcing: Towards best practice guidelines. Presented at the LREC 2014, Ninth International Conference on Language Resources and Evaluation, European Language Resources Association (ELRA), Reykjavik, Iceland.

Schmidt, M., Klettner, S., Steinmann, R., Häusler, E., 2012. The impact of the contributor in VGI projects, in: Workshop on Map Creation from User Generated Data, Hannover, Germany, 8 October 2012. Available at: http://www.visualisierung.dgfk.net/docs/Schmidt-etal.pdf [Last accessed 16 May 2017].

Sedano, E., 2016. "Sensor"ship and Spatial Data Quality. *Urban Planning*, 1(2), 75–87. DOI: https://doi.org/10.17645/up.v1i2.608

See, L., Mooney, P., Foody, G., Bastin, L., Comber, A., Estima, J., Fritz, S., Kerle, N., Jiang, B., Laakso, M., Liu, H.-Y., Milčinski, G., Nikšič, M., Painho, M., Pődör, A., Olteanu-Raimond, A.-M., Rutzinger, M., 2016. Crowdsourcing, citizen science or Volunteered Geographic Information? The current state of crowdsourced geographic information. *ISPRS International Journal of Geo-Information* 5, 55. DOI: https://doi.org/10.3390/ijgi5050055

Skopeliti, A, Antoniou, V and Bandrova, T. 2017. Visualisation and Communication of VGI Quality. In: Foody, G, See, L, Fritz, S, Mooney, P, Olteanu-Raimond, A-M, Fonte, C C and Antoniou, V. (eds.) *Mapping and the Citizen Sensor*. Pp. 197–222. London: Ubiquity Press. DOI: https://doi.org/10.5334/bbf.i.

Turner, A., 2006. Introduction to Neogeography. O'Reilly, Sebastopol, Calif.

van Exel, M., Dias, E., Fruijtier, S., 2010. The impact of crowdsourcing on spatial data quality indicators, in: Wallgrün, J.O., Lautenschütz, A.K. (eds.), Proceedings of the GIScience 2010 Doctoral Colloquium, Zurich, Switzerland, 14–17 September 2010. Available at: http://www.giscience2010.org/pdfs/paper_213.pdf [Last accessed 16 May 2017]

Viglino, J.-M., 2009. Managing partners feedbacks through the geoweb, in: Streilein, A. & Kellenberger, T. (eds.), 2010. Proceedings of the EuroSDR Workshop 'Crowd Sourcing for Updating National Databases' held from 20th to 21st August 2009 in Wabern, Switzerland.

Yanenko, O., Schlieder, C., 2014. Game principles for enhancing the quality of user-generated data collections. Presented at the AGILE 2014 Workshop on Geogames, Castellón, Spain, pp. 1–5.

CHAPTER 14

Opportunities for Volunteered Geographic Information Use in Spatial Planning

Matej Nikšič*, Michele Campagna†, Pierangelo Massa†, Matteo Caglioni‡, Thomas Theis Nielsen§

*Urban Planning Institute of the Republic of Slovenia, Ljubljana, Slovenia, matej.niksic@uirs.si

†DICAAR, Università di Cagliari, Cagliari, Italy

‡Université Côte d'Azur, CNRS, ESPACE, France

§University of Roskilde, Roskilde, Denmark

Abstract

This chapter highlights two types of georeferenced User-Generated Content (geo-UGC) that show considerable potential for fruitful usage in spatial planning in practice: Volunteered Geographic Information (VGI) and Social Media Geographic Information (SMGI). By describing selected case studies, the chapter illustrates how geo-UGC can be used at different stages of spatial planning processes, supporting a more pluralist understanding of places, fostering the collaboration between decision-makers and contributing to a more participatory practice in spatial planning. The Geodesign approach is used as the framework for underpinning the discussion. Selected case studies developed by the authors are presented showing how geo-UGC can be beneficial for building knowledge on current urban and territorial dynamics, for identifying possible alternative futures and for finding agreement on preferable future developments. In all the selected cases, large numbers of users were involved

How to cite this book chapter:

Nikšič, M, Campagna, M, Massa, P, Caglioni, M, Nielsen, T T. 2017. Opportunities for Volunteered Geographic Information Use in Spatial Planning. In: Foody, G, See, L, Fritz, S, Mooney, P, Olteanu-Raimond, A-M, Fonte, C C and Antoniou, V. (eds.) *Mapping and the Citizen Sensor.* Pp. 327–349. London: Ubiquity Press. DOI: https://doi.org/10.5334/bbf.n.

in collecting volunteered content. The findings are also interpreted within the Smart Cities paradigm, where participation is an essential factor for building successful smart communities.

Keywords

VGI, SMGI, urban planning, urban design, Geodesign, Smart Cities

1 Introduction

Spatial planning, as an interdisciplinary practice of managing the development of space in its physical, functional and socio-economic dimensions, aims to provide efficient, economically viable, just and sustainable space arrangements. It is traditionally a competence of a state, regional or local authority, and usually involves a number of actors and institutions.

In the last few decades a stronger emphasis has been placed on the involvement of the community and the users of space in urban planning procedures. In part this has arisen from the general democratisation of the processes in contemporary societies in many Western countries, but it has also emerged out of a need to avoid conflicts between opposing parties, which often have contrary interests in space (Arnstein, 1969; European Commission, 2003; McTague and Jakubowski, 2013; Cerar, 2014).

Prior to the widespread diffusion of new Information and Communication Technologies (ICT), public participation was largely understood as a form of public commenting on already prepared plans, while emerging technologies have opened up new and innovative ways of realising the active involvement of the wider public in spatial planning (Bizjak, 2012). Opportunities have arisen in different fields, e.g. improving the communication between authorities and citizens, providing more accurate and up-to-date databases on the current state of territorial conditions, and collecting the ideas and visons for future developments of different stakeholders (Berntzen et al., 2005; Brabham, 2009; Seltzer and Mahmoudi, 2013).

As a dynamic and complex socio-technical process, spatial planning may entail multi-faceted paradigms originating in a variety of workflows in practice. The aim of this chapter is to use the concept of Geodesign (Steinitz, 2012), which is one of many possible ways of approaching spatial planning, to explore the opportunities for exploiting georeferenced User-Generated Content (geo-UGC) in spatial planning. We can differentiate between two main categories of geo-UGC of particular interest in spatial planning, either as an information resource or as a communication platform, or both: Volunteered Geographic Information (VGI), which is geo-UGC purposely collected by a group of users for a given purpose (e.g. OpenStreetMap.com); and Social Media Geographic

Information (SMGI), which is geo-UGC collected passively (e.g. Twitter.com; instagram.com) or actively (e.g. fixmystreet.org; projectnoah.org; carticipe.net) on social networking platforms. In the next section, the Geodesign approach is outlined, along with the opportunities for effective use of geo-UGC. This is followed by a set of case studies from the authors, which illustrate how geo-UGC has been used in planning, relating these examples to different stages in the Geodesign approach. Finally, we consider how VGI and SMGI can support 'smart cities' initiatives.

2 The Geodesign Approach: Opportunities Arising from VGI and SMGI

In the last decade, the term Geodesign has gained popularity among a growing number of spatial planners, landscape architects and Geographic Information Systems (GIS) scholars, formalising an innovative approach to planning and design deeply rooted in geographic analysis and at the same time able to foster collaboration in decision-making. Geodesign may be defined as an integrated process, informed by environmental sustainability appraisal, that aims to address complex problems related to territorial and environmental issues and to social and economic matters (Dangermond, 2010). The main novelty in the Geodesign approach is the extensive use of digital spatial data and processing and of communication resources such as ICT and GIS, aimed at easing the integration of societal and scientific knowledge in planning, design and decision-making (Ervin, 2011). Current technologies may be considered mature enough to exploit ICT support in spatial planning processes, overturning the barriers that in the past limited the use of new technologies in practice (Göçmen and Ventura, 2010). Additionally, ICT, the Internet and, more recently, Web 2.0 technologies are increasingly channeling digital Geographic Information (GI) into the daily lives of a growing number of users. This phenomenon is leading to a paradigmatic shift in the contents and characteristics of GI, as well as in its modes of production and dissemination (Elwood et al., 2012). In the spatial planning domain, this unprecedented wealth of digital GI provides great opportunities for advances in methodologies such as Geodesign, fostering opportunities for supporting design, analysis and decision-making processes. Most of the opportunities arising for innovation emerge from the avalanche of spatial big data, which Web 2.0 technologies are making available to the wider public.

In the last two decades, developments in Spatial Data Infrastructures (SDIs) have enabled access to digital GI produced and maintained by public or private institutions for public or business purposes. In Europe, the implementation of Directive 2007/02/CE, establishing a shared Infrastructure for Spatial Information in Europe (INSPIRE), fostered the development of National and Regional SDIs in the Member States, allowing the public access and reuse of available official information, or Authoritative Geographic Information (A-GI), according to common data, technology and policy standards. Secondly, several platforms, continuously flourishing through the Internet as a result of Web 2.0 technologies, are supporting the production and diffusion of User-Generated Content (UGC), which often has a geographic reference embedded, potentially transforming the Web into a big warehouse of spatial data (Elwood et al., 2012). Spatial UGC is commonly labelled as VGI, emphasising the voluntary activities of users to collect and contribute information related to the geographic world (Goodchild, 2007). In spatial planning, VGI may supply both experiential knowledge from local communities and expert knowledge from professionals in a bottom-up approach, e.g. through citizen science initiatives. SMGI, which is a subset of UGC (Campagna, 2014), is spatial information produced and shared through social network sites, and may allow for the collection of quantitative GI related to a study area but also of qualitative information concerning the perceptions of users about phenomena in space and time. Indeed, SMGI is different from traditional common vector spatial datasets such as A-GI supplied by institutional SDIs, which exclusively feature spatial and thematic information: the SMGI data model features spatial, temporal and multimedia dimensions (i.e. image, text, video and audio), as well as a user dimension, including specific information about the user profiles. Furthermore, in certain cases, the SMGI data model also includes a preference dimension, i.e. SMGI appreciation expressed by the social network community by means of scores, stars or likes/dislikes, thus widely expanding the range of analytical opportunities for planners and analysts (Campagna et al., 2015). A comparison between the SMGI and traditional A-GI data models is shown in Figure 1.

The general SMGI data model may foster advances in spatial planning methodologies and may be a valuable complement to traditional A-GI that can support several stages of the Geodesign process. To formalise the Geodesign approach, Steinitz (2012) proposed a methodological framework that relies on six models: representation, process, evaluation, change, impact and decision models. These are iteratively implemented to design future development

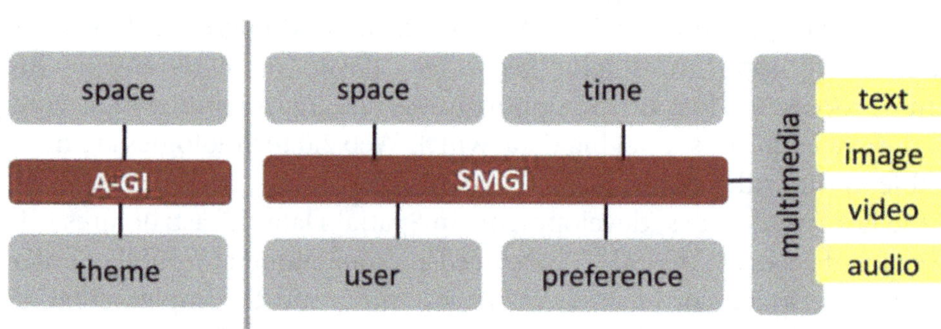

Fig. 1: Comparison between the A-GI data model and the SMGI data model (Adapted from Campagna, 2016).

alternatives and to identify their potential consequences by means of a territorial context description, an analysis of the dynamics and an evaluation of the impacts. The first three models describe the present situation of the territorial context considering (1) the environmental system, and (2) explaining its evolution, mainly focusing on (3) opportunities and threats that may arise from the current situation. Conversely, the last three models define potential alternatives for (4) transforming the system, (5) assessing the transformation alternatives' potential beneficial or dangerous impacts on environmental and human systems, and eventually (6) supporting stakeholders during the decision-making process.

VGI and SMGI may thus be used to complement the availability of official information for the implementation of all the Geodesign models, supplying useful societal data. In the representation model, SMGI may be used to facilitate the description of a geographic context, providing experiential knowledge that is usually dismissed in official information and integrating A-GI with a pluralist vision of geographic phenomena, which may be used to identify social and cultural dynamics affecting the area. For example, SMGI from several Location-Based Social Networks (LBSNs) has been used to identify the most appreciated Points of Interest (POIs) and landmarks in a study area (Jankowski et al., 2010), the pedestrian paths in the historical centre of a city, the neighbourhoods featuring the lowest number of services and the different land uses in an urban environment (Frias-Martinez et al., 2012), and to classify urban areas (Noulas et al., 2011).

Regarding the development of process models, SMGI may be used to investigate how detected phenomena evolve over time thanks to the real-time supply of information, which may be used for monitoring and to feed predictive models for studying future trends and dynamics. SMGI may also be extracted and analysed for different periods from different social networks, investigating first whether current phenomena were already present in the past and secondly if the potential factors affecting these phenomena persist, in order to evaluate the future situation. Similarly, users' preferences about urban mobility or cultural dynamics may be elicited from SMGI with the aim of feeding agent-based models that can simulate individual behaviours.

In the evaluation model, SMGI may be used to assess the current situation of the geographic area, due to the preferences, opinions and behaviours of users, which are embedded in this source of information. For instance, SMGI may be extracted for studying the movements of users in urban environments (Jankowski et al., 2010), the utilisation rates of public spaces (Torres and Costa, 2014) and the neighbourhood perceptions of users (Massa and Campagna, 2014), as well as the dynamics of different population groups (Longley et al., 2015).

Furthermore, social networks, representing a means to gain useful insights about the social and cultural dynamics of an area, may support the development of alternative scenarios in the Geodesign change model, and, at the same time, they may be used to actively involve local communities during planning and design (Eräranta et al., 2015). In addition, SMGI may be useful in the Geodesign impact model to assess the potential alternative effects on the territory, due to the possibility to present change scenarios to the local community and to collect feedback using a participatory planning approach (Rantanen and Kahila, 2009).

Finally, despite the difficulties in transposing the experiential knowledge of local communities into practice (Nonaka and Takeuchi, 1995), SMGI might be used to foster a communicative process among participants in the decision model, wherein the mutual integration of expert and experiential knowledge is a crucial step (Khakee et al., 2000) to build a shared, sustainable and democratic development process for the territory. Commonly, a local community's experiential knowledge is considered exclusively an opinion in planning processes (Fischer, 2000); however, the technical knowledge of experts may not be sufficient to properly guide decision-making processes (Lindblom, 1990). Hence, the integration of A-GI and SMGI may support the decision model, and may foster the development of more transparent, pluralist and democratic decision-making.

In the next section, selected case studies that we carried out will be briefly outlined to demonstrate the value of SMGI at different stages of the planning process, using the Geodesign framework as a reference.

3 Case Studies on the Value of VGI and SMGI in Spatial Planning and Design

3.1 Representation Model

Representation of geographic information is extremely important for planners and citizens. Both of them use visualisation methods to explore the real world and as a basis for analysing different scenarios based on spatial data. Visualisation is one of the possible representations for VGI, and probably the most powerful one. Geovisualisation explores geospatial information and supports decision-making processes in spatial planning.

One innovative example of representation is the interactive visualisation of OpenStreetMap (OSM), which allows users to upload quantitative and qualitative data in a Web-based GIS, as was the case in the GeoCampPACA event. GeoCampPACA2016 was a mapping party organised by OSM France, the Provence-Alpes-Côte d'Azur (PACA) French region and the region's centre for geoinformation, CRIGE (Figure 2). The aim of this event was to make a survey related to different modes of transport, such as pedestrian, bicycle, car, bus, tram and train routes, including infrastructure, equipment, services, etc., and to represent the information in cartographic form. This two-day event was a real participatory mapping operation, open to all students in geography and GIS of the PACA French region. The first day was dedicated to OSM protocols

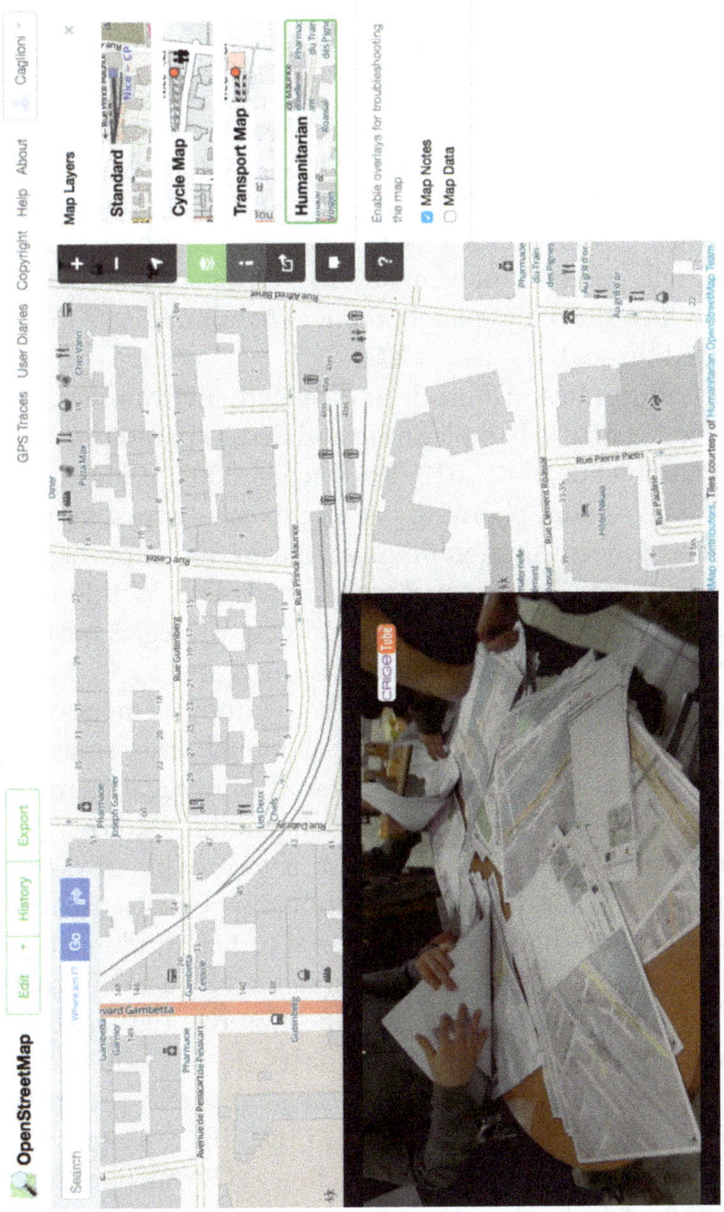

Fig. 2: Images from the GeoCampPACA2016 web mapping party organised by OpenStreetMap in France. ©OpenStreetMap contributors.

and basic notions of crowdsourcing and GIS, while the second day was devoted to practical and field activities in the different main train stations of the region. The event facilitated the creation of open data available on the OSM portal, while allowing participants to gain a better understanding of their surrounding environment.

3.2 Process Model

As mentioned earlier, the Geodesign process model concerns the understanding of current territorial dynamics. This model will be illustrated with two examples. The first is a case study of volunteered urban cycling information via GPS devices, which demonstrates how VGI can help planners monitor current behaviour and preferences in movement and transport dynamics. The second case study shows how the daily spatial practices of homeless people can be better comprehended through the use of VGI.

Rising motorisation rates in Europe and related environmental issues have created a demand for new urban planning and design paradigms in relation to urban transportation (Eurostat, 2012; Knoflacher, 2007; Zubelzu and Fernández, 2016). The new spatial planning paradigms are advocating for a change in the proportion of means of mobility in favour of non-motorised and public transportation to account for personal motorised traffic. Within these endeavors, urban cycling is gaining momentum, and new strategies have been developed to accommodate urban cycling into existing cities.

One of the related urban planning issues is the improvement of the existing and provision of new cycling infrastructures. Contemporary smart approaches, however, do not deal with the infrastructure as a physical element, but deal with it solely in relation to perceptual and behavioural patterns, i.e. how people tend to perceive and use it; the main aim is to provide infrastructure that will be efficient and safe and to encourage enough people to use it regularly. A wide range of approaches have been developed to help understand what kind of cycling infrastructure is preferred and demanded by users in contemporary cities, and VGI is playing an increasingly important role in these developments (Latham and Wood, 2015; Yeboah and Alvanides, 2015; Winters et al., 2016).

Such an attempt has been made with CyCity, a research programme by the Swedish governmental agency Vinnova, with the aim to improve the knowledge on urban cyclists' preferences in route choices (Envall and Koucky, 2013). Through a combined technique of using GPS devices and online questionnaires, each participating urban cycler has provided valuable information for the planning and (re)design of cycling path networks in the cities of implementation (Ljubljana in Slovenia and Linköping in Sweden). For a limited time, participants were given user-friendly GPS devices and asked to record every biking route they made in the city, as well as filling out a questionnaire regarding qualitative data on the cycling routes (Tominc et al., 2012). Even though the GPS technology proved to be not very precise and accurate (e.g. the mapped polylines overlapped with built blocks, etc.), the research revealed a big potential to fulfil the needs of urban planning (Figure 3), namely in the following aspects:

- The appropriate amount of mapped cycling tracks clearly indicates where in the city the cycling trips densify, as well as where they are non-existent. The densely cycled areas may be regarded as potential locations to place and develop programmes that appeal to cyclists, which may generate new urban activities, much longed for in urban regeneration processes.
- Areas that have no records of tracks at all should be observed in detail to determine the reasons why and the possible solutions for increasing cycling opportunities.
- The cross-interpretation of GPS tracks and qualitative data offers an exclusive insight into how different sections of the cycling network are perceived by users and what their preferences are when choosing their cycling routes.

Urban transportation, as one of the most dynamic and changeable features of urban settlements, is certainly a planning sector that can greatly benefit from the usage of VGI, where urban cycling is just one example. As the main mission of urban settlements is to provide settings for human interactions and exchanges, it is important to reveal people's perceptions, expectations and desires in various fields of urban life. In this respect, the CyCity initiative showed that VGI can provide a valuable source of direct information.

Another example of how VGI has been used to shed light on the spatial practice of local communities is one launched in 2014 in Denmark. In the city of Odense, a project was initiated whereby the homeless population in the city was invited to participate in monitoring their daily spatial practices using portable GPS technology. Homeless people and other vulnerable groups are under-represented in the planning and political apparatus of the modern city, so the physical planning of the city is not influenced by these groups, despite the fact that group members are often very present in the city, and often with no place else to turn to than the streets.

Much of the research to date has investigated homelessness and homeless mobility in the city (e.g. Wolch et al., 1993; Cloke et al., 2008), as well as in the countryside (Cloke et al., 2003). The spatial practice of homeless people has also been the topic of numerous studies. Some studies have focused on homelessness among immigrant groups in Europe (e.g. Pezzoni, 2011) while others have focused on gender issues (e.g. Crystal, 1984) involved in homelessness. However, only very few studies, if any, can be identified that utilise contemporary location technology in relation to monitoring the spatial practice of homeless groups.

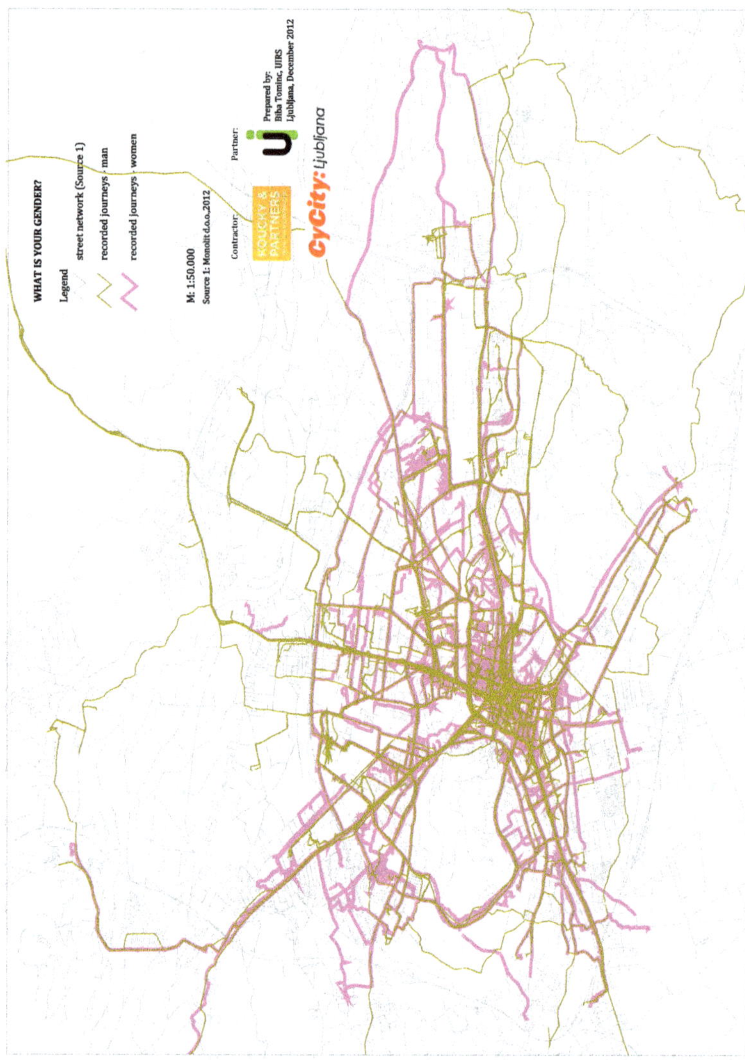

Fig. 3: An example of analyses of the cycled tracks as recorded during the CyCity research in Ljubljana.

In the Odense project, data are collected twice a year. A number of GPS devices are left in one of the shelters operated by the Blue Cross NGO in collaboration with the municipality. The homeless people are encouraged to put a GPS device in their pockets and to hand the GPS back the next day. It is, to some extent, a leap of faith for the homeless to participate in such an enterprise, as many doubts and fears about the use of the data can be raised; here, the close collaboration with officials from the municipality and high ethical standards (F. Harvey, 2013) are paramount, as the data contributors have to be assured that data on their spatial patterns are not revealed to any third party. After one day of carrying, the GPS units are collected and the data are gathered and analysed.

To date, the project has implemented three data collection routines, and already the results are being used by officials in the municipality as part of the planning process. Data on mobility patterns have revealed new bottlenecks in the spatial practices of the homeless; confluences of mobility have been identified, and places for resting and meeting up have been confirmed or investigated as part of the data analysis. The results from these analyses and the new insights into homeless mobility are further being used in the physical planning of the city of Odense in order to identify places to erect new structures such as shelters and roofed open spaces for the homeless and other vulnerable groups. The results are also being considered whenever new projects are initiated in the city.

As such, the Odense project highlights the fact that locational data on vulnerable groups can be collected in a volunteered data collection regime and can be used very effectively as a means to give voice to a group of citizens that does not traditionally get heard in the physical planning of the city. This type of information, and empowerment, would not be possible without data being provided by contemporary techniques; users volunteering the data; and ethical procedures and analysis protocols to structure the understanding and use of the results in a manner that, on the one hand, meets the requirements of the planning organs of the municipality while, on the other hand, makes sense to the vulnerable groups volunteering the data.

3.3 *Evaluation Model*

Another example of the considerable value of VGI for urban planning is in the field of the (re)design and (re)establishment of the quality of open urban public spaces. Open public spaces are the most contested spaces of contemporary cities, as they are common spaces and different users and interest groups have different conceptions and aspirations related to them. At the same time they are the places that connect the urban population in real space and time and play a crucial role in the socio-economic dynamics of cities (Madanipour et al., 2014).

In order to reveal people's spatial perceptions on urban public spaces, various techniques have been developed, from traditional mental mapping techniques inspired by Lynch (1960)'s work to a variety of contemporary IT-supported community techniques (Davis, 2007; Evans-Cowley, 2010; Bizjak, 2012).

The perceptual dimension of space, namely emotions related to concrete spatial arrangements, proves to be rather difficult to grasp in a form that could effectively support the processes of spatial planning; it is personally conditioned and varies greatly among individuals. Nevertheless, as technically supported VGI allows large samples to be collected, this aspect of urban planning may well find a way onto urban-planning agendas of the future, if the communication tools are adjusted to the knowledge and skills of the general public. A concrete example is the project outlined in Healey and Ramaswamy (2016), which explores possibilities to estimate and visualise sentiments through text mining methods, starting from short, incomplete text snippets on Twitter. Collections of real-time tweets are visualised in various ways: by sentiments, by topic, by location, by frequent terms and their co-occurrence, etc. Another very appropriate medium to reveal one's perception of space is photography and the descriptions attached to photographs. An example that has revealed the attitudes and perceptions of inhabitants regarding their immediate living environment through photography is the Human Cities (2016) online project (Figure 4). One of its many activities is a participatory collection of urban neighbourhood photographs. The project is based on a conviction that it is important to reveal the shared values that local inhabitants have to propose sensible urban design improvements to neighbourhoods. The Human Cities (2016) online photograph contest runs as a web-blog as well as a mobile phone app and has been organised with pre-defined thematic categories, e.g. Most pleasant place in my neighbourhood; Professions in my neighbourhood; My neighbour; Borders of my neighbourhood; Shared values in my neighbourhood. By analysing the photographs in each category and their subtitles, planners are given a deeper insight into the otherwise hidden layer of local environments, i.e. the interpretations of local places by users, which would not traditionally be taken into consideration in urban (re)design processes or would have to be undertaken through time-consuming interviewing.

3.4 Change, Impact and Decision Models

According to Simon (1969), any design process entails devising courses of action aimed at changing existing situations into preferred ones. In order to achieve a design, Simon (1969) proposes a three-tier iterative workflow of *intelligence* (i.e. the knowledge base is created), *design* (i.e. the alternative possible future courses of action are devised) and *choice* (where the preferable option is selected for implementation). These definitions and this approach can be considered applicable to the majority of spatial planning (and Geodesign) processes.

Fig. 4: An entry page to the Human Cities-generated portal for collection of photographs from inhabitants with subtitles to reveal local perceptions of living environments.

While previous case studies gave evidence of how VGI and SMGI can be used as information resources in the *intelligence* phase (i.e. the representation, process, and evaluation models in Geodesign), the following example shows how a Web-based collaborative platform with social networking features can be used to involve a large number of users in collecting volunteered content about *design* and *choice* (i.e. the change, impact and decision models in Geodesign).

While social media have been acknowledged as a potentially powerful means for engineering design and communication (Gopsill et al., 2013) and for supporting design studio work (Güler, 2015), until recently there have not been many Web-based platforms that were available to support collaborative planning and design. One example of such a platform is the geodesignhub.com platform developed by Ballal and Steinitz (2015), which implements the Steinitz Geodesign Framework (Steinitz, 2012). This platform, which has been successfully applied in a growing number of Geodesign workshops (Rivero et al., 2015; Nyerges et al., 2016; Campagna et al., 2016), allows for crowdsourcing of spatial data diagrams (i.e. georeferenced lines and polygons) representing design options (i.e. projects and policies) by a number of users (usually, but not necessarily, around 30). After the project and policy diagrams are collected (see Figure 5 for examples), the users can combine them in complex design syntheses that can be compared and evaluated against an impact model highlighting positive and negative impacts as well as costs (Figure 6). The platform also features a number of tools supporting negotiations so that the users participating in a workshop (which can be virtual and of same/different place/time types) can eventually find consensus on a common shared design.

The data stored in the project geodatabase of geoidesignhub.com can be considered as a design stemming from VGI. In addition, the data feature SMGI characteristics for design diagrams, i.e. they have spatial, temporal, user and preference dimensions, which can be further used to analyse the overall design process and participant behaviours. This demonstrates a novel approach in making value of crowdsourced design contents in spatial planning and (geo) design processes.

4 VGI and SMGI to Support Smart Cities Initiatives

The examples in the previous section aimed to support the idea that the increasing wealth of digital GI, made freely available through the Internet to analysts, planners and practitioners, may affect the current practices in spatial planning. While this process may still be at an early stage, it is likely that it may foster the development of 'smart city' strategies in the future. These strategies rely not only on the development of intelligent technologies but also on smart governance models according to which strategic and management decisions

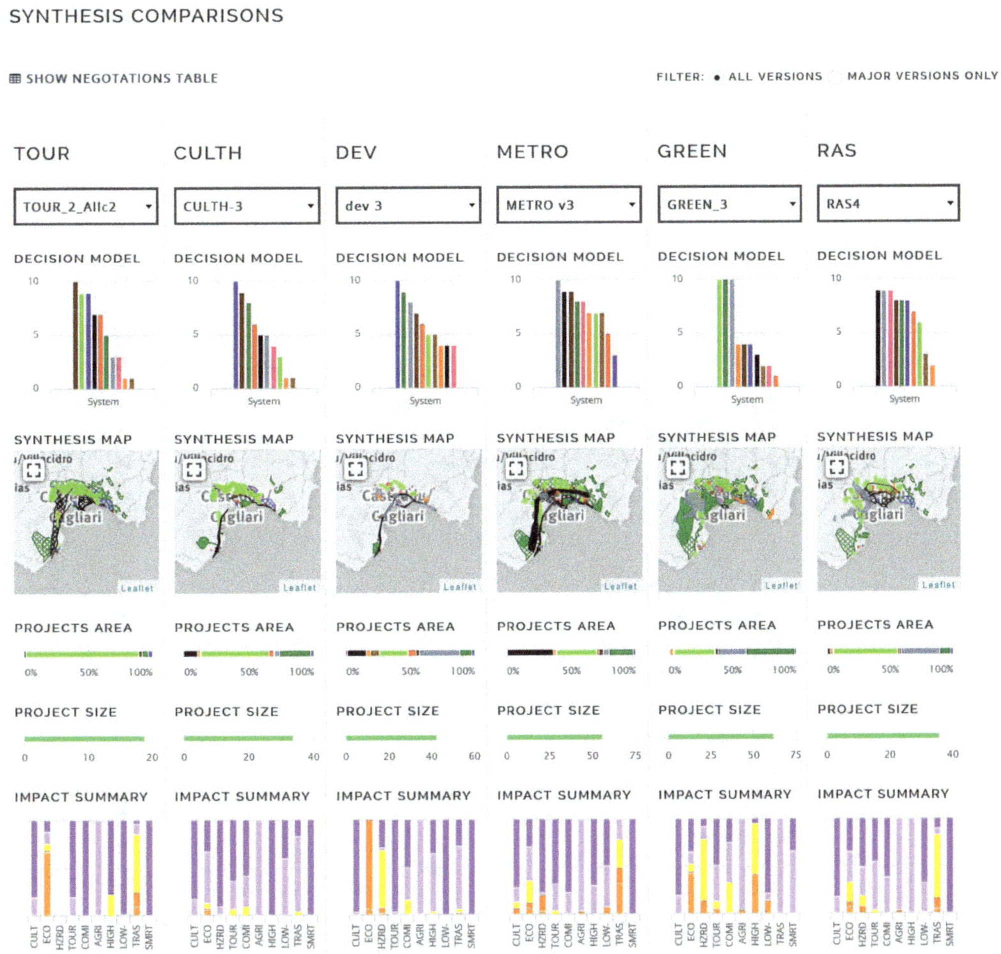

Fig. 5: Project and policy diagrams of the Cagliari (Italy) metro area crowdsourced at a Geodesign workshop in 2016 with geodesignhub.com. Each diagram in the matrix represents a project or a policy proposed by the participants during the crowdsourcing design exercise.

are informed by the real concerns and preferences of local communities as a result of real-time monitoring of needs, requirements and movements in urban environments.

In recent years, the label 'smart city' emerged as a broad term for identifying not only technology and smart infrastructure issues, but also strategies suitable to address societal problems generated by uncontrolled urbanisation and population growth in cities. Smart city strategies rely upon the Internet and Web 2.0 technologies to deal with several challenges, such as urban welfare, quality of life, societal participation and environmental sustainability (Schaffers et al., 2010). In the literature, many other smart city definitions may be found concerning different elements that contribute to the success of such initiatives. ICT represents the fundamental element to improve urban

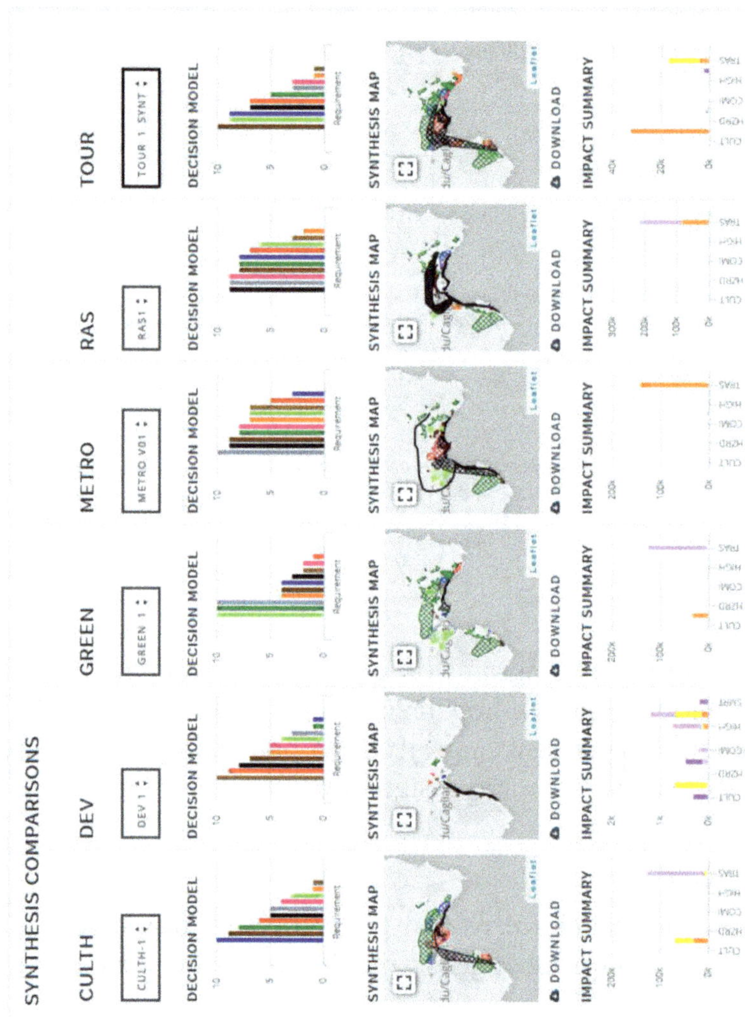

Fig. 6: Comparing Geodesign syntheses in the Cagliari (Italy) metro area at a Geodesign workshop in 2016 with geodesignhub.com.

livability and sustainability, as well as to ensure the integration, efficiency and connections in the network of urban infrastructure and services (Washburn and Sindhu, 2009). However, technology is also intended to foster the spatial enablement of citizens by improving the access to, and the sharing and integration of, spatial data within urban services (Roche et al., 2012).

Nevertheless, the technological advances offered by ICT are not the only key elements leading to the success of smart city strategies, which also depends on the managerial, political and contextual dimensions of a city (Nam and Pardo, 2011). Several factors of the political dimension, such as governance, policy and local community, may play a central role in the development of such strategies. Indeed, many stakeholders are involved in the implementation of smart city strategies, and tight relationships between these actors are fundamental to ensure the exchange of knowledge in order to avoid the failure of projects (Scholl et al., 2009). At the same time, local communities play a fundamental role in defining smart city strategies by taking into account their own needs and opinions in order to guarantee transparency, democracy and pluralism while avoiding negative effects on their quality of life.

In light of the above considerations, the participation of local actors and people should represent an essential factor for tailoring successful smart city initiatives. In this regard, the unprecedented wealth of digital GI, namely SMGI and VGI, supplies insights not only about opinions, needs, perceptions and movements of local communities in the urban environment but also about design requirements and strategies, and may result in unprecedented opportunities for leading the development of smart city strategies, taking into account the real requirements of multiple stakeholders and of the local community and the people living in a place.

5 Conclusions

To conclude, let us remind ourselves of the concept of the *Right to the city*, addressed by D. Harvey (2008: 23) as follows: 'The right to the city is far more than the individual liberty to access urban resources: it is a right to change ourselves by changing the city. It is, moreover, a common rather than an individual right since this transformation inevitably depends upon the exercise of a collective power to reshape the processes of urbanisation. The freedom to make and remake our cities and ourselves is, I want to argue, one of the most precious yet most neglected of our human rights'.

As shown in this chapter, it is realistic to foresee broader and pluralist knowledge of the places enclosed in VGI and SMGI in the near future. This knowledge might be proficiently used by developing advanced technological solutions that integrate official and experiential information with an urban sensor data infrastructure, fostering the implementation of strategies informed and supported by local communities in a bottom-up approach.

Such an approach must not be seen as beneficial only for citizens, but also equally for the authorities at different levels, and in particular for the decision-makers who may one day rely upon VGI and SMGI to discriminate among different alternatives, paying specific attention to the concerns of users and selecting among the solutions that will satisfy the requirements of involved stakeholders. VGI and SMGI may also foster scenarios where city planners are able to listen to the local community's concerns and preferences, eventually interacting with the community through new technologies and communication channels to design alternative projects and to assess future development options through a constructive and participatory dialogue. This may sound rather like a distant promise, but it represents a possible future development in spatial and urban planning and design, thus contributing to finally making the concept of the *right to the city* a realised one.

Reference list

Arnstein, S.R., 1969. A ladder of citizen participation. *Journal of the American Institute of Planners* 35, 216–224.

Ballal, H., Steinitz, C., 2015. A Workshop in Digital GeoDesign Synthesis, in: Buhmann, E., Ervin, S., Pietsch, M. (Eds.), *Proceedings of Digital Landscape Architecture 2015*. Herbert Wichmann Verlag, Berlin, Germany, pp. 400–407.

Berntzen, L., Steinmann, R., Krek, A., 2005. Innovative use of Geographic Information Systems to facilitate collaboration between the government and citizens, in: Cunningham, P., Cunningham, M. (Eds.), *Innovation and the Knowledge Economy: Issues, Applications, Case Studies*. IOS Press, Amsterdam, Netherlands, pp. 746–753.

Bizjak, I., 2012. Improving public participation in spatial planning with Web 2.0 tools. *Urbani izziv* 112–124. DOI: https://doi.org/10.5379/urbani-izziv-en-2012-23-01-004

Brabham, D.C., 2009. Crowdsourcing the public participation process for planning projects. *Planning Theory* 8, 242–262. DOI: https://doi.org/10.1177/1473095209104824

Campagna, M., 2016. Social Media Geographic Information: Why social is special when it goes spatial? in: Capineri, C., Haklay, M., Huang, H., Antoniou, V., Kettunen, J., Ostermann, F., Purves, R. (Eds.), *European Handbook of Crowdsourced Geographic Information*, pp. 45–54. London: Ubiquity Press. DOI: https://doi.org/10.5334/bax.d

Campagna, M., 2014. The geographic turn in social media: Opportunities for spatial planning and Geodesign, in: Murgante, B., Misra, S., Rocha, A.M.A.C., Torre, C., Rocha, J.G., Falcão, M.I., Taniar, D., Apduhan, B.O., Gervasi, O. (Eds.), *Computational Science and Its Applications – ICCSA 2014*. Springer International Publishing, Cham, pp. 598–610.

Campagna, M., Floris, R., Massa, P., Girsheva, A., Ivanov, K., 2015. The role of Social Media Geographic Information (SMGI) in spatial planning, in: Geertman, S., Ferreira, Jr, J.F., Goodspeed, R., Stillwell, J. (Eds.), *Planning Support Systems and Smart Cities, Lecture Notes in Geoinformation and Cartography*. Springer International Publishing, pp. 41–60.

Campagna, M., Mourão Moura, A.C., Borges, J., Cocco, C. 2016. Future scenarios for the Pampulha region: A Geodesign workshop, *Journal of Digital Landscape Architecture* 1, 292–301. DOI: https://doi.org/10.14627/537612033

Cerar, A., 2014. From reaction to initiative: Potentials of contributive participation. *Urbani izziv* 25, 93–106. DOI: https://doi.org/10.5379/urbani-izziv-en-2014-25-01-002

Cloke, P., May, J., Johnsen, S., 2008. Performativity and affect in the homeless city. *Environment and Planning D: Society and Space* 26, 241–263. DOI: https://doi.org/10.1068/d84j

Cloke, P., Milbourne, P., Widdowfield, R., 2003. The complex mobilities of homeless people in rural England. *Geoforum* 34, 21–35. DOI: https://doi.org/10.1016/S0016-7185(02)00041-6

Crystal, S., 1984. Homeless men and homeless women: The gender gap. *Urban and Social Change Review* 17, 2–6.

Dangermond, J., 2010. GeoDesign and GIS – Designing our Futures, in: Buhmann, E., Pietsch, X, Kretzler, X. (Ed.). Peer Reviewed Proceedings of Digital Landscape Architecture 2010. Anhalt University of Applied Sciences, Bernburg, Germany, pp. 502–514.

Davis, T.S., 2007. Mapping patterns of perceptions: A community-based approach to cultural competence assessment. *Research on Social Work Practice* 17, 358–379. DOI: https://doi.org/10.1177/1049731506295103

Elwood, S., Goodchild, M.F., Sui, D.Z., 2012. Researching volunteered geographic information: Spatial data, geographic research, and new social practice. *Annals of the Association of American Geographers* 102, 571–590. DOI: https://doi.org/10.1080/00045608.2011.595657

Envall, P., Koucky, M., 2013. *Varför behöver man beskriva trafiknätets cykelvänlighet? En rapport inom CyCity*. Cycity Bikeroute, Stockholm, Sweden.

Eräranta, S., Kahila-Tani, M., Nummi-Sund, P., 2015. Web-based public participation in urban planning competitions. *International Journal of E-Planning Research* 4, 1–18. DOI: https://doi.org/10.4018/ijepr.2015010101

Ervin, S., 2011. A system for GeoDesign, in: Proceedings of Digital Landscape Architecture. Anhalt University of Applied Science, pp. 145–154.

European Commission, 2003. Directive 2003-4-ES on providing for public participation in respect of the drawing up of certain plans and programmes relating to the environment. Brussels: Official Journal of EU 156-14.

Eurostat, 2012. Motorisation rate. Available at http://ec.europa.eu/eurostat/web/products-datasets/-/tsdpc340 [Last accessed 9 May 2016].

Evans-Cowley, J., 2010. Planning in the real-time city: The future of mobile technology. *Journal of Planning Literature* 25, 136–149. DOI: https://doi.org/10.1177/0885412210394100

Fischer, F., 2000. *Citizens, Experts, and the Environment: The Politics of Local Knowledge*. Duke University Press, Durham, NC.

Frias-Martinez, V., Soto, V., Hohwald, H., Frias-Martinez, E., 2012. Characterizing urban landscapes using geolocated tweets, in: Proceedings of the 2012 ASE/IEEE International Conference on Social Computing and 2012 ASE/IEEE International Conference on Privacy, Security, Risk and Trust, SOCIALCOM-PASSAT '12. IEEE Computer Society, Washington, DC, USA, pp. 239–248. DOI: https://doi.org/10.1109/SocialCom-PASSAT.2012.19

Göçmen, Z.A., Ventura, S.J., 2010. Barriers to GIS use in planning. *Journal of the American Planning Association* 76, 172–183. DOI: https://doi.org/10.1080/01944360903585060

Goodchild, M.F., 2007. Citizens as sensors: the world of volunteered geography. *GeoJournal* 69, 211–221. DOI: https://doi.org/10.1007/s10708-007-9111-y

Gopsill, J.A., McAlpine, H.C., Hicks, B.J., 2013. A social media framework to support engineering design communication. *Advanced Engineering Informatics* 27, 580–597. DOI: https://doi.org/10.1016/j.aei.2013.07.002

Güler, K., 2015. Social media-based learning in the design studio: A comparative study. *Computers & Education* 87, 192–203. DOI: https://doi.org/10.1016/j.compedu.2015.06.004

Harvey, D., 2008. The right to the city. *New Left Review* 53, 23–40.

Harvey, F., 2013. To volunteer or to contribute locational information? Towards truth in labeling for crowdsourced geographic information, in: Sui, D., Elwood, S., Goodchild, M. (Eds.), *Crowdsourcing Geographic Knowledge*. Springer Netherlands, Dordrecht, Netherlands, pp. 31–42.

Healey, C.G., Ramaswamy, S.S., 2016. Visualisation Twitter Sentiment. Available at https://www.csc.ncsu.edu/faculty/healey/tweet_viz/ [Last accessed 22 May 2016].

Human Cities, 2016. Fotozgodba naše soseske / Photo-story Our Neighbourhoods. Available at http://humancities.uirs.si/Domov.aspx [Last accessed 8 June 2016].

Jankowski, P., Andrienko, N., Andrienko, G., Kisilevich, S., 2010. Discovering landmark preferences and movement patterns from photo postings. *Transactions in GIS* 14, 833–852. DOI: https://doi.org/10.1111/j.1467-9671.2010.01235.x

Khakee, A., Barbanente, A., Borri, D., 2000. Expert and experiential knowledge in planning. *The Journal of the Operational Research Society* 51, 776. DOI: https://doi.org/10.2307/253959

Knoflacher, H., 2007. Success and failures in urban transport planning in Europe—understanding the transport system. *Sadhana* 32, 293–307. DOI: https://doi.org/10.1007/s12046-007-0026-6

Latham, A., Wood, P.R.H., 2015. Inhabiting infrastructure: exploring the interactional spaces of urban cycling. *Environment and Planning A* 47, 300–319. DOI: https://doi.org/10.1068/a140049p

Lindblom, C.E., 1990. *Inquiry and Change: The Troubled Attempt to Understand and Shape Society*. Yale University Press, New Haven, Conn.

Longley, P.A., Adnan, M., Lansley, G., 2015. The geotemporal demographics of Twitter usage. *Environment and Planning A* 47, 465–484. DOI: https://doi.org/10.1068/a130122p

Lynch, K., 1960. *The Image of the City*. MIT Press, Cambridge, MA, USA.

Madanipour, A., Knierbein, S., Degros, A., 2014. A moment of transformation, in: Madanipour, A., Knierbein, S., Degros, A. (Eds.), *Public Space and the Challenges of Urban Transformation in Europe*. Routledge, London UK, pp. 1–8.

Massa, P., Campagna, M., 2014. Social Media Geographic Information: Recent findings and opportunities for smart spatial planning. *TeMA Journal of Land Use, Mobility and Environment* INPUT 2014 Special Issue, 645–658. DOI: https://doi.org/10.6092/1970-9870/2500

McTague, C., Jakubowski, S., 2013. Marching to the beat of a silent drum: Wasted consensus-building and failed neighborhood participatory planning. *Applied Geography* 44, 182–191. DOI: https://doi.org/10.1016/j.apgeog.2013.07.019

Nam, T., Pardo, T.A., 2011. Smart city as urban innovation: focusing on management, policy, and context, in: Proceedings of the 5th International Conference on Theory and Practice of Electronic Governance. ACM Press, pp. 185–194. DOI: https://doi.org/10.1145/2072069.2072100

Nonaka, I., Takeuchi, H., 1995. *The Knowledge-creating Company: How Japanese Companies Create the Dynamics of Innovation*. Oxford: Oxford University Press.

Noulas, A., Scellato, S., Mascolo, C., Pontil, M., 2011. Exploiting semantic annotations for clustering geographic areas and users in location-based social networks, in: *Proceedings of the Social Mobile Web Workshop held during the 2011 ICWSM Conference,* Barcelona, Spain, 21 July 2011, pp. 32–35.

Nyerges, T., Ballal, H., Steinitz, C., Canfield, T., Roderick, M., Ritzman, J., Thanatemaneerat, W., 2016. Geodesign dynamics for sustainable urban watershed development. *Sustainable Cities and Society* 25, 13–24. DOI: https://doi.org/10.1016/j.scs.2016.04.016

Pezzoni, N., 2011. The city and its vision. The collapse of urban representation: Migrants mapping Milan, in: *My Ideal City. Scenarios for the European City of the 3rd Millennium*. Università Iuav di Venezia, pp. 190–200.

Rantanen, H., Kahila, M., 2009. The SoftGIS approach to local knowledge. *Journal of Environmental Management* 90, 1981–1990. DOI: https://doi.org/10.1016/j.jenvman.2007.08.025

Rivero, R., Smith, A., Balla, H., Steinitz, C., 2015. Promoting collaborative Geodesign in a multidisciplinary and multiscale environment: Coastal Georgia 2050, USA, in: Buhmann, E., Ervin, S., Pietsch, M. (Eds.), *Proceedings of Digital Landscape Architecture 2015*. Herbert Wichmann Verlag, Berlin, Germany, pp. 1–20.

Roche, S., Nabian, N., Kloeckl, K., Ratti, C., 2012. Are "Smart Cities" smart enough? Presented at the GSDI World Conference (GSDI 13), Global Spatial Data Infrastructure Association, Québec City, Canada, pp. 215–235.

Schaffers, H., Guzman, J.G., Navarro, M., Merz, C., 2010. *Living Labs for Rural Development. Results from the C@ R Integrated Project*, TRAGSA and FAO, Madrid, Spain.

Scholl, H.J., Nahon, K., Ahn, J.-H., Re, B., 2009. E-Commerce and E-Government: How do they compare? What can they learn from each other? Presented at the 42nd Hawaii International International Conference on Systems Science (HICSS-42 2009), IEEE, Waikoloa, Big Island, HI, USA, pp. 1–10. DOI: https://doi.org/10.1109/HICSS.2009.169

Seltzer, E., Mahmoudi, D., 2013. Citizen participation, open innovation, and crowdsourcing: Challenges and opportunities for planning. *Journal of Planning Literature* 28, 3–18. DOI: https://doi.org/10.1177/0885412212469112

Simon, H.A., 1969. *The Sciences of the Artificial*, 1st ed. MIT Press, Cambridge, Mass.

Steinitz, C., 2012. *A Framework for Geodesign: Changing Geography by Design*, 1st ed. Esri Press, Redlands, Calif.

Tominc, B., Bizjak, I., Mladenovič, L., Nikšič, M., 2012. *CyCity Ljubljana. Cyclists' route choice and route preferences*. Urban Planning Institute of the Republic of Slovenia, Ljubljana, Slovenia.

Torres, Y.Q.A., Costa, L.M.S.A., 2014. Digital narratives: Mapping contemporary use of urban open spaces through geo-social data. *Procedia Environmental Sciences* 22, 1–11. DOI: https://doi.org/10.1016/j.proenv.2014.11.001

Washburn, D., Sindhu, U., 2009. Helping CIOs Understand "Smart City" Initiatives, Making Leaders Successful Every Day. Forrester Research, Inc., Cambridge, MA, USA. Available at http://public.dhe.ibm.com/partnerworld/pub/smb/smarterplanet/forr_help_cios_und_smart_city_initiatives.pdf [Last accessed 21 Mar 2014].

Winters, M., Teschke, K., Brauer, M., Fuller, D., 2016. Bike Score®: Associations between urban bikeability and cycling behavior in 24 cities. *International Journal of Behavioral Nutrition and Physical Activity* 13, 18. DOI: https://doi.org/10.1186/s12966-016-0339-0

Wolch, J.R., Rahimian, A., Koegel, P., 1993. Daily and periodic mobility oatterns of the urban homeless. *The Professional Geographer* 45, 159–169. DOI: https://doi.org/10.1111/j.0033-0124.1993.00159.x

Yeboah, G., Alvanides, S., 2015. Route choice analysis of urban cycling behaviors using OpenStreetMap: Evidence from a British urban environment, in: Jokar Arsanjani, J., Zipf, A., Mooney, P., Helbich, M. (Eds.), *OpenStreetMap in GIScience*. Springer International Publishing, Cham, pp. 189–210.

Zubelzu, S., Fernández, R.Á., 2016. *Carbon Footprint and Urban Planning: Incorporating Methodologies to Assess the Influence of the Urban Master Plan on the Carbon Footprint of the City*. Springer.

CHAPTER 15

Citizen Science and Citizens' Observatories: Trends, Roles, Challenges and Development Needs for Science and Environmental Governance

Hai-Ying Liu, Sonja Grossberndt and Mike Kobernus

NILU-Norwegian Institute for Air Research, Kjeller 2027, Norway

Email address for corresponding author: hai-ying.liu@nilu.no

Abstract

This chapter explores growing and important trends within citizen sensing, especially those linked to major initiatives that form citizens' observatories and address novel ways to engage citizens in science and environmental policy-making. On the basis of providing an overview of existing and planned citizen science and citizens' observatories programmes, this chapter identifies areas where citizen science and citizens' observatories have actively contributed to, and can be expected to see further development in, the formation of various policies in Europe. Furthermore, this chapter considers the motivations for developing citizen science and citizens' observatories and how these initiatives can contribute to awareness raising and decision support systems. We address key challenges and development needs for policy- and decision-making within the context of widespread and accessible citizen science and of the activities of citizen observatories.

How to cite this book chapter:

Liu, H-Y, Grossberndt, S and Kobernus, M. 2017. Citizen Science and Citizens' Observatories: Trends, Roles, Challenges and Development Needs for Science and Environmental Governance. In: Foody, G, See, L, Fritz, S, Mooney, P, Olteanu-Raimond, A-M, Fonte, C C and Antoniou, V. (eds.) *Mapping and the Citizen Sensor.* Pp. 351–376. London: Ubiquity Press. DOI: https://doi.org/10.5334/bbf.o. License:

Keywords

Awareness raising; Citizens' observatories; Citizen science; Decision support systems; Environmental policy-making

1 Citizen Science and Citizens' Observatories: A Growing and Important Trend to Engage Citizens in Science and Environmental Policy-making

The participation of citizens in environmental monitoring and related scientific activities has a long tradition, dating back at least two centuries (Silvertown, 2009; UWE, 2013). The present digital era facilitates people's easy access to advanced Information and Communication Technology (ICT) systems (e.g., social media platforms, mobile Internet, online gaming or smartphone apps, etc.), enabling the public to participate in (scientific) projects on issues relevant to their local environment and to easily access data and information about the state of those data. The collaborative power of these advanced ICT systems is enormous, and can leverage a collective intelligence that has the potential to change the way environmental policy-making and monitoring is performed, as well as more effectively raise citizens' awareness of environmental issues. Numerous collaborative and co-design approaches have been developed and tested during the last decades. In this chapter, we will focus on two methodologies that are well suited to be applied in the context of 'Mapping and the citizen sensor': Citizen Science (CS) and Citizens' Observatories (COs), which both have applicability in the acquisition of spatial data through Volunteered Geographic Information (VGI).

In this section, we first define our terms (CS and CO) and discuss how these methodologies have become increasingly vital within science and policy-making (Sections 1.1 and 1.2). We then distinguish between CS and COs in general, and also especially in relation to major CS and CO initiatives that engage citizens in science and environmental policy-making.

1.1 Citizen Science: old wine in new bottles

Before diving directly into the world of CS, let us first review its definition. Generally, the term describes the activities of non-scientist citizens that contribute to scientific research. In the Oxford dictionary, we find the following definition: 'scientific work undertaken by members of the general public, often in collaboration with or under the direction of professional scientists and scientific institutions' (OED, 2014). CS approaches are also described as Public Participation in Scientific Research (PPSR). PPSR describes all efforts of lay people directed towards their involvement into scientific research activities (Shirk et al., 2012);

it includes CS, but augments it with a broader definition of participation, not only limited to collecting scientifically relevant data. However, these definitions do not provide any information on the extent to which citizens are involved in the scientific work, whether they are only collecting data or whether they also participate in the creation of the study. Based on relevant literature, we have created an overview of the most prominent categories of public participation in scientific research (Figure 1, adapted from Bonney et al., 2009) and visualised a range of popular terms that are used in this context in a cloud tag (Figure 2).

Why do we need CS? CS offers many advantages. Due to restricted time and limited monetary resources, scientists cannot always collect large amounts of data or cover big geographic areas for both data collection and documentation (Dickinson et al., 2010; Tulloch et al., 2013). For this reason, the help of volunteers in collecting data can be extremely valuable. For example, since the US Weather Service did not have enough resources to set up a countrywide meteorological measuring network, they made use of volunteers all over the country to help in the data collection. The resultant data were one of the most important long-term datasets in the history of North America and have been used for essential work within climate research, agriculture and development planning

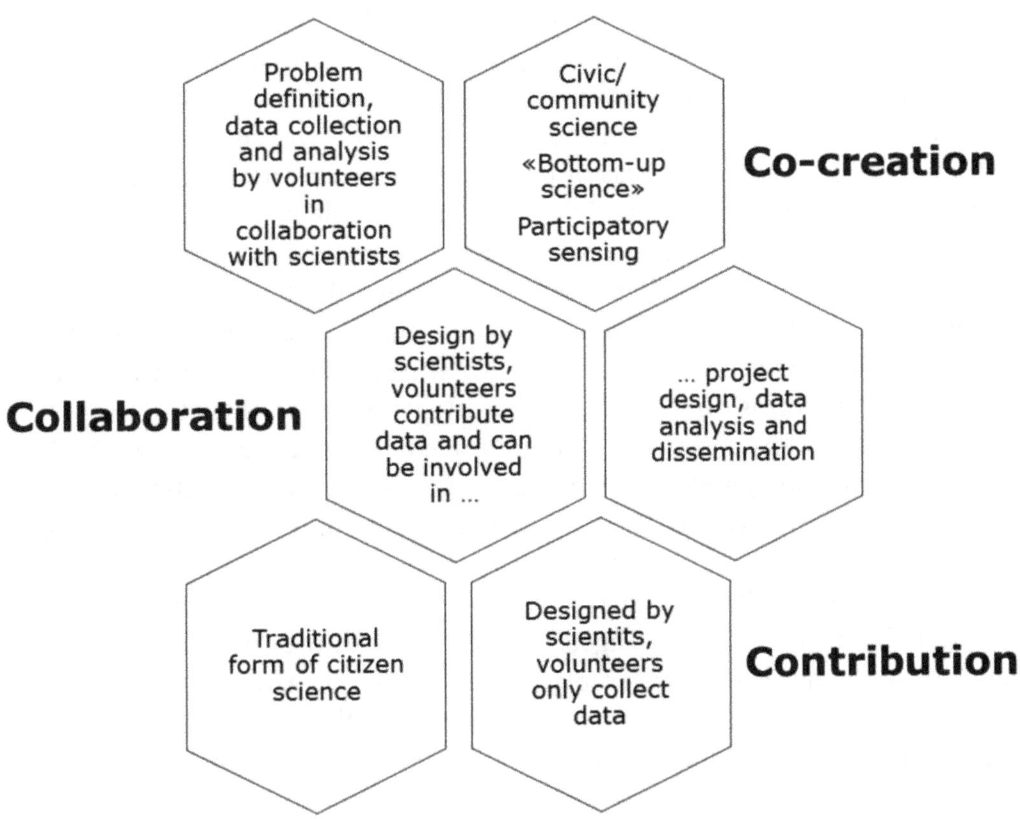

Fig. 1: Categories of citizen science. Modified from Grossberndt and Liu (2016).

Fig. 2: Cloud tag visualising terms related to citizen science.

(Vetter, 2011). This example shows that the collection of data over many decades has led to the compilation of long-term data series, which are extremely valuable for the work of modern science (Miller-Rushing et al., 2012).

Another reason for the application of CS and other participatory approaches is to increase citizens' awareness of problems related to their immediate environment. In some cases, the activities can also result in greater interest and increased engagement in these issues. Engaging citizens can also have educational effects and increase science literacy (Haklay, 2015).

One would think that CS was a rather novel invention, considering that many scientists prefer to keep to themselves in their ivory towers and the concept of public participation is only gradually making its way into their thinking. Surprisingly enough, the roots of CS can be traced at least as far back as the 18th century. At this time, a Norwegian bishop engaged a large number of clergymen throughout the whole country and assigned them with the task of collecting observations and natural objects from all over Norway in order to assist him in his research (Brenna, 2011). Throughout the centuries, non-scientists/laypeople have often been engaged in assisting scientists in the collection of data. Another more recent example is the traditional Christmas Bird Count in the USA, Canada and other Western countries that began in 1900: in the 2014/15 season, more than 72,000 volunteers participated in that programme (LeBaron, 2015).

Nowadays, a large number of CS activities have been initiated and are still ongoing, covering many different fields (see Section 2.2). The list of CS programmes is endless, and, during the last decades, CS activities have sprung up like mushrooms all around the globe. What has caused this phenomenon? There are several reasons. First and foremost, there have been rapid changes within ICT; for example, easy Internet access, the emergence of Web 2.0 systems and the rise of social media have enabled increased engagement with the public. Another aspect is the improvement and simplification of the collection, management and storage of data. More and more people have access to easy-to-use devices like smartphones and other mobile devices with GPS positioning technology; this facilitates the involvement and connection of citizens around the world. Collecting data or taking a picture and sending it to a data server with the exact time and geographic position now takes split seconds, not hours or days. A second important reason for the emergence of CS initiatives is the changes in society. At least in Western countries, the level of education amongst the public has been increasing. More leisure time and a growing understanding of scientific concepts, as well as increased technical skills, even for the youngest in society, are contributing factors to CS initiatives. Thirdly, scientists have become more aware of the fact that citizen participation in the collection of scientific data can also be beneficial, due to resource limitations, as mentioned above. Recent study results indicate that savings in labour cost per project can reach up to US$200,000 over the project's first 180 days, depending on the project (Sauermann and Franzoni, 2015).

1.2 Citizens' Observatories: A New Concept

As early as the 1970s, P.K. Feyerabend suggested that it was time for a democratisation of science; he claimed that 'everywhere science is enriched by unscientific methods and unscientific results' (Feyerabend, 1970). Essentially, he believed that the monopolisation of research by universities, corporations and other large institutions was contrary to the best interest of science, which, as we have seen, has a long history of public participation. However, in spite of his attempts to redress the lack of citizens or non-scientists within research, amateur participation was declining. This deficit was eventually recognised, and, in order to promote a more active participation from the public, the EU first commissioned the SOCIENTIZE project (2012–2014), to create a common forum for cooperation between e-Infrastructure providers and CS infrastructure providers, including any end user with an interest in contributing to the scientific process (Socientize, 2012–2014). The project produced the Green Paper on Citizen Science, which helped to create a 'roadmap' for CS in Europe. This led to a series of further initiatives where CS was incorporated in some form, especially within the development of the new concept of the CO (see Section 2.1).

The term CO was first addressed in the EU FP7 Topic ENV.2012.6.5-1: 'Developing community-based environmental monitoring and information systems using innovative and novel earth observation applications' (EC, 2014).

It is a term that is applied to a framework that combines participatory community monitoring with monitoring by policy-makers, scientists and other stakeholders. Typically, this is achieved via a technological system that may include web portals, mobile technologies and sensors (Liu et al., 2014). The term was further developed within five projects that were funded within the EU FP7 Topic ENV.2012.6.5-1 (see Section 2.1). For example, in the CITI-SENSE project, a CO for supporting community-based environmental governance has been defined as 'the citizens' own observations and understanding of environmentally related problems and in particular as reporting and commenting on them within a dedicated ICT platform' (Liu et al., 2014) and was tested in nine cities in the field of air quality. In the WeSenseIt project, Ciravegna et al. (2013) defined a CO as 'a method, an environment and an infrastructure supporting an information ecosystem for communities and citizens, as well as emergency operators and policymakers, for discussion, monitoring and intervention on situations, places and events.' The CO in the WeSenseIt project is therefore seen as an environment for implementing collaboration, as infrastructure to validate the CO concept and as a method to demonstrate the applicability of its outcome (Lanfranchi et al., 2013).

There is no doubt that the term CO has become popular in CS programmes (especially EU-funded ones), and many new CO-related initiatives have been created at different levels. Accordingly, this new term represents a growing and important trend in both science and policy-making.

In practice, all CO projects typically share a similar model, including the main aspects needed to develop COs as a method for data collection. These include engaging the participation of citizens in data collection, data interpretation and information delivery. Alternatively, the CO model (Figure 3) combines (i) sequential aspects, (ii) interaction with citizens and other stakeholders, (iii) data collection tools, and (iv) an ICT infrastructure that underlies the CO framework and supports effective citizen participation.

A set of sequential aspects (the pyramid within Figure 3) has been identified by Liu et al. (2014) as follows: A) identifying what citizens want and what citizens can offer; B) exploring what products and services a CO can provide for the citizens; C) recruiting and retaining citizens to participate in and contribute to environmental governance; D) providing tools that support citizens to report their observations, inferences and concerns; and E) supplying tools to access/receive information on the environment in a manner that is both easily understood and useful, for citizens and other stakeholders, including policymakers.

The essential aspects of the interaction with citizens and other stakeholders (who are represented by the five circles along the bottom outer open edge in Figure 3) have been addressed in all existing CO models. A CO includes observations from not just professionals and scientists, but also citizens. An effective CO shall enable a two-way communication between citizens

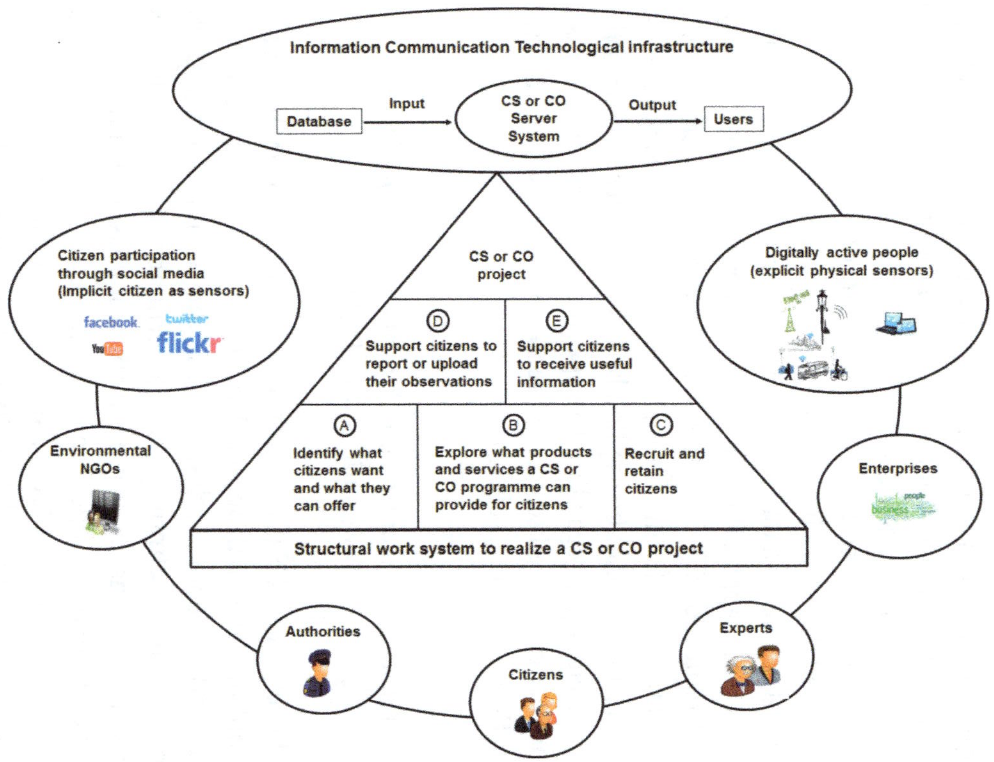

Fig. 3: A common model for citizens' observatories programmes in practice. Modified from Grossberndt and Liu (2016).

and other stakeholders, potentially resulting in profound changes to local environmental management processes, and, as such, shall engage in social innovation processes and outcomes (Wehn and Evers, 2015). For example, the WeSenseIt project used social media and co-design approaches, exploring citizens' needs and providing a framework in which authorities and citizens cooperate in sharing collective intelligence and participate in planning, decision-making and governance regarding the water environment, including flood risk (WeSenseIt, 2012–2016).

The data collection tools (the two ovals along the outer open edge of Figure 3) are highlighted in the existing CO models as well. For example, the CITI-SENSE project engages citizens to use low-cost micro-sensors to monitor air quality in their surroundings (hard layer of data collection), and interacted with citizens via various social media and mobile apps (soft layer of data collection; CITI-SENSE, 2012–2016).

The ICT infrastructure (the large oval at the top of Figure 3) is an essential part of the CO model that includes boundary services with sensors and apps, data management services, data storage support and the reusable visualisation widgets used for both apps and web portals. Currently, existing CO projects are building all required ICT infrastructure towards a systematic, simple and reusable method to facilitate the setting up of new COs in various environmental fields, a method which can be applied by communities and organisations to overcome their challenges regarding the specific technical ICT skills and programming knowledge needed to create the necessary server infrastructure and mobile applications (Zaman et al., 2014).

1.3 Citizen Science and Citizens' Observatories – Commonalities and Differences

As mentioned previously, CS is a novel take on an old approach and is generally described as 'public participation in scientific research'. COs are a new concept that evolved from EU policy circles, defining the combination of participatory community monitoring, technology and governance structures that are needed to monitor, observe and manage an environmental issue (Haklay, 2015).

Both CS and COs involve citizens in scientific research or various monitoring programmes, help citizens to play an active role in the data collection process and enable them to exchange data/information and knowledge, to reach the expert who can answer questions about various issues that are being addressed, and to disseminate information to further the understanding of such issues. The Chinese proverb 'Tell me and I'll forget; show me and I may remember; involve me and I'll understand' is an apt quotation in this context, since both CS and COs have great potential to be a suitable instrument to raise awareness, increase citizen participation and support community-based environmental decision-making.

Whereas CO approaches focus very much on a two-way communication between citizens and other stakeholders, such as scientists, this may not always be the case for CS: here, the degree of participation can vary from only collecting data to participating in the study design and data analysis. In addition, CS usually refers to science/scientific projects, whereas COs include a broad range of stakeholders, including authorities or policy-makers. However, the combination of both top-down and bottom-up approaches makes COs a more complex tool, especially as they require an ICT infrastructure, which is not necessarily required for CS initiatives.

2 Current Citizen Science and Citizens' Observatory Programmes in Europe

2.1 Citizens' Observatory Projects

In recent years, there have been many ongoing COs projects in Europe. For example, the European Commission (EC) has seen the possibility of empowering European citizens in environmental monitoring, with the consequent increase in observational possibilities. The EC has provided funding through their Seventh Framework Programme for five projects (i.e., Citclops, CITI-SENSE, COBWEB, OMNISCIENTIS and WeSenseIt) with the aim of building COs in the various environmental fields. For example, OMNISCIENTIS has combined the active participation of citizens with the implementation of innovative technologies for improving the governance of odour nuisance (OMNISCIENTIS, 2012–2014). Other projects that emphasise the need for citizens' participation are COBWEB, which aimed at creating a test-bed environment that would enable citizens living within Biosphere Reserves to collect environmental data using mobile devices (COBWEB, 2012–2016; Higgins et al., 2016); Citclops, which aimed at developing an observatory based on CS applications for bio-optical monitoring of coast and ocean (Ceccaroni et al., 2016; Citclops, 2012–2015); and WeSenseIt, which puts emphasis on enabling citizens to become active stakeholders in information capturing, evaluation and communication for the marine environment, including flood risk (WeSenseIt, 2012–2016). Finally, CITI-SENSE aimed at empowering citizens to participate in environmental governance by developing various CO supporting services related to outdoor air quality, indoor air quality in schools and environmental perception in public spaces (CITI-SENSE, 2012–2016). These five CO projects were designed independently of each other; however, they had considerable similarities in terms of their structure, operation and methodology for communication with the public (Liu et al., 2014). Furthermore, there has been cross-project collaboration amongst these five projects to (i) facilitate data, knowledge and success sharing amongst the projects, and (ii) establish common methodologies and standards for crowdsourcing/citizen science within GEOSS and aligned with INSPIRE and Copernicus[1].

In addition, four projects have been funded under the EC H2020 topic SC5-17-2015, 'Demonstrating the concept of "Citizen Observatories"' (EC, 2015–2016), that aim to scale up, demonstrate, deploy, test and validate, under real-world conditions, the concept of CO and the effective transfer of environmental knowledge for policy, industrial, research and societal use, with a focus on the domain of land cover/land use, both in rural and urban areas. The EC H2020 topic CSA-2017 ('Coordination of Citizens' Observatories initiatives'; EC, 2016–2017b) aims at bringing existing CO and related communities together, and also the EC H2020 topic RIA-2017 ('Novel in-situ observation systems') will further develop ICTs and test them in various CO activities (EC, 2016–2017a).

Furthermore, with an increasing number of CO-based initiatives, the EU H2020 Work Programme 2016–2017 (Topic in SC5-19-2017) issued a call for the coordination of citizens' observatories initiatives (EC, 2016–2017b) to create a CO knowledge base in Europe across disciplines in order to avoid duplication, ensure interoperability, create synergies and facilitate the gradual uptake of this knowledge base by environmental authorities.

There are more existing and planned CO-related activities supported by the EC programmes and calls, for example:

- CAPS – Collective Awareness Platforms for Sustainability and Social Innovation (ICT Calls; 34 existing projects (EC, 2016));
- The new calls in 2016–2017 (EC, 2016–2017c) and Pilots and Coordination and Support Actions;
- Integrating Society in Science and Innovation (RRI) (EC, 2017);
- MYGEOSS (EC, 2015); and
- RIA – Novel in-situ observation systems (EC, 2016–2017a), etc.

2.2 Citizen Science Projects

In recent years, there has been a boom in CS projects, with many now harnessing new technologies, such as mobile Internet and smartphone apps, to increase accessibility and remote participation. For example, more than 1,600 formal and informal research projects, tools and events are listed on SciStarter and the number is increasing rapidly (SciStarter 2017). Some of the best known projects were and are run by the previous Zooniverse team, now Citizen Science Alliance, which launched the Galaxy Zoo galaxy-classifying project in 2007 (Zooniverse, 2013), and whose crowdsourcing model has been adopted by many other groups. However, there are many more examples of CS projects, which include, but are certainly not limited to, topics such as biological monitoring (e.g., the Cornell Lab of Ornithology, www.birds.cornell.edu; the Great Backyard Bird Count[2]; the big butterfly count[3],), geography (e.g., OpenStreetMap[4]), air quality (e.g., Air Quality Egg[5]), and others that encompass different models of CS; within the environmental sciences, these span a diverse range of subjects.

The CS activities can differ in focus, approach or technique. Various reviews indicate that the most prominent topics for CS are biology, conservation and ecology, with citizens assisting in the collection and classification of data (Kviner, 2012; Science Communication Unit, University of the West of England, 2013; Liu et al., 2014; Grossberndt and Liu, 2016). Another main cluster is geographic information research, with citizens collecting geographic data; as the third most prominent group of CS topics, the study identified research involving the public in relation to environmental and health issues (Kullenberg and Kasperowski, 2016). There are also 'higher level' initiatives, like the Open Air Laboratories (OPAL) for CS initiatives focused on nature[6], Geo-Wiki for projects addressing global land cover issues[7] or Zooniverse, serving as a hub for projects from different fields[8].

In Europe, CS has grown in scale and scope, and is therefore receiving increasing attention from scientists and policy-makers at local, national and international levels. Some of the well known European CS projects are ENERGIC[9], EmoMap[10] and EveryAware[11]. Gradually, CS has been considered as an independent discipline. For example, there are academic groups and collaborations (Science Communication Unit, University of the West of England, 2013), including the Citizen Cyberlab[12], a Swiss partnership involving CERN, the UN Institute for Training and Research and the University of Geneva; and OPAL[13];). Furthermore, there are large-scale experiments at JRC (EC JRC, 2014) to (i) assess the quality of social network data of 2010–2012 (by comparison with official data from EFFIS); (ii) map CS and Smart Cities projects; (iii) develop the typology of CS, set up facilities for social media data analysis and develop analytical tools; (iv) set up a framework for hosting citizen science project data (e.g. CitObs, EveryAware), websites and code after the end of project; (v) develop interoperability protocols and integration with official data sources (INSPIRE, Copernicus); (vi) develop partnerships with relevant stakeholders (e.g. ECSA, 2016); and (vii) explore the use of citizen-generated content to develop new indicators of quality of life in urban areas, with comparison to official sources (e.g. Eurobarometer).

3 Citizen Science and Citizens' Observatories for Policy and Decision-Making

The increasing numbers of CS activities and the rise of COs in recent years demonstrates one key fact: science needs public participation. We have already stated that the involvement of volunteers in the collection of observations and data can be beneficial for scientists who suffer from a constraint of resources. Another advantage that we inevitably come across is the fact that the participation of citizens in science will also serve the purpose of awareness raising, i.e. that people become more aware of problems or issues related to their direct environment and are consequently more likely to be in turn interested in the initiative and to be more willing to participate (Evans et al., 2005; Haklay, 2015). Several reviews of CS and CO projects indicate that the involvement of volunteers in science offers added value to science literacy and education effects (Kviner, 2012; Science Communication Unit, University of the West of England, 2013; Haklay, 2015; Grossberndt and Liu, 2016). A review of more than 230 "citizen science" projects concluded that volunteers have proven to provide information that has 'high value to research, policy, and practice' (Tweddle et al., 2012).

Although public participation has been given more attention in environmental governance processes recently, in most places it is still in its infancy. In 1998, the Aarhus Convention strengthened public participation through the establishment of 'the right to know', i.e., the access to environmental information, public participation in environmental decision-making and access to justice (UNECE, 1998). The EC Directive 2003/35/EC was adopted in 2003 to provide for public participation and thus implement the Aarhus Convention in the Member States of the EU (EC, 2003).

Involving citizens, and not only scientific experts, in environmental governance processes creates new opportunities. The EC published a White Paper in 2001 (EC, 2001), where they called upon different actors for cooperation within the whole process of environmental governance. The White Paper points to decision-makers and scientists as actors of such governance, but also requests explicitly the inclusion of representatives from civil society. In 2014, the EU project Socientize developed a White Paper on Citizen Science for Europe (Socientize, 2012–2014), which aimed to support policy-makers at the European, national and regional levels to set up future strategies of civic engagement.

Both CS and COs can provide scientists with important and reliable data, enabling authorities to carry out informed policy-making, while providing citizens with opportunities to address issues affecting them at different scales. As citizens develop an increased scientific and environmental understanding, they may begin to influence decision-making and policy through activities such as petitions, public debate and advocacy, e.g., for identifying new policy issues, generating policy options, lobbying, supporting joined-up governance, etc. (Walters et al., 2000). An example of participatory monitoring impacting policy can be seen in Cambodia, where the Committee for Free and Fair Elections uses voter scorecards and volunteers with mobile phones to monitor if elected representatives keep their election promises. These examples have a direct impact on local policy and are the direct result of citizen participation and observation (Bottomley, 2014). However, many CS and CO programmes have yet to be evaluated for these impact attributes.

As addressed earlier in this chapter (see Section 1.2), the CO as a new concept that considers the wider implications of CS has evolved in EU policy circles. The existing and planned CO projects, and the results of their preliminary testing in practice, indicate that COs have a great potential to complement in-situ observation networks and to contribute to European policies covering areas from water management and air quality protection to biodiversity conservation.

In the 'Citizen Science and Policy: A European Perspective' report (Haklay, 2015), the following three policy dimensions are distinguished: (1) level of geography; (2) policy domains; and (3) level of engagement and type of CS activity. CS initiatives can influence policy decisions in a specific geographic area, i.e. local, regional, national or international. Usually, problems that affect the direct environment lead to more engagement, since people are more concerned (Haklay, 2015). This increased awareness can be leveraged to engage local people to contribute to CS initiatives. Local CS is often linked to environmental activism and supports community management by working towards effective and meaningful management planning and stewardship (Conrad and Hilchey, 2011). Local CS can also apply the so-called community-based monitoring (CBM) approach. CBM describes a process where concerned citizens, public authorities and further stakeholders collaborate to monitor, track and respond to issues that arise from common community concerns (Whitelaw et al., 2003).

There is an increasing need for communities to fall back on CS approaches (or on CBM ones) and include different stakeholders with their diverse knowledge and experience into decision-making processes (Conrad and Daoust, 2008). In addition to potential savings in time and money for decision-making bodies, the societal benefits of CBM will be to create environmental democracy, social capital, and an increase in scientific literacy and inclusion in local issues (Conrad and Hilchey, 2011).

Policy areas can be manifold and partially overlapping. For example, city-scale policy includes public transport, environmental quality, education, infrastructure and public health. Thus, cities can be a canvas for a potpourri of local monitoring activities, originating from different concerns but using the accumulated data to see the bigger picture. Moving CS projects to the regional, national or even international level is likely to meet even more challenges than there already are. Since bottom-up initiatives usually dispose of limited budgets only, it will be less likely to find community science approaches with an active involvement of citizens in all parts of the participation cycle, i.e. citizens will instead only be asked to share observations or viewpoints on certain issues. Nevertheless, national and even international initiatives including CS are possible and do exist. Projects funded by the EC and formations of international organisations like the European Citizen Science Association[14] provide frameworks for national initiatives and NGOs to create synergies to promote CS on larger scales and to call on international institutions such as the European Environment Agency (EEA) to promote citizen participation at the international level as well (Haklay, 2015).

At present, there are still relatively few CS and CO examples that demonstrate where such projects have had a clear and distinct impact on both policy- and decision-making. However, this is dependent on how one perceives the 'level' of impact. Monitoring projects may not bring about immediate policy change, but their usefulness in building up evidence bases is invaluable. For example, the UK Biodiversity Indicators rely directly on the long-term data that NGOs and their volunteers collect for species such as birds and butterflies. These biodiversity indicators feed directly into wider UK and global policy, such as the Convention on Biological Diversity Strategic Plan for Biodiversity 2011–2020. Other projects that focus on observing and identifying invasive species, for example PlantTracker and the Harlequin Ladybird Survey, are valuable and will become increasingly relevant to policies in this area, such as the recently proposed EU Regulation on Invasive Alien Species and the developing of tree health policies within the UK (British Ecological Society, 2013).

Both CS and COs have an extremely important role to play in today's environmental science and research, and, through modern technology, innovative projects and new partnerships, the involvement of the public will only increase. The role of CS and CO projects in policy is relatively hard to gauge, but they are invaluable for building up evidence bases and directing change – especially those projects that are linked to some pressure groups (i.e. a group that tries to influence public policy in the interest of a particular cause) or that address environmental issues at the population level. Equally, given the educational values that citizen projects can provide, such projects may be influencing people's mindsets, which in turn could influence policy decisions in ways that are more abstract. As such, people really are power, not just for science but for policy-making too (British Ecological Society, 2013).

4 Challenges and Development Needs

As we have seen in this chapter so far, the idea of citizens participating in environmental governance is found not only in citizens' initiatives, but also at the international level, with e.g. the EU or UN as driving forces. However, there is still a discrepancy between theory and practice, owing to different circumstances and challenges. We shall now look a bit closer into the challenges that are connected to the implementation of CS and COs in environmental governance.

In this section, we distinguish between four different categories of challenges:

- Technologies and data;
- Citizen engagement;
- Policies and framework;
- Additional requirements for COs.

4.1 Technologies and Data

CS approaches and COs require strict data management. In both cases, volunteers who do not necessarily possess the required skills for the collection process can still gather large amounts of data; however, the obtained data often contain errors and bias. It takes time and resources to train the volunteers to enable them to collect data in the manner and of a quality that is useful for scientists, decision-makers and other stakeholders (Conrad and Hilchey, 2011; Dickinson et al., 2010; Engelken-Jorge et al., 2014; Goodchild and Li, 2012; Hanahan and Cottrill, 2004). An insufficient experimental design can hence lead to undesired outcomes (Conrad and Hilchey, 2011). Another requirement is the management and analysis of the continuously increasing volume, variety and velocity of the data that are collected throughout the whole course of the initiative (Zikopoulos et al., 2011). One option to deal with this issue is to build networks with other existing projects or initiatives to use already existing datasets and combine them with newly obtained data (Dickinson et al., 2010). However, special attention must be paid to accuracy and uncertainty, especially when comparing crowdsourced with referenced data. The same applies for the interpretation of qualitative data; indicators such as 'quality of life' or 'wellbeing' should be developed together with more quantitative data. In addition, data security and privacy are important issues that require special attention. Especially when using smartphones and/or mobile sensing devices, it has to be ensured that the data from the volunteers are anonymised and treated according to national and international data protection laws and standards. In addition, ethical restrictions may apply (Liu et al., 2014). Increasing the amount of data requires progressive technologies and data analysis methods that reduce measurement uncertainties through real-time, reliable and fast quality assurance/quality control tools. Furthermore, there is an urgent need to explore and develop technologies for data collection and analysis by building the technical capacity required to combine environmental monitoring with the exchange and integration of different types of data, then visualise and communicate the results to end users (Liu et al., 2014; DFID, 2008).

The evaluation of citizen science and especially of CO approaches is another topic that requires further research. Indicators for evaluation and value proposition have to be developed to facilitate the comparison of initiatives from different fields and their effectiveness/efficiency, especially regarding engagement and participation.

4.2 Citizen Engagement

Engaging with volunteers to participate in any form of activity related to CS or COs can be quite challenging. The most crucial task is to raise the interest of the volunteers to actively participate and continue until the end of the initiative.

If there is no interest, there will be no data. In addition, few people will spend their spare time and resources for nothing; the volunteers must clearly know what to expect in return, i.e. what is in it for them. Thus, it is essential to implement various tailor-made tools to recruit and sustain citizen participation in environmental monitoring activities (Fernandez-Gimenez et al., 2008; Conrad and Hilchey, 2011). One of the preconditions for successful involvement of volunteers in CS activities is their level of interest in the research itself. Nevertheless, many volunteers seem to contribute very little at the beginning of data collection activities, leaving a rather small amount of volunteers contributing the most (Sauermann and Franzoni, 2015). Thus, keeping the volunteers' interest through fun activities seems to bear potential for a higher contribution rate. So-called 'gamification' for this purpose seems to show positive results; however, this is very much dependent on the project type and the volunteers and requires further research (Prestopnik et al., 2014). Immediate and continuous feedback of results in a visually attractive and easy to understand manner is also important. Social media can also be a good way to keep in contact with the volunteers (Gottschalk Druschke and Seltzer, 2012). Furthermore, it is very helpful to engage and to retain citizens by clearly addressing the positive aspects of their participation, for example the benefits they can gain, such as improved health, knowing which areas are polluted and how to avoid exposure (in the case of air quality) or personal recognition (e.g. through a leader board in the community). Being able to access data from other volunteers and to compare them to the data collected by oneself, as well as dashboard and analytical tools accessible to the volunteers, etc., are all useful methods to engage citizens.

4.3 Policies and Framework

Even though participative approaches in environmental governance have been repeatedly promoted at an international level, this does not mean that these approaches are automatically followed up at national, regional or local levels. Next to the obvious willingness of decision-makers, their level of readiness is a crucial precondition for success. In this context, funding opportunities play an important role (Conrad and Hilchey, 2011; Litke and Day, 1998). CS and COs represent powerful and usually low-cost solutions to address existing gaps in environmental governance. These platforms can allow authorities to obtain evidence and provide citizens with opportunities to address environmental concerns. However, often, citizens participating in environmental governance are considered a 'threat' rather than a resource to decision-makers, since they are deemed to be in opposition to the plans of the authorities or industries. Citizen participation should rather be considered as a means to make environmental governance more transparent so that the citizens' trust in the conclusions of experts will increase. Here, the challenge lies in integrating CS in environmental decision-making in a manner that enhances the process by enabling it to deal with issues concerning the community appropriately and that at the same time takes into consideration the risks and opportunities that go along with these practices (Hakley, 2015).

4.4 Additional Requirements for Citizens' Observatories

Additionally to the challenges that have been mentioned so far, the establishment of COs is accompanied by a number of further development needs. COs usually have a similar structure; however, when starting a new CO, the whole infrastructure and data flow have to be installed from scratch (Liu et al., 2014). So far, there are no systematic, easy and reusable methods to do so. This causes an unsurmountable hurdle for institutions and organisations, as they usually lack the specific technical ICT and programming knowledge to create the required server infrastructure and mobile applications. As a result, organisations can fall back on old-fashioned, non-technological methods (which can take longer to implement) or spend tremendous amounts of their often limited budget on external ICT and programming experts (D'Hondt et al., 2014; Zaman et al., 2014).

Liu et al. (2014) have identified the following development needs to ensure a functional and operational CO with the active involvement of citizens:

A. The adequate promotion of a CO platform, including tools and activities for capacity building, awareness raising, recruiting and maintaining the participation of citizens;
B. A good understanding of the current and future societal demography in order to create COs that meet the actual and future needs of the population;
C. Building a long-lasting infrastructure, including open source software with the following requirements: use of open standards, easy exploitation through an open Application Programming Interface (API), and the ability to be widely accessed, extended and maintained. A CO should be seen rather as a generic environmental enabler than as a project-specific outcome;
D. Addressing and evaluating both citizens' views on certain environmental issues and their related actions ('Citizens' Voice') and the accountability of the governments for their environmental actions ('Accountability') in the social and political context of each CO (Fernandez-Gimenez et al., 2008). These two concepts should be actively promoted as important dimensions of good environmental governance, and that also in relation to the improvement of social justice (DFID, 2008; Kamar et al., 2012);
E. Developing tailor-made channels and mechanisms to enable citizens to actually influence environmental governance processes.

5 Conclusions

Engaging citizens in science and environmental observations is a challenging task. While many scientists are cautious about using data from volunteered observations, others believe that the quality of such data is sufficient to allow them to either use or publish the data while admitting that further work may be required before applying such data in other ways. However, we cannot say much about the quality of data from COs, as further research is still needed. The need for further research also applies to the validation processes, data integration and quality management. Merging citizen data with authoritative data and integration with other existing data may also be considered. Another way to improve data quality is to pay attention to the composition of the volunteer groups. In order to avoid imbalances and biases in the observations, the volunteers should be representative of different groups (e.g. different age, gender or cultural background groups, etc.). Applying co-design approaches in the design of the study/initiative can also be a useful way to maximise outputs of the observation process.

In order for citizens to participate in CS and CO initiatives, we have to create activities with low barriers and with incentives for citizens to both start participating and continue to do so. To succeed, we (the scientists) have to respect every volunteer and the role they play, manage their expectations and be transparent in our plans and actions. In addition, we must ensure to protect private data and create secure solutions. To the same degree, we have to respect and deal with the expectations, concerns and fears of public authorities in the same open and transparent manner. It is important to include and engage public authorities, where applicable, from the start to increase the chances of sustainable outcomes and solutions, and to influence their policies.

More can be done to promote citizen participation in environmental governance. With its latest Framework Programme for Research and Innovation, Horizon 2020, the EC is strongly promoting citizen engagement. Aiming to deepen the relationship between science and society and to reinforce public confidence in science, Horizon 2020 should foster the informed engagement of citizens and civil society in research and innovation by promoting science education, making scientific knowledge more accessible and developing responsible research, as well as innovation agendas that meet the actual concerns and expectations of citizens. In order to facilitate the participation of citizens in Horizon 2020, the engagement of citizens and civil society should be coupled with public outreach activities to generate and sustain public support for Horizon 2020 and beyond. Furthermore, EU research in this area often consists of top-down prescribed CO and CS programmes, which would need to be compatible with the existing bottom-up networks and the true data needs of citizens. Together, these top-down and bottom-up approaches allow us to minimise the differences and maximise the similarities among multiple systems, enabling both individual-case-study data analysis and integrated data analysis to be performed (Liu et al., 2014).

The growth in Web-based CS and COs and the use of mobile phones have opened many new opportunities for instrumental observations that can enhance the abilities of analysts to use this information for decision-making processes. Overall, policy-makers and government officials need to be aware that CS and COs, in the latter's new incarnation, are a phenomenon that will continue to grow and impact all levels of government. Each CS and CO activity will always involve trade-offs between inclusion of people, education, awareness of science and contribution to scientific research; the emerging examples from Europe show that, with appropriate multidisciplinary teams, it is possible, however, to achieve several of these goals in any given activity.

Another opportunity within COs is the potential for social innovation, novel partnerships and creating new opportunities for SMEs. This would meet the need for more cross-cutting and transdisciplinary activities that again would result in the creation of synergies and the facilitation of interoperability and coordination.

Whereas CS initiatives have had the chance to learn and undergo different changes through the course of the last decades, the concept of CO is rather young. Initiatives following this approach are still at an early stage and an honest discussion about their risks and opportunities needs to be carried out with citizens, scientists, authorities and other potential stakeholders in order to determine the full potential and areas of application of COs; only the future will show if our efforts were worth it.

Acknowledgements

The ideas presented here evolved from work undertaken in the context of studies funded under the CITI-SENSE and hackAIR projects. CITI-SENSE was a collaborative project partly funded by the EU FP7-ENV-2012 under Grant Agreement No. 308524. hackAIR is supported through the EU programme on 'Collective Awareness Platforms for Sustainability and Social Innovation' and funded through the Horizon 2020 Research and Innovation Programme under Grant Agreement No. 688363. We would like to thank, Steffen Fritz, the referee for this chapter, for his very valuable comments.

Previous publication

Figures 1 and 3 in this chapter have been modified from the following work: Citizen Participation Approaches in Environmental Health. In: Jozef M. Pacyna and Elisabeth G. Pacyna (eds.) *Environmental Determinants of Human Health* 2016, pp. 225–248, Sonja Grossberndt and Hai-Ying Liu.

Notes

[1] http://citizen-obs.eu/
[2] http://gbbc.birdcount.org/
[3] http://www.bigbutterflycount.org/
[4] http://www.openstreetmap.org
[5] http://airqualityegg.com/
[6] http://www.opalexplorenature.org/aboutOPAL and http://www.imperial.ac.uk/opal/
[7] http://www.geo-wiki.org
[8] https://www.zooniverse.org/
[9] http://vgibox.eu/
[10] http://cartography.tuwien.ac.at/emomap/
[11] http://www.everyaware.eu/
[12] http://www.citizencyberscience.net/
[13] http://www.opalexplorenature.org/aboutOPAL and http://www.imperial.ac.uk/opal/
[14] https://ecsa.citizen-science.net/

Reference list

Bonney, R., Ballard, H., Jordan, R., McCallie, E., Phillips, T., Shirk, J., Wilderman, C. C., 2009. *Public Participation in Scientific Research: Defining the*

Field and Assessing Its Potential for Informal Science Education. A CAISE Inquiry Group Report. Washington, D.C.: Centre for Advancement of Informal Science Education (CAISE).

Bottomley, R., 2014. The Role of Civil Society in Influencing Policy and Practice in Cambodia: Report for Oxfam Novib. Available at https://www.worldcitizenspanel.com/assets/The-Role-of-Civil-Society-in-Influencing-Policy-and-Practice-in-Cambodia.pdf [Last accessed 16 December 2016].

Brenna, B., 2011. Clergymen Abiding in the Fields: The Making of the Naturalist Observer in Eighteenth-Century Norwegian Natural History. *Science in Context* 24(2): 143–166.

British Ecological Society, 2013. Is there a role for the citizen scientist in policy making? Available at http://www.britishecologicalsociety.org/blog/2013/12/11/is-there-a-role-for-the-citizen-scientist-in-policy-making/ [Last accessed 30 May 2016].

Ceccaroni, L., Velickovski, F., Blaas, M., Wernand, M.R., Blauw, A., Subirats, L., 2016. Citclops Data Explorer: exploring water quality in the Wadden Sea, in: Berre, A.J., Schade, S., Piera, J. (Eds.) Proceedings of the Workshop "Environmental Infrastructures and Platforms 2015 - Infrastructures and Platforms for Environmental Crowd Sensing and Big Data", co-located with the European Citizen Science Association General Assembly 2015 (ECSA GA'2015). Barcelona, Spain, October 28–30 October, 2015.

Ciravegna, F., Huwald, H., Lanfranchi, V., Wehn de Montalvo, U., 2013. Citizen observatories: the WeSenseIt Vision, in: Proceeding of the Infrastructure for Spatial Information in the European Community (INSPIRE 2013). Florence, Italy, 23–27 June, 2013, 1–2.

Citclops, 2012–2015. Citizens' observatory for coast and ocean optical monitoring. Available at http://www.citclops.eu. [Last accessed 20 October 2015].

CITI-SENSE, 2012–2016. Development of sensor-based citizens' observatory community for improving quality of life in cities. Available at http://www.citi-sense.eu [Last accessed 20 October 2015].

COBWEB, 2012–2016. Citizen Observatory web. Available at http://cobweb-project.eu [Last accessed 20 October 2015].

Conrad, C.C., Hilchey, K.G., 2011. A review of citizen science and community-based environmental monitoring: issues and opportunities. *Environmental Monitoring and Assessment* 176: 273–291.

Conrad, C.T., Daoust, T., 2008. Community-based monitoring frameworks: increasing the effectiveness of environmental stewardship. *Environmental Management* 41: 358–366.

Department for International Development (DFID), 2008. *Citizens' voice and accountability evaluation. Mozambique country case study (final report). Evaluation report EV688*. Available at https://www.gov.uk/government/publications/citizens-voice-and-accountability-evaluation-mozambique-country-case-study-final-report-ev688 [Last accessed 1 May 2017].

D'Hondt, E., Zaman, J., Philips, E., Gonzalez Boix, E., De Meuter, W., 2014. Orchestration Support for Participatory Sensing Campaigns, in: Proceedings of the 2014 ACM International Joint Conference on Pervasive and Ubiquitous Computing (UbiComp 2014). ACM, New York, USA, 13 September, 2014, pp., 727–738.

Dickinson, J.L., Zuckerberg, B., Bonter, D.N., 2010. Citizen Science as an Ecological Research Tool: Challenges and Benefits. *Annual Review of Ecology, Evolution, and Systematics* 41: 149–172.

EC (European Commission), 2001. European Governance – A White paper. 2001/C 287/01. Available at http://europa.eu/rapid/press-release_DOC-01-10_en.htm [Last accessed 1 May 2017].

EC (European Commission), 2003. Directive 2003/35/EC of the European Parliament and of the Council providing for public participation in respect of the drawing up of certain plans and programmes relating to the environment and amending with regard to public participation and access to justice Council Directives 85/337/EEC and 96/61/EC. Available at http://eur-lex.europa.eu/resource.html?uri=cellar:4a80a6c9-cdb3-4e27-a721-d5df1a0535bc.0004.02/DOC_1&format=PDF [Last accessed 1 May 2017].

EC (European Commission), 2014. CORDIS – Programmes: Developing community-based environmental monitoring and information systems using innovative and novel earth observation applications. Available at http://cordis.europa.eu/programme/rcn/18152_en.html [Last accessed 10 May 2016].

EC (European Commission), 2015. What is MYGEOSS. Available at http://digitalearthlab.jrc.ec.europa.eu/mygeoss/ [Last accessed 12 May 2016].

EC (European Commission), 2015–2016. TOPIC – Demonstrating the concept of 'Citizen Observatories'. Available at https://ec.europa.eu/research/participants/portal/desktop/en/opportunities/h2020/topics/sc5-17-2015.html [Last accessed 12 May 2016].

EC (European Commission), 2016. CAPS projects. Available at https://ec.europa.eu/digital-single-market/node/66639 [Last accessed 12 May 2016].

EC (European Commission), 2016–2017a. TOPIC – Novel in-situ observation systems. Available at http://ec.europa.eu/research/participants/portal/desktop/en/opportunities/h2020/topics/sc5-18-2017.html [Last accessed 12 May 2016].

EC (European Commission), 2016–2017b. TOPIC – Coordination of citizens' observatories initiatives. Available at http://ec.europa.eu/research/participants/portal/desktop/en/opportunities/h2020/topics/sc5-19-2017.html [Last accessed 12 May 2016].

EC (European Commission), 2016–2017c. TOPIC – Collective Awareness Platforms for Sustainability and Social Innovation. Available at http://ec.europa.eu/research/participants/portal/desktop/en/opportunities/h2020/topics/5070-ict-11-2017.html [Last accessed 12 May 2016].

EC (European Commission), 2017. TOPIC – Integrating Society in Science and Innovation – An approach to co-creation. Available at http://ec.europa.eu/research/participants/portal/desktop/en/opportunities/h2020/topics/2264-swafs-13-2017.html [Last accessed 12 May 2016].

EC JRC (European Commission Joint Research Centre), 2014. Citizen Science and Smart Cities. Available at http://publications.jrc.ec.europa.eu/repository/bitstream/JRC90374/lbna26652enn.pdf [Last accessed 12 May 2016].

Engelken-Jorge, M., Moreno, J., Keune, H., Verheyden, W., Bartonova, A., 2014. CITISENSE consortium: Developing citizens' observatories for environmental monitoring and citizen empowerment: challenges and future scenarios, in: Parycek P, Edelmann N. (eds.) Proceedings of the Conference for E-Democracy and Open Government (CeDEM14). Danube University Krems, Austria 21–23 May 2014, 49–60.

European Citizen Science Association (ECSA), 2016. Available at http://ecsa.citizen-science.net/ [Last accessed 12 May 2016].

Evans, C., Abrams, E., Reitsma, R., Roux, K., Salmonsen, L., Marra, P.P., 2005. The neighborhood nestwatch program: participant outcomes of a citizen-science ecological research project. *Conservation Biology* 19:589–594.

Fernandez-Gimenez, M.E., Ballard, H.L., Sturtevant, V.E., 2008. Adaptive management and social learning in collaborative and community-based monitoring: a study of five community-based forestry organizations in the western USA. *Ecology and Society* 13:4.

Feyerabend, P.K., 1970. Against method: online of an anarchistic theory of knowledge. *Minnesota Studies in the Philosophy of Science* 4:17–130.

Goodchild, M.F., Li, L., 2012. Assuring the quality of volunteered geographic information. *Spatial Statistics* 1:110–120.

Gottschalk Druschke, C., Seltzer, C.E., 2012. Failures of Engagement: Lessons Learned from a Citizen Science Pilot Study. *Applied Environmental Education & Communication* 11(3–4): 178–188. DOI: https://doi.org/10.1080/1533015X.2012.777224.

Grossberndt, S., Liu, H.-Y., 2016. Citizen participation approaches in environmental health. In Pacyna, J., Pacyna, E. (Eds.), *Environmental Determinants of Human Health*. London, UK: Springer, pp. 225–248.

Haklay, M., 2015. *Citizen science and policy: a European perspective*. Washington, DC: Woodrow Wilson International Centre for Scholars.

Hanahan, R.A., Cottrill, C., 2004. A comparative analysis of water quality monitoring programs in the southeast: lessons for Tennessee. Available at http://isse.utk.edu/wrrc/programsprojects/pdfs/mainbook.pdf [Last accessed 12 May 2016].

Higgins, C.I., Williams, J., Leibovici, D.G., Simonis, I., Davis, M.G., Muldoon, C., O'Grady, M., 2016. Generic Infrastructure Platform to Facilitate the Collection of Citizen Science data for Environmental Monitoring. *International Journal of Spatial Data Infrastructures Research* 11, 20–48.

Kamar, E., Hacker, S., Horvitz, E., 2012. Combining human and machine intelligence in large-scale crowdsourcing, in: Proceedings of the 11th International Conference on Autonomous Agents and Multiagent Systems (AAMAS 2012). Valencia, Spain, 4–8 June 2012, pp. 467–474.

Kullenberg, C., Kasperowski, D., 2016. What Is Citizen Science? – A Scientometric Meta-Analysis. *PLoS ONE* 11(1): e014. 7152. DOI: https://doi.org/10.1371/journal.pone.0147152.

Kviner, M., 2012. Review highlights role of citizen science projects. Available at http://www.bbc.com/news/science-environment-20445296 [Last accessed 12 May 2016].

Lanfranchi, V., Wrigley, S.N., Ireson, N., Ciravegna, F., Wehn, U., 2013. Citizens' observatories for situation awareness in flooding, in: Hiltz, S.R., Pfaff, M.S., Plotnick, L., Shih, P.C. (Eds.) *Proceedings of the 11th International ISCRAM Conference*. University Park, Pennsylvania, USA, 19–22 May 2014, pp. 145–154. Available at: http://www.iscram.org/legacy/ISCRAM2014/papers/p40.pdf [Last accessed 16 May 2017].

LeBaron, G.S., 2015. The 115th Christmas bird count. Available at https://www.audubon.org/news/the-115th-christmas-bird-count-0 [Last accessed 12 May 2016].

Litke, S., Day, J.S., 1998. Building local capacity for stewardship and sustainability: the role of community-based watershed management in Chilliwack, British Colombia. *Environments* 25: 91–110.

Liu, H-Y., Kobernus, M., 2017. Citizen science and its role in the sustainable development. In Ceccaroni, L., Piera, J. (Eds.), *Analysing the Role of Citizen Science in Modern Research. Advances in Knowledge Acquisition, Transfer, and Management*. Hershy, PA, USA: IGI Global, pp. 147–167.

Liu, H-Y., Kobernus, M., Broday, D., Bartonova, A., 2014. A conceptual approach to a citizens' observatory – supporting community-based environmental governance. *Environmental Health* 13:107.

Miller-Rushing, A., Primack, R., Bonney, R., 2012. The history of public participation in ecological research. *Frontiers in Ecology and the Environment* 10(06):285–290, DOI: https://doi.org/10.1890/110278.

Omniscientis, 2012–2014. Odour monitoring and information system based on citizen and technology innovative sensors. Available at http://www.omniscientis.eu/ [Last accessed 1 May 2017].

Oxford English Dictionary (OED), 2014. Third Edition. Available at http://www.oed.com/view/Entry/33513?redirectedFrom=citizen+science#eid316619123 [Last accessed 12 May 2016].

Prestopnik, N., Crowston, K., Wang, J., 2014. Exploring Data Quality in Games with a Purpose, in: *Proceedings of the iConference 2014*, Berlin, Germany, 4–7 March 2014, pp. 213–228. DOI: https://doi.org/10.9776/14066. Available at: https://www.ideals.illinois.edu/bitstream/handle/2142/47311/066_ready.pdf?sequence=2&isAllowed=y [Last accessed 12 May 2017].

Sauermann, H., Franzoni, C., 2015. Crowd science user contribution patterns and their implications. *PNAS* 2015 112 (3) 679–684; published ahead of print January 5, 2015, DOI: https://doi.org/10.1073/pnas.1408907112.

Science Communication Unit, University of the West of England, Bristol, 2013. *Science for Environment Policy In-depth Report: Environmental Citizen Science.* Report produced for the European Commission DG Environment, December 2013. Available at: http://ec.europa.eu/science-environment-policy [Last accessed 12 May 2016].

SciStarter, 2017. https://scistarter.com/about. [Last accessed 04 July 2017].

Shirk, J. L., Ballard, H. L., Wilderman, C. C.,Phillips, T., Wiggins, A., Jordan, R., McCallie, E., Minarchek, M., Lewenstein, B. V.,Krasny, M. E., Bonney, R., 2012. Public participation in scientific research: a framework for deliberate design. *Ecology and Society* 17(2): 29. http://dx.doi.org/10.5751/ES-04705-170229. Accessed 16 December 2016.

Silvertown, J., 2009. A new dawn for citizen science. *Trends in Ecology & Evolution* 24 (9): 467 – 471.

Socientize, 2012–2014. Society as e-Infrastructure through technology, innovation and creativity. Available at http://www.socientize.eu/ [Last accessed 12 May 2016].

Tulloch, A.I.T., Possingham, H.P., Joseph, L.N., Szabo, J., Martin, T.G., 2013. Realising the full potential of citizen science monitoring programs. *Biological Conservation* 165:128–138.

Tweddle, J. C., Robinson L. D., Pocock, M. J., Roy, H. E., 2012. *Guide to citizen science: developing, implementing and evaluating citizen science to study biodiversity and the environment in the UK.* Natural History Museum and Centre for Ecology and Hydrology.

United Nations Economic Commission for Europe (UNECE), 1998. Convention on Access to Information, Public Participation in Decision-Making and Access to Justice in Environmental Matters. Available from: https://www.unece.org/fileadmin/DAM/env/pp/documents/cep43e.pdf [Last accessed 16 May 2017].

University of the West of England (UWE), 2014. *Science for Environment Policy In-depth Report: Environmental Citizen Science.* Report produced for the European Commission DG Environment, February 2014. Available at: http://ec.europa.eu/environment/integration/research/newsalert/pdf/IR10_en.pdf [Last accessed 12 May 2016].

University of the West of England (UWE), 2013. *Science for Environment Policy In-depth Report: Environmental Citizen Science.* Report produced for the European Commission DG Environment, December 2013. Available at: http://ec.europa.eu/science-environment-policy [Last accessed 12 May 2016].

Vetter, J., 2011. Lay observers, telegraph lines, and Kansas weather: the field network as a mode of knowledge production. *Science in Context* 24:259–280.

Walters, L.C., Aydelotte, J., Miller, J., 2000. Putting more public in policy analysis. *Public Administration Review* 60(4):349–359.

Wehn, U., Evers, J., 2015. The social innovation potential of ICT-enabled citizen observatories to in-crease eParticipation in local food risk management. *Technology in Science* 2015 42: 187–198.

WeSenseIt, 2012–2016. Citizen water observatories. Available at http://www.wesenseit.com [Last accessed 20 October 2015].

Whitelaw, G., Vaughan, H., Craig, B., Atkinson, B., 2003. Establishing the Canadian community monitoring network. *Environmental Monitoring and Assessment* 88:409–418.

Zaman, J., D'Hondt, E., Gonzalez Boix, E., Philips, E., Kambona, K., De Meuter, W., 2014. Citizen-Friendly Participatory Campaign Support, in: Proceedings of 2014 IEEE International Conference on Pervasive Computing and Communications Work in Progress (PerCom WiP'14). Budapest, Hungary, 24–28 March, 2014, Pages 232 - 235.

Zikopoulos, I.B.M.P., Eaton, C., Zikopoulos, P., 2011. *Understanding Big Data: Analytics for Enterprise Class Hadoop and Streaming Data.* McGraw-Hill Professional, New York, USA.

Zooniverse, 2013. Purpose. Zooniverse. Available at http://www.zooniverse.org/about [Last accessed 1 May 2017].

CHAPTER 16

The Future of VGI

Vyron Antoniou*, Linda See†, Giles Foody‡, Cidália Costa Fonte§, Peter Mooney¶, Lucy Bastin‖,**, Steffen Fritz§, Hai-Ying Liu††, Ana-Maria Olteanu-Raimond‡‡ and Rumiana Vatseva§§

*Hellenic Army General Staff, Geographic Directorate, PAPAGOU Camp, Mesogeion 227-231, Cholargos, 15561, Greece, v.antoniou@ucl.ac.uk
†International Institute for Applied Systems Analysis (IIASA), Schlossplatz 1, 2361 Laxenburg, Austria
‡School of Geography, University of Nottingham, UK
§Department of Mathematics, University of Coimbra, 3001-501 Coimbra, Portugal / INESC Coimbra, Rua Sílvio Lima, Pólo II, 3030-290 Coimbra, Portugal
¶Maynooth University, Maynooth, Ireland
‖European Commission, Joint Research Centre, Ispra, Italy
**Aston University, Birmingham UK
††Norwegian Institute for Air Research (NILU), Kjeller 2027, Norway
‡‡Paris-Est, LASTIG COGIT, IGN, ENSG, F-94160 Saint-Mande, France.
§§National Institute of Geophysics, Geodesy and Geography, Bulgarian Academy of Sciences, Bulgaria

Abstract

In this final chapter, we speculate on future developments in the field of Volunteered Geographic Information (VGI); we focus on how VGI will be affected by future technological developments, but we also consider issues such as VGI

How to cite this book chapter:
Antoniou, V, See, L, Foody, G, Fonte, C C, Mooney, P, Bastin, L, Fritz, S, Liu, H-Y, Olteanu-Raimond, A-M and Vatseva, R. 2017. The Future of VGI. In: Foody, G, See, L, Fritz, S, Mooney, P, Olteanu-Raimond, A-M, Fonte, C C and Antoniou, V. (eds.) *Mapping and the Citizen Sensor.* Pp. 377–390. London: Ubiquity Press. DOI: https://doi.org/10.5334/bbf.p.

quality, the relationship of VGI with science and citizens, and the impact of VGI in future cities and societies.

Keywords

Future of VGI, technology, Digital Earth, Smart Cities, citizen science, legal and ethical concerns

1 Introduction

Katherine is a typical citizen of the future. The year is 2030. Like most mornings, Katherine gets up and goes for a run, wearing sensors embedded in her clothes. These sensors monitor her vital signs and communicate with her smartphone, alerting her of anything unusual. With her permission, the sensors also send the data to many different places, including to her medical records, her health insurance company and a vast supercomputing facility, which uses her Volunteered Geographic Information (VGI), along with that of millions of other citizens, to uncover behavioural and health patterns that can be used to provide doctors with preventative health care advice. Before going to work, Katherine controls the environment of her house using her smartphone; this VGI gets sent to her gas and electricity companies, who use the data to bill her, but also to determine customer behaviour so that they can optimise their tariffs and provide customers like Katherine with advice on how to save money while being environmentally friendly. Katherine's driverless electric car takes her to work, where she is a spatial data quality expert at the National Mapping Agency (NMA) in her country. She is responsible for the quality assurance and quality control of the NMA's spatial databases. Today she is focused on doing some routine quality assurance on the main topographic database, which is a dynamically updated set of layers that takes in changes from a range of users, including citizens. She does some checks to ensure that the automated quality assurance procedures are filtering out data that do not meet the minimum requirements for the database and determines where to send field surveyors to confirm any critical changes. Today is Friday and Katherine is looking forward to attending a weekend mapping party, which will focus on helping another country build up their own, quality assured topographic database with seamless input from experts like her, interested citizens, businesses and non-governmental organisations on the ground.

This vision of a future world in which Katherine lives is not that far away and many of these things are already happening, even if only on a small scale at present. Although providing longer term predictions about VGI is a challenge because VGI is heavily reliant on rapidly changing technologies, it is clear that the role of citizen sensors is likely to become much more prominent than it is today. It is anticipated that citizen-derived data will grow considerably and be used in increasingly diverse ways in the near future. The amount of spatial data available is increasing exponentially (Craglia and Shanley, 2015), and the diversity of data sources and types is also increasing, e.g. through current trends such as Digital Earth (Craglia et al., 2012), Smart Cities (Batty et al., 2012), Citizen Science (Bonney et al., 2009; Silvertown, 2009), the Internet of Things (IoT; Ashton, 2009) and Data Analytics (Kitchin, 2013). Thus, this chapter will attempt to examine the relationship between VGI and a number of these current technological trends. We also consider VGI quality, which will continue to be one of the most important obstacles for the future diffusion of VGI, as well as legal and ethical concerns.

2 Technology

VGI has been heavily based on advances in the information and communication technology (ICT) domain. Web 2.0 applications (O'Reilly, 2007), GPS-enabled devices and the open availability of very fine spatial resolution satellite sensor imagery, sensor-equipped portable devices and smartphones have all been growth drivers for crowdsourced spatial data. Thus, it is expected that future advances in these areas will continue to play a major role in the future of VGI.

As an initial technological consideration, it can be noted that the basic infrastructure, such as Internet availability, bandwidth and processing power, has an important role to play; such infrastructure examples are all expected to evolve considerably and thus to greatly affect both the number of people online and the quality of connectivity and communication. Based on what we have experienced during the last few decades, it is safe to say that the way in which people are connected online will move to a totally new level.

The continuing developments in location-aware, data capturing devices are likely to impact greatly on the future of VGI. The removal of the selected availability of the GPS signal (Clinton, 2000) has led to the proliferation of GPS-enabled sensors in even low-cost everyday devices. Thus, location-enabled devices are now everywhere, from smartphones and cameras in our pockets to cars, airplanes and ships around the world. However, there is a clear distinction to be made: on the one hand, there are human-controlled devices that collect data in relation to an individual's activity, while, on the other hand, there are sensors that constantly collect and transmit location-aware data about a phenomenon. Regarding the former, our generation has witnessed the appearance of mobile phones, which then evolved into smartphones and have now been transformed into location-capturing devices; when combined with web applications and social networking, the volumes of data created are immense. There are many examples of Web-based applications, such as Facebook, Flickr, Foursquare, etc., where the data come from the conscious use of these applications but the geographic information (GI) is generated implicitly by the users without the original aim of actually creating geospatial datasets. This can be distinguished from the proliferation of all kinds of sensors that passively collect spatial data, mostly in an urban context. From high-end sensors to do-it-yourself, low-cost devices based on hardware platforms such as Raspberry Pi and Arduino, the flow of sensor-recorded location data is expected to increase. All these connected sensors are part of the vision of the IoT. Widespread sensor networks may dominate the urban fabric initially, but then expansion to a global-wide sensor network would be a natural continuation of this trend in sensor technology.

While the human-controlled and sensor network data sources of GI have, up until now, been working in a complementary way, this situation could also change in the future. A key question is whether developments in ubiquitous sensing will lead to a decline in human-collected VGI. For example, to know how people are moving inside a city, will it be necessary to tap into data from wearable technology if we can use sensors to automatically count the number of people crossing every street in every city? Will we need people to measure air quality (Goodchild, 2007) or make noise-maps (Foerster et al., 2010) if we have low-cost air and noise sensors located on every street corner? Moreover, sensor-collected data will not suffer from some of the quality issues or biases that usually accompany human-collected VGI. Some technologies may, however, rely on VGI to function properly or to realise their full potential. Take, for example, smart thermostats, which are intended to learn over time and make adjustments that improve the efficiency of heating/cooling systems while maximising the comfort of users. Such connected devices or sensors of the IoT require some active human intervention and thus will always involve some form of VGI. Many more electronic devices of this nature are expected to emerge in the near future.

Technological trends also cover advances in software and algorithms. It is likely that the technology for handling large and complex datasets will advance in ways that will more fully exploit the use of VGI. Data quality is a major issue related to VGI at present, so it is likely that in the future we will develop new, sophisticated algorithms to address biases and quality issues that arise from the spatial distribution of participation (see e.g. Haklay, 2010; Antoniou, 2011; Barron et al., 2014). This will reveal the areas and feature types that suffer more in terms of quality and thus need more directed attention from volunteers. Just imagine a map with the following stated differences in scale, and hence in positional accuracy, due to heterogeneous citizen contributions: 'in urban areas roads are of scale 1:5,000, buildings are of scale 1:25,000 and land cover is of scale 1:50,000, but in rural areas land cover is of scale 1:10,000, roads are of scale 1:25,000 and buildings are of scales 1:10,000; urban areas are more complete than rural ones'. One could imagine similar caveats regarding thematic accuracy. It is, therefore, anticipated that VGI projects, based on this algorithmic evaluation of quality, will want to guide their contributors to specific areas or spatial feature types in order to counterbalance any recorded biases (see for example how Geograph[1] informs its contributors). However, it is uncertain what this 'algorithmic management' (Lee et al., 2015) will do to VGI. On one hand, it may greatly enhance the quality and thus the acceptance by a broader audience of VGI. On the other hand, if this results in removing features such as the freedom of expression, fun and intuitiveness from the contribution process, this may severely curtail VGI as a phenomenon in the future.

In summary, technology will continue to evolve, and VGI will certainly continue to leverage technological advances. Strong indications of what the near future will bring are already visible. Indoor positioning and mapping devices (see for example Google's Tango project[2]) will bring VGI into built-up areas. Drones are becoming increasingly popular and we are still exploring their potential as a source of data for many different fields, from humanitarian applications to land cover and elevation mapping. Finally, wearable technology, which is still at an early stage, is expected to become ubiquitous and will vastly multiply the amount of spatial data on the Web. These are just a few examples of what the future holds, and they have the potential to vastly influence and shape the field of VGI.

3 VGI, Smart Cities and Digital Earth

Both the growth of VGI and the evolution of technology have pushed forward the initiatives of Smart Cities and Digital Earth. The transformation of our living environment into a smart, interconnected place will lead to a more detailed recording, and hence a better understanding, of the spatial-temporal pattern of human activity. As Roche (2014) points out, the future of smart cities will probably be spatially enabled and develop new spatial skills. Thus, if we better understand the structure of future cities and of the human activities taking place within them, we will also be better placed to understand the role of VGI within them.

Spatially enabling our cities is easier said than done but will very soon prove to be a priority. According to the United Nations Environment Program (n.d.), while cities will cover only 3% of the Earth's inhabited land area by 2050, almost 80% of the population on the globe will live in cities, which will account for 75% of the total energy consumed and 60–80% of Greenhouse Gas (GHG) emissions. It is easy for anyone to understand that sustainability is one of the most important, yet elusive, societal concerns. However, if we do not want to lower our living standards, then improvements in urban functions will become a necessity. To this end, geospatial data and particularly VGI can be a valuable input. Urban planners, authorities, local administrations, NGOs and active communities can benefit from detailed, up-to-date, timely and freely available GI. A list of examples of how VGI is used by governments and authorities is provided in Haklay et al. (2014), where the added value of using VGI alone or in combination with authoritative data to improve resource allocation, efficiency and transparency is presented.

While technology will continue to play an important role in Smart Cities, human capital is equally fundamental to city intelligence. Spatially literate citizens are needed both to embrace new developments and to push for innovative solutions. To this end, VGI has much to offer now, and even more so in the future. Ubiquitous crowdsourced spatial information can serve as the baselayer on top of which all future 'smart' functionalities of a city could develop.

4 VGI Quality

Although VGI has been a growing phenomenon for over a decade now (Capineri et al., 2016; See et al., 2016), one of the major factors that hinders the more widespread diffusion and uptake of VGI is the lack of a robust and standardised way to evaluate data quality, as outlined in Chapter 7 by Fonte et al. (2017). VGI could both facilitate and accelerate the transition to Smart Cities and Digital Earth if it were credible enough to trust and hence use in applications that require accurate GI. However, this quest for trust, fitness-for-purpose and usability of VGI data comes down to implementing or devising tangible ways of measuring and reporting VGI quality. Without concrete knowledge of the state of a VGI dataset, its use might end up being a leap of faith that no serious stakeholder is willing to take. Yet if the quality requirements for VGI are too stringent in terms of data specifications, precision, update cycles, spatial coverage or metadata, then we may end up discouraging volunteers. At the same time, we need to avoid the situation whereby VGI is considered to be 'laypeople's data' of de-facto inferior quality, full of biases, with no metadata and only occasional respect for protocols and best practices; such a development would disrupt the momentum and the dynamic that VGI has developed so far and will mark this kind of data out as marginal or as a cheap and untrustworthy replacement for authoritative datasets. It is important to note that VGI is already sometimes as good as, if not superior to, authoritative data and can even exceed the quality requirements of NMAs for common mapping applications (Olteanu-Raimond et al., 2017).

For these reasons, the evaluation of VGI data quality has been a hot topic in academia (see e.g. Haklay et al., 2010; Bégin et al., 2013; Antoniou and Skopeliti, 2015; Foody et al., 2015; Senaratne et al., 2016; Fonte et al., 2017), and research on this topic will continue in the future, not least because improving the methods for reporting quality could end up becoming a catalyst for the widespread diffusion of VGI in mainstream geomatics engineering. Well established methods for spatial data quality evaluation (e.g. ISO specifications), while still valid, need to be supplemented with additional evaluation tools that take the specific nature of VGI into account (Antoniou and Skopeliti, 2015; Fonte et al., 2017). If adequate quality assurance tools and algorithms fail to materialise, then the future uses of VGI might not expand much beyond what we see today. That said, VGI is highly interdisciplinary, combining underlying social, economic and technological factors within the geospatial domain; the result is the recording of space and phenomena based on what citizens perceive to be important. Thus, uncertainty, biases and noise in the data might never be fully eliminated. Instead, we need to understand, model and handle these issues so that VGI can be used effectively.

Future efforts might focus on data harmonisation, which can play an important role in the era of big data since it may enable data comparison, allowing the application of the law of large numbers, i.e. the tendency to arrive at the expected value by averaging the results obtained from repeating an experiment a large number of times (Kuhn, 2007), and contribute to an automated and fast preliminary data quality assessment and even data conflation. To address the availability of multiple sources that may potentially be useful, methodologies need to be developed to assist users in choosing the right dataset or the right combination of datasets for each application. Decisions such as these will be aided by the provision of information about the data, and hence metadata are likely to become increasingly important accompaniments of citizen-derived datasets. Given the huge amount of VGI foreseen in the future, it is likely that there will be a focus on the development of approaches that are more automated for the assessment of VGI quality; this development will be challenging given the greatly varied nature of the data, which can be unstructured and heterogeneous, but is nevertheless of high potential value.

5 VGI in Science

Despite VGI quality being an obstacle to the larger diffusion of crowdsourced data in everyday applications, there has been considerable use of VGI in scientific research, in particular in citizen science projects. Citizen science typically refers to the involvement of citizens in scientific research, either in collaboration with or under the direction of professional scientists (Silvertown, 2009). A considerable number of such projects actively use geospatial or geotagged data. Citizens usually use smartphones, cheap do-it-yourself devices or more advanced purpose-built sensors to observe or measure a phenomenon associated with geographic information on a volunteered basis.

Large-scale scientific projects that need a regional or even global-wide spatial coverage are now feasible via the power of the crowd. In fact, any project of such scale needs to seek assistance from the crowd in order to collect the volumes of data needed for research. Examples include the Christmas Bird Count[3], Asteroid Zoo[4] or iNaturalist[5]. Apart from simple data collection, people participating in citizen science projects might get more involved in the analysis of the data or in the interpretation of the results; for an analysis on the typology of participation see Haklay (2013). This increasing trend in citizen participation in citizen science projects will most likely continue in the future, particularly given the success of many different citizen science projects and the active interest shown by authorities such as the European Union in building citizen observatories. This trend is also an important development for VGI on many levels. First, as more and more citizens get actively involved in scientific projects at a local or global scale, collaboration and volunteerism will become stronger. Also, involvement in science has much to teach enthusiastic but untrained contributors of VGI. If we start considering VGI observations and measurements as scientific ones, then following rigorous data protocols for production and evaluation, explicitly documenting measurements with metadata, and the ability to replicate results may become more important for VGI projects; in some cases it may even become obligatory, as with many current citizen science projects.

6 VGI, Citizens and Societies

Throughout the book, it has been repeatedly shown that the driving force of VGI is volunteers and their modes of engagement. Although technological advancements provide the means for novel ways of ubiquitous data capturing, what transforms the technological means into a global-wide phenomenon that challenges the fundamentals of the geospatial domain is the role of citizens and their engagement with volunteered contributions of location-based data. Consequently, the future of VGI is closely related to the future of social trends and social evolution.

Crowdsourcing, volunteerism, active communities, citizen science and social enterprises are early formations that can take the lead in the sustainable production of VGI. If such social initiatives evolve further, gain momentum and become commonplace, then the bottom-up production of geotagged data will rise to entirely new levels. For example, it is worth noting how online communities in citizen science projects address real-world problems. Similar examples exist in the VGI sphere, and can be found in the efforts of the Humanitarian OpenStreetMap Team (HOT), which mobilises volunteers in mapping areas that have been hit by natural disasters. Interestingly, such grassroots collaboration overcomes societal barriers and enables citizens to participate in the management and improvement of quality of life, a common goal of visions such as Digital Earth and Smart Cities.

A really intriguing, and equally interesting, future development might arise if we consider location and spatial information as common goods (Roche et al., 2012) that are mainly produced and maintained by people. What changes will this generate in our society? What will be the benefits to and responsibilities of the citizens and the authorities? For instance, we will need to steer future societies into geospatial crowdsourcing, understand its value, its benefits, its potential and the steps that we need to take in order to create and sustain spatial infrastructures. Consequently, citizens should be initiated and trained into the world of geospatial information from the early years of their education. Geography curricula and lessons should be redesigned to include the collection of geotagged information in a volunteered and collaborative mode. There are already excellent examples available to provide initial best practice. These include the activities of the Finnish Environment Institute and the Finnish National Land Survey Agency, which have introduced citizen science and crowdsourced data collection in elementary schools, the Muséum National d'Histoire Naturelle in France, which introduced collaborative science on biodiversity into French schools, or the positive experiences of the Dutch Kadaster, which introduced a new curriculum on crowdsourcing and mapping in elementary schools.

It should be noted, however, that future developments in citizen sensing may require greater consideration of the citizen as well as the end use of the data generated. A greater understanding of citizen sensors is required as there is a two-way dialogue between those using and contributing the VGI, especially as citizens may also be the source of very useful ideas. Feedback to citizen contributors is likely to become much more important, especially in developing the citizens' skills and maintaining motivation. A new reality in which the role of geospatial information is highlighted, which renders its collection and maintenance a common responsibility, might prove a very efficient way to secure the motivation and long-term engagement from large parts of the population that is needed to support global-wide geospatial data collection.

7 Understanding the True Value of VGI

Much of the literature on VGI is about understanding this phenomenon. The subjects examined range from the motivation behind volunteered contributions, the quality of the data obtained or the biases that VGI datasets might possess to the integration of VGI with other sources of data. Little has been written about the true value of VGI. By 'true value', we refer to what VGI has offered not only to the geomatics domain but also to people and society as a whole.

The bottom-up production of VGI has democratised the production and use of GI. VGI has changed a landscape where spatial data creation was once the responsibility and privilege of a few governmental agencies or large corporations (e.g. NMAs), and where the access to spatial information was limited and usually very expensive for the public. What VGI did, and probably will continue to do in the future, was to create a closer relationship between the public, on the one hand, and geography, cartography, web mapping and geospatial applications, on the other hand; in a sense, the public have been introduced to the value of GI. The omnipresence of GI in everyday devices and the multiple applications and services offered today that are based on spatial data would not have been possible without this new, enlightened relationship. Moreover, there is a constantly increasing demand for more GI, both in terms of quantity and detail. As VGI has, in a sense, spatially enabled our societies, the need for more data of this nature will only intensify in the future. Now, for the first time, it is possible to have a tangible picture of how people understand space, what matters to them and what they think needs to be on a map. The horizon of what GI should cover has been considerably broadened, ranging from the mapping of litter[6], noise pollution (Maisonneuve et al., 2010) and other relevant urban problems[7] to the support of Smart Cities and a wealth of other applications. This information is valuable for understanding how societies function and what we need to do in the future to help improve them.

8 Future Legal and Ethical Concerns

The importance of legal and ethical issues has already been raised in Chapter 6 by Mooney et al. (2017), but much more attention will need to be given to these issues in the future. It is anticipated that VGI will increasingly be harvested from diverse sources including social media and wearable devices. While potentially yielding vast amounts of useful VGI, including information about human location, movement and behaviour, this comes with a suite of data privacy, ethical and legal concerns. These are complex issues, since legislation tends to lag behind advances in technology and also differs from country to country. There are also serious concerns with the reuse of VGI; in many instances, especially when it is mined from open resources, VGI may be used for different applications than the original purpose of data collection, which some volunteers may be uncomfortable with. As the ability to integrate and fuse together greater numbers of complex and disparate datasets increases, it is of crucial importance that the issue of data reuse be addressed. Data reuse also links to legal concerns; for example, if the VGI was acquired by digitising from a map or image without the relevant permissions, what are the implications for those that reuse the VGI? Equally important are possible cases of vandalism. Intentional deterioration of the quality of a VGI dataset or the insertion of false data could have considerable ramifications if the data are then used in decision-making or policy implementation. It is anticipated that in the future, as VGI gains momentum, there will be a need to better safeguard the integrity and objectivity of this data source.

9 The Final Word

This is a time of very rapid change – in the last decade the geomatics domain has witnessed unprecedented growth. GI has moved from the control of a few producers to the hands of many, who now have the power to produce and update many different spatial data repositories. At the same time, demand for timely, free and accurate GI is multiplying. Whether from the move to a digitised environment or from the frequent use of map-based applications, the value of GI has been widely recognised by many. VGI has been a catalyst for these changes, but we are currently standing at a very important crossroads: either VGI will move to a new level in which it will be the key enabling factor for future developments or it will remain at current levels of acceptance, running the danger of being overtaken by developments in other domains, and possibly even decline or decay. The responsibility for what happens is, at least partially, in the hands of GI professionals as well as citizens. Fortunately, networks such as COST Action TD1202[8], out of which this book has arisen, are succeeding in bringing together an interdisciplinary community including professionals from NMAs. By working together to address VGI quality issues and potential dangers to the field of VGI, we will strive to ensure that VGI has a strong and exciting future.

Notes

1. http://www.geograph.org.uk/
2. https://get.google.com/tango/
3. http://www.audubon.org/conservation/science/christmas-bird-count
4. https://www.asteroidzoo.org/
5. http://www.inaturalist.org/
6. http://www.litterati.org/
7. https://www.fixmystreet.com/
8. http://www.citizensensor-cost.eu/

Reference list

Antoniou, V., 2011. User Generated Spatial Content: An Analysis of the Phenomenon and its Challenges for Mapping Agencies. Unpublished PhD Thesis. University College London (UCL), London, UK.

Antoniou, V., Skopeliti, A., 2015. Measures and indicators of VGI quality: An overview, in: ISPRS Annals of the Photogrammetry, Remote Sensing and Spatial Information Sciences. Presented at the ISPRS Geospatial Week 2015, ISPRS Annals, La Grande Motte, France, pp. 345–351.

Ashton, K., 2009. That "internet of things" thing. *RFiD Journal* 22, 97–114.

Barron, C., Neis, P., Zipf, A., 2014. A comprehensive framework for intrinsic OpenStreetMap quality analysis. *Transactions in GIS* 18, 877–895. DOI: https://doi.org/10.1111/tgis.12073

Batty, M., Axhausen, K.W., Giannotti, F., Pozdnoukhov, A., Bazzani, A., Wachowicz, M., Ouzounis, G., Portugali, Y., 2012. Smart cities of the future. *The European Physical Journal Special Topics* 214, 481–518. DOI: https://doi.org/10.1140/epjst/e2012-01703-3

Bégin, D., Devillers, R., Roche, S., 2013. Assessing Volunteered Geographic Information (VGI) quality based on contributors' mapping behaviours, in: *International Archives of the Photogrammetry, Remote Sensing and Spatial Information Sciences, XL-2/W1*. Hong Kong, China, pp. 149–154.

Bonney, R., Cooper, C.B., Dickinson, J., Kelling, S., Phillips, T., Rosenberg, K.V., Shirk, J., 2009. Citizen science: A developing tool for expanding science knowledge and scientific literacy. *BioScience* 59, 977–984. DOI: https://doi.org/10.1525/bio.2009.59.11.9

Capineri, C., Haklay, M., Huang, H., Antoniou, V., Kettunen, J., Ostermann, F., Purves, R., 2016. Introduction, in: Capineri, C., Haklay, M., Huang, H., Antoniou, V., Kettunen, J., Ostermann, F., Purves, R. (Eds.), *European Handbook of Crowdsourced Geographic Information*. Ubiquity Press, London, UK, pp. 1–11. DOI: https://doi.org/10.5334/bax

Clinton, W., 2000. *Improving the civilian global positioning system (GPS)*. Washington, D.C., USA.

Craglia, M., de Bie, K., Jackson, D., Pesaresi, M., Remetey-Fülöpp, G., Wang, C., Annoni, A., Bian, L., Campbell, F., Ehlers, M., van Genderen, J., Goodchild, M., Guo, H., Lewis, A., Simpson, R., Skidmore, A., Woodgate, P., 2012. Digital Earth 2020: towards the vision for the next decade. International *Journal of Digital Earth* 5, 4–21. DOI: https://doi.org/10.1080/17538947.2011.638500

Craglia, M., Shanley, L., 2015. Data democracy – increased supply of geospatial information and expanded participatory processes in the production of data. *International Journal of Digital Earth* 8, 679–693. DOI: https://doi.org/10.1080/17538947.2015.1008214

Foerster, T., Jirka, S., Stasch, C., Pross, B., Everding, T., Bröring, A., Jürrens, E.H., 2010. Integrating human observations and sensor observations—The example of a noise mapping community, in: A. Devaraju, A. Llaves, P. Maué, C. Keßler (eds.) *Proceedings of the Workshop Towards Digital Earth: Search, Discover and Share Geospatial Data 2010*, Münster, Germany, 20 Sep 2010. Available at: http://ceur-ws.org/Vol-640/paper1.pdf [Last accessed 17 May 2017].

Fonte, C C, Antoniou, V, Bastin, L, Estima, J, Arsanjani, J J, Bayas, J-C L, See, L and Vatseva, R. 2017. Assessing VGI Data Quality. In: Foody, G, See, L, Fritz, S, Mooney, P, Olteanu-Raimond, A-M, Fonte, C C and Antoniou, V. (eds.) *Mapping and the Citizen Sensor*. Pp. 137–163. London: Ubiquity Press. DOI: https://doi.org/10.5334/bbf.g.

Foody, G., See, L., Fritz, S., Velde, M. van der, Perger, C., Schill, C., Boyd, D.S., Comber, A., 2015. Accurate attribute mapping from Volunteered Geographic Information: Issues of volunteer quantity and quality. *The Cartographic Journal* 52, 336–344. DOI: https://doi.org/10.1080/00087041.2015.1108658

Goodchild, M.F., 2007. Citizens as sensors: the world of volunteered geography. *GeoJournal* 69, 211–221. DOI: https://doi.org/10.1007/s10708-007-9111-y

Haklay, M., 2013. Citizen science and volunteered geographic information: Overview and typology of participation, in: Sui, D., Elwood, S., Goodchild, M. (Eds.), *Crowdsourcing Geographic Knowledge*. Springer Netherlands, Dordrecht, Netherlands, pp. 105–122.

Haklay, M., 2010. How good is volunteered geographical information? A comparative study of OpenStreetMap and Ordnance Survey datasets. *Environment and Planning B: Planning and Design* 37, 682–703. DOI: https://doi.org/10.1068/b35097

Haklay, M., Antoniou, V., Basiouka, S., Soden, R., Mooney, P., 2014. *Crowdsourced Geographic Information Use in Government. Report to GFDRR (World Bank)*. UCL-Geomatics, London.

Haklay, M., Basiouka, S., Antoniou, V., Ather, A., 2010. How many volunteers does it take to map an area well? The validity of Linus' Law to volunteered geographic information. *The Cartographic Journal* 47, 315–322.

Kitchin, R., 2013. Big data and human geography Opportunities, challenges and risks. *Dialogues in Human Geography* 3, 262–267. DOI: https://doi.org/10.1177/2043820613513388

Kuhn, W., 2007. Volunteered geographic information and GIScience. Position Paper for the NCGIA and Vespucci Workshop on Volunteered Geographic Information. Santa Barbara, CA, 13–14 Dec 2007. Available at: http://ncgia.ucsb.edu/projects/vgi/docs/position/Kuhn_paper.pdf [Last accessed 16 May 2017]

Lee, M.K., Kusbit, D., Metsky, E., Dabbish, L.A., 2015. Working with machines: The impact of algorithmic and data-driven management on human workers – semantic scholar, in: Proceedings of the 33rd Annual ACM Conference on Human Factors in Computing Systems. ACM, pp. 1603–1612.

Maisonneuve, N., Stevens, M., Ochab, B., 2010. Participatory noise pollution monitoring using mobile phones. *Information Polity* 15, 51–71.

Mooney, P, Olteanu-Raimond, A-M, Touya, G, Juul, N, Alvanides, S and Kerle, N. 2017. Considerations of Privacy, Ethics and Legal Issues in Volunteered Geographic Information. In: Foody, G, See, L, Fritz, S, Mooney, P, Olteanu-Raimond, A-M, Fonte, C C and Antoniou, V. (eds.) *Mapping and the Citizen Sensor*. Pp. 119–135. London: Ubiquity Press. DOI: https://doi.org/10.5334/bbf.f.

Olteanu-Raimond, A.-M., Hart, G., Foody, G., Touya, G., Kellenberger, T., Demetriou, D., 2017. The scale of VGI in map production: A perspective of European National Mapping Agencies. *Transactions in GIS* 21, 74–90. DOI: https://doi.org/10.1111/tgis.12189

O'Reilly, T., 2007. What is Web 2.0: Dcsign patterns and business models for the next generation of software. *Communications & Strategies* 65, 17–37.

Roche, S., 2014. Geographic Information Science I: Why does a smart city need to be spatially enabled? *Progress in Human Geography* 38, 703–711. DOI: https://doi.org/10.1177/0309132513517365

Roche, S., Nabian, N., Kloeckl, K., Ratti, C., 2012. Are "smart cities" smart enough? in: Rajabifard, A., Coleman, D. (Eds.), *Spatially Enabling Government, Industry and Citizens: Research and Development Perspectives*. GSDI Association Press, Needham, pp. 215–236.

See, L., Mooney, P., Foody, G., Bastin, L., Comber, A., Estima, J., Fritz, S., Kerle, N., Jiang, B., Laakso, M., Liu, H.-Y., Milčinski, G., Nikšič, M., Painho, M., Pődör, A., Olteanu-Raimond, A.-M., Rutzinger, M., 2016. Crowdsourcing, citizen science or Volunteered Geographic Information? The current state of crowdsourced geographic information. *ISPRS International Journal of Geo-Information* 5, 55. DOI: https://doi.org/10.3390/ijgi5050055

Senaratne, H., Mobasheri, A., Ali, A.L., Capineri, C., Haklay, M. (Muki), 2016. A review of volunteered geographic information quality assessment methods. *International Journal of Geographical Information Science* 1–29. DOI: https://doi.org/10.1080/13658816.2016.1189556

Silvertown, J., 2009. A new dawn for citizen science. *Trends in Ecology & Evolution* 24, 467–471. DOI: https://doi.org/10.1016/j.tree.2009.03.017

United Nations Environment Program (UNEP), n.d. *Global Initiative for Resource Efficient Cities*. Available at: http://www.resourceefficientcities.org/ [Last accessed 16 May 2017].

www.ingramcontent.com/pod-product-compliance
Lightning Source LLC
LaVergne TN
LVHW082110200926
848484LV00008B/1936
9781013288715